BIOTERRORISM, PREPAREDNESS, ATTACK AND RESPONSE

ADVANCES IN HEALTH CARE MANAGEMENT

Series Editors: John D. Blair, Myron D. Fottler and
Grant T. Savage

Recent volumes:

CONTENTS

PART II: CHAOS, COMPLEXITY AND CHANGE

PART III: ORGANIZATIONS RESPOND . . . OR NOT

**PART IV: DEFENDING THE HOMELAND: CHANGES
AND CHALLENGES**

LIST OF CONTRIBUTORS

James W. Begun	Carlson School of Management, University of Minnesota, USA
John D. Blair	Rawls College of Business, Texas Tech University, USA
George W. Buck, Jr.	College of Public Health, University of South Florida, USA
Nancy M. Dodge	Rawls College of Business, Texas Tech University, USA
Mari K. Eder	United States Army, Office of the United States Secretary of Defense, USA
Myron D. Fottler	College of Health and Public Affairs, University of Central Florida, USA
Leonard Friedman	Department of Public Health, Oregon State University, USA
Robert S. Fry	Office of Assistant Secretary Defense, TRICARE Management Activity, USA
David N. Gans	Practice Management Resources, Medical Group Management Association, USA
Brian J. Gerber	Department of Political Science, Texas Tech University, USA
James J. Hoffman	Rawls College of Business, Texas Tech University, USA

Cynthia A. Holubik	Rawls College of Business, Texas Tech University, USA
E. L. Hunter	Department of Sociology, University of Maryland, USA
H. Joanna Jiang	Center for Delivery, Organization and Markets, Agency for Healthcare Research and Quality, USA
Robert K. Keel	Rawls College of Business, Texas Tech University, USA
Ryan Kelty	Department of Sociology, University of Maryland, USA
Meyer Kestnbaum	Department of Sociology, University of Maryland, USA
Donna Malvey	College of Health and Public Affairs, University of Central Florida, USA
Peter Marghella	The Joint Staff, Department of Defense, USA
Reuben R. McDaniel, Jr.	McCombs School of Business, University of Texas at Austin, USA
Timothy W. Nix	Rawls College of Business, Texas Tech University, USA
Reid M. Oetjen	College of Health & Public Affairs, University of Central Florida, USA
Thomas P. Peterson	Idaho Emergency Physicians, Medical Group Management Association, USA
Kourtney Scharoun	College of Health & Public Affairs, University of Central Florida, USA

David R. Segal   Department of Sociology, University of Maryland, USA

Steven R. Tomlinson   Rawls College of Business, Texas Tech University, USA

K. Wade Vlosich   Rawls College of Business, Texas Tech University, USA

Paul S. Westney   Tindal & Foster, LLC, USA

Carlton J. Whitehead   Rawls College of Business, Texas Tech University, USA

Richard O. Wightman, Jr. United States Army, Office of the United States Secretary of Defense, USA

Lawrence F. Wolper   President, L. Wolper, Inc., Great Neck, USA

Albert C. Zapanta   Reserve Forces Policy Board, Office of the United States Secretary of Defense, USA

REVIEW BOARD MEMBERS

REVIEWERS

Diane Brannon	Department of Health Policy and Administration, Pennsylvania State University, USA
Leonard Friedman	Department of Public Health, Oregon State University, USA
Cynthia Holubik	Rawls College of Business, Texas Tech University, USA
Pete Marghella	The Joint Staff, United States Department of Defense, USA
Reuben McDaniel	Department of Management Science & Information Systems, University of Texas at Austin, USA
Keith Provan	School of Public Administration and Policy, University of Arizona, USA
Jorge Ramirez	School of Law, Texas Tech University, USA
Carlton J. Whitehead	Rawls College of Business, Texas Tech University, USA

NOTES FROM THE EDITORS

A Note from the Editors of *Bioterrorism, Preparedness, Attack and Response*

This thematic volume on preparing for bioterrorism, bioterrorist attacks, and the response to such attacks is, in part, the collective response of both the authors of specific chapters as well as that of the three editors to the events of September 11, 2001.

As scholars, we looked at the existing literature on bioterrorism preparedness, attack, and response and concluded that we could make a new set of contributions. Each of the authors challenges were to think outside or beyond the existing paradigm and to contribute their best thinking on the certainty of our uncertain future, which is facing us all.

Because of its wide-availability already, we have not included clinical information on bioterrorism agents or their treatment in the volume itself. However, we have included a resource guide to assist in finding such information as well as specific and changing information on preparedness and response activities.

Instead, we have focused on the managerial and organizational perspectives, which were underdeveloped in the research literature. We invited an extremely interdisciplinary set of scholars, policy makers, and practitioners to try to provide an unusual set of perspectives to be included in one volume.

Perspectives focused the attempts to defend the homeland and provide for its security we included with analyses of terrorist strategies and tactics – from the perspective of terrorists seeking the most effective forms of attack. Examination of the chaos and complexity facing healthcare organizations and perspectives on their needed change were included with analyses of how organizations have, could and should respond. This collective group of authors, who focus on the specific areas involved in understanding, often do not have a chance to bring their unique insights together in one place.

We hope that the reader will find these research contributions of value whether they are:

- Other researchers in multiple areas of scholarship or
- Faculty members who are preparing future healthcare, military or emergency preparedness leaders or
- Local, state and federal civilian and military policymakers who must attempt to craft appropriate policy and find efficient by effective ways to fund the programs needed to make these policies real, or

- Executives who are required daily to confront bioterrorism uncertainty, which is now a certainty to continue to exist.

We hope these authors and others will continue to help develop new intellectual, policy and practice insights into confronting the potential horror of the bioterrorism formula, which we discuss in the first chapter.

John D. Blair, Myron D. Fottler and Albert C. Zapanta
Editors

A Note from the *Advances in Health Care Management* Series Editors

This volume was conceived after a series of discussions among the volume editors and with Tom Clark and other Elsevier publishing executives. Previous editions (Volumes 1, 2, and 3) of *Advances in Health Care Management* have been traditionally eclectic, as one would find in an academic journal, with several different subtopics emerging from the combinations of: (a) commissioned papers; (b) manuscripts submitted in response to several "Call for Papers" and from the "Best Papers" from the Healthcare Management Division presented at the national meetings of the Academy of Management.

For marketing purposes, the Elsevier publishers felt "thematic" volumes would make it easier to target and reach specific audiences who might have an interest in specific topics. After much discussion, we, the Series editors (John Blair, Myron Fottler, and Grant Savage), agreed to change the strategic direction of the Series to both address the marketing issues but also to draw on the intellectual synergy ideally found in special issues of journals.

We decided to publish the first thematic volume on "Bioterrorism" (Volume 4). We (Blair and Fottler) would edit this volume with our colleague Albert Zapanta, Chair of the Reserve Forces Policy Board, which is one of two congressionally mandated boards and headquartered in the Office of the Secretary of Defense.

A second thematic volume on "International Health Care Management" (Volume 5) will be edited by Grant Savage, Jon Chilingerian, and Michael Powell, which will appear in 2004 and will follow this bioterrorism volume. The sixth volume edited by Blair and Fottler, will have its thematic focus on "Strategic Thinking and Entrepreneurial Action in the Health Care Industry," and will appear in 2005.

John D. Blair, Myron D. Fottler and Grant T. Savage
Series Editors

PART I:
BIOLOGICAL AGENTS AND
TERRORIST AGENTS

THE BIOTERRORISM FORMULA: FACING THE CERTAINTY OF THE UNCERTAIN FUTURE

John D. Blair, Myron D. Fottler and Albert C. Zapanta

ABSTRACT

This paper presents an overview of the articles used in this edition of Advances in Health Care Management. The beginning of the article gives the reader a history of bioterrorist activity within the United States, and how these events have led to current situations. It also provides a model for health care leaders to follow when looking at a bioterrorist attack. The model includes descriptions of how the articles within this book relate to an overall bioterrorist formula. Through this, the reader shall be able to deduce which individual article fits into the vastness of healthcare research pertaining to bioterrorism.

In 1996, Usama bin Laden declared war on the United States (bin Laden, 1996). On the second anniversary of September 11, 2001, many analyses of the whats and whys of 9/11 have been made (Baxter & Downing, 2002; Carr, 2002; Fielding & Fouda, 2003; Friedman, 2003; Gunaratna, 2002; Hoge & Rose, 2001; Kean & Hamilton, 2003; Miller & Stone, 2002; Posner, 2003). These analyses provide rich insights into the nature of the terrorist threat to the United States and to other countries in the "West," at this time. The bulleted information below is simply to remind us that the threat had been real for some time before the events of September 2001, and continues to be a threat.

Bioterrorism, Preparedness, Attack and Response
Advances in Health Care Management, Volume 4, 3–24
© 2004 Published by Elsevier Ltd.
ISSN: 1474-8231/doi:10.1016/S1474-8231(04)04001-7

- In 1999, the "Encyclopedia of the Afghan Jihad: Training Manual" for al Qaeda was found and published, together with detailed information about the tactics to be used.
- In 2001, public documents and the growing threat from a number of smaller attacks on U.S. interests were not fully heeded.
- In 2001, radical Islamists attacked the World Trade Center and the Pentagon, and they attempted an additional strike that ended in a Pennsylvania field. This followed a number of smaller attacks on American interests overseas.
- In 2001 (October and November), a still unknown terrorist, individual or group, used the U.S. Postal Service to disseminate Anthrax.
- In 2002, records from al Qaeda sites in Afghanistan and subsequent investigations by counter-terrorist experts reveal the desire and efforts of al Qaeda to acquire and then use biological agents against the West, particularly the United States.
- In 2003, the War in Iraq (Operations Iraqi Freedom) has failed to locate the stocks of biological agents known to exist, and whose destruction was never recorded.

The United States has made significant efforts to assess national, state, and local vulnerabilities to terrorist attacks (Anderson, 2003; Ridge, 2003b). The United States Government issued several major strategy statements in homeland security (2002), national security (2002), terrorism (2002), and cyber security (2003). The Department of Homeland Security formed to improve security at home. In a complementary move, the Defense Department has formed a new Northern Command with the explicit mission to defend the homeland.

As Laqueur (1999, 2001) and others (Byman et al., 2001; Gurr & Cole, 2000; Hoffman, 1998; Hudson, 1999; Perl, 1997; Pillar, 2001; Rappaport, 1988; Stern, 1999) have repeatedly reminded us, terrorism is not a new phenomenon. Although, 9/11 clarifies the current type of threat from terrorists (Benjamin & Simon, 2002; Berkowitz, 2003; Laqueur, 2003; Moore, 2003; Scheitzer, 2002; U.S. Department of State, 2003). Terrorists have applied "Asymmetric war" concepts because these concepts focus on how the weak can defeat the strong (Gray, 2002; Mack, 1975; McKensie, 2002; Metz & Johnson, 2001; Staten, 1999). Businesses are concerned with "business continuity" in light of the impact on companies housed fully or partially in the World Trade Center (DeCarlo, 2002; Kuong, 2002; Smith, 2002).

Collectively, the United States public sector, private sector, and non-profit sector improved the capabilities we have to prevent, prepare for, respond to, and manage the consequences of terrorist attacks, including bioterrorism (Ahiquist & Burns, 2002; Cosgrove, 2002; Heinrich, 2001; Henderson, 2002; Institute of Medicine National Research Council, 2002; Jackson et al., 2002; Salvucci, 2001). The growing list of potential types of weapons, some of which were earlier recognized and some that are only recent concerns, is increasingly becoming real (Beelman, 2002; Bolkcom & Elias, 2003; Boyd & Sullivan, 1997; CDI, 2003; Central Intelligence Agency, 2003; Dolnik, 2003; Karasik, 2002; Lugar, 1999; Macintyre et al., 2000; Miller, 2002; Tonat & Barbera, 2000).

Non-fiction scenarios can be both alarming, but sometimes instructive. In the same year that bin Laden declared war on the United States, Tom Clancy detailed a fictional bioterrorism attack launched against the United States by Radical Islamists (Clancy, 1996). He had described a jetliner full of fuel that was used as a self-propelled bomb with a high explosive yield to attack the U.S. political and military leadership in Washington, DC (Clancy, 1994).

The actual use of biological agents against an enemy is a very old phenomenon (Wheelis, 2002). True biological warfare implies that the attacker is a state using biological agents against its enemies. Bioterrorism is now perceived as a real threat because of those anthrax attacks, but a bioterrorist attack should not have been a surprise (Alibek, 1999; Bardi, 1999; Bartlett, 1999; CDC, 1994; Cieslak & Eitzen, 2000; Cordesman, 2001; Henderson, 1998, 2002; Inglesby, 1999; Inglesby, Grossman & O'Toole, 2001; Inglesby, Henderson & Barlett, 1999; Miller, Engleberg & Broad, 2001; O'Toole, Mair & Inglesby, 2002; Preston, 1997).

After the anthrax attack, there has been considerable focus on preparedness and response. For example, some key problems in bioterrorism preparedness and response have been identified, resources levels and types have been changed, plans have been revised, and new exercises have been planned and implemented (Anser Institute, 2002; Bozzette et al., 2003; Ridge, 2003a; *Yahoo News*, 2003). Agroterrorism, a special case of bioterrorism, is also a growing concern (Segarra, 2002).

The results of the Anthrax attack of 2001 have further stimulated many health care facilities creating emergency operations plans. It is very questionable if all of our health care facilities can effectively respond to a bioterrorist attack (Fauci, 2002; Grow & Rubinson, 2003; Joint Commission on the Accreditation of Health Care Organizations, 2003; Leavitt, 2003; Macintyre & Deatley, 2001; Preston, 2002; U.S. GAO, 2003). However, a number of reports have continued to show that public health, healthcare facilities, police and fire departments, and local and state governments are not ready for a bioterrorist event. For example, the most recent one on hospital preparedness by the U.S. Government Accounting Office (2003) came to that conclusion for urban hospitals. Realistically, even less preparedness from rural hospitals is a reasonable expectation that should not be ignored.

The interface between the public health system and the actual provider and care delivery system is clearly tenuous, if not truly broken. There are significant differences in professional perspectives from comparable clinicians in both groups, as well as between the administrators of the two types of organizations. It is not uncommon for one to also see is clear professional jealousy or antagonism between physicians in the two groups. To assume that they will be able to work together without conflict over appropriate and relative authority is unrealistic.

Still incomplete and very difficult to achieve is a functioning multi-agency, inter-organizational communication, and functional collaborative structures. Given a myriad of unrealistic or incomplete emergency operation plans and still limited and uncoordinated resources, the federal, state, and local governments are not prepared. Although, these key stakeholders are taking active roles in trying to adequately prepare for any new bioterrorist event. The most reasonable assumption is that we have a long road to travel yet (Blair, Holubik, Keel, Roberson & Tomlinson, 2003; Fricker, Jacobson & Davis, 2002; Peters, 2003).

Developing formal plans that are put on the shelf or in required accreditation or licensing reports is not enough. Even the best emergency preparedness plan can never cover all possibilities of mass casualty, and responding to a bioterrorism event is extremely difficult. The current financial crisis of the United States health care system directly affects preparedness and response for mass casualty incidents in a number of ways. It is argued that the lack of proper funding to hospitals and response facilities causes an adequate response to lack in equipment, personnel, and training even if a preparation plan is devised by the facility; a plan cannot be implemented if the financial institutions do not exist in order to implement the plan. The lack of financial infrastructure in current health care facilities is only one hindrance to adequate responding to a bioterrorist event.

Perhaps the most disquieting information is that hospitals may be more prepared than before, but physicians' offices and ambulatory clinics are not (Wolper, Gans & Peterson, 2003). Although generally small in size and limited in complexity, there are many unprepared settings for the fewer than five thousand emergency rooms that must prepare.

Why is this a problem? Most biological agents result in flu-like symptoms. It is far more likely that in early stages for infected or contaminated victims to go to see their family primary care physicians, than to go straight to the emergency room with an illness that does not appear to be emergent. Physician offices and ambulatory clinics do not routinely have any significant protection for others, who may become infected or contaminated. This includes patients in waiting rooms as well as the clinic or office staff, including the physicians, nurses, receptionists, and cashiers. Most emergency drills and simulations include a limited number of physicians in addition to healthcare and other executives outside the health care industry, but those physicians are included in their roles within the hospitals, not in their own offices or clinics.

The specific paper's and this entire volume's objectives are to facilitate the reader's:

(1) Understanding of the broader issues of terrorism and the context it puts around the potential of bioterrorism preparedness, attack and response the "new reality" that creates certainty in the uncertainty facing them in the future.

(2) Understanding of bioterrorism-specific issues and appreciation for how terrorists can use biological weapons.

(3) Recognition and acknowledgement of homeland vulnerabilities, homeland defense and homeland security, with their societal and organizational implications.

(4) Understanding a realistic appreciation for what preparedness will require and what response must be able to do, with particular emphasis on health care delivery organizations and their administrative as well as clinical leaders.

(5) Insight into what political, military, and civilian (private sector and public sector) decision makers must face up to in their own organizations and in systems, of which their organization is a part.

We will begin in this paper by providing a way to think of how multiple factors and their related variables come together to limit or enhance the consequences of an attack. We will focus on what goes into to preventing or mitigating the outcomes of terrorists' actions as well as what factors result in facilitating terrorists' ability to create an even greater environmental jolt to organizations, health care or not.

Understanding the terrorists and their tools has not been in the life experience of most of the readers, whether scholars, governmental officials, or practitioners. In the past, that was a good thing. Now, lack of understanding limits creativity and imagination in improving their respective organizations, policies, and systems or in formulating and conducting cutting-edge conceptual and empirical research that will help address these issues. As the authors of this paper as well as editors of this unique volume, we believe that such insights are essential to enhance effective preparedness and meaningful response.

Later in this paper, we will provide also a systematic overview of the structure of the volume and, more importantly, the contributions our colleagues and we have made to the understanding of the phenomena. It is these phenomena we will next describe in the form of a conceptual formula.

THE BIOTERRORISM FORMULA: THE BROADER CONTEXT OF PREPAREDNESS

In this volume, there are several complex hypothetical scenarios reported on, or new ones presented. These provide some key observations about bioterrorist attacks, including the nature of the agents used by terrorists and exposure of wide-range vulnerabilities in healthcare organizations during response to such an attack. It is also necessary to understand the larger picture or context of terrorist "warfare," in general and bioterrorism in particular.

The Nature of Terrorism

There are a number of domestic and/or international social conditions that can influence the development of a perceived need for change, which potentially lead to war and terrorism. These conditions may be ideological, political, economic, religious, and ethnic concerns, and often, terrorist groups in the Middle East are reacting to a combination of these social conditions. The conditions can be within a country or may cross international borders and be greatly facilitated by the ready availability of cyberspace, both for organizing and communicating as well as serving as a potential weapon itself (Arquilla & Ronfeldt, 2001; Richardson, 2003; Vatis, 2001).

Due to social conditions that often create poverty, lack justice, or are inconsistent with the beliefs of many people, it is likely that change-oriented organizations will emerge (Gamson, 1975). These may be "traditional" political parties or interest groups. They may be either non-violent or violent in their ideology and philosophy (Epstein, 2002). They may operate overtly or covertly and may be found in the military or among civilians. When a desired change is frustrated, sometimes change-oriented organizations become terrorist-like (Byman, 2001; Krebs, 2002; Thompson, 1989).

Many strategies for change may also emerge. Conventional strategies use elections, political influence, or even legal action. Sometimes conventional war is used by organizations in power, or guerrilla war by those who are not. Terrorism always remains an option to introduce change (Boot, 2002).

The current worldwide battle between the Radical Islamists and the West increasingly appears to be part of the "clash of civilizations" predicted by Samuel Huntington (1996). The religious bases of this clash have been examined in detail subsequently to the 9/11 attacks (Benjamin & Simon, 2002; Emerson, 2002; Williams, 2002).

An example of the extent of the clash appears in a *fatwa*, or religious directive, from Usama bin Laden and others calling themselves the World Islamic Front, issued in 1998. Their *fatwah* instructed Muslims throughout the world to kill Americans, either combatant or noncombatant, to the extent possible, anywhere in the world (World Islamic Front, 1998). An excerpt from the statement:

> . . . in compliance with God's order, we issue the following *fatwa* to all Muslims:
>   The ruling to kill the Americans and their allies – civilians and military – is an individual duty for every Muslim who can do it in any country in which it is possible to do it . . . . This is

in accordance with the words of Almighty God, "And fight the pagans all together as they fight you together . . .

We – with God's help – call on every Muslim who believes in God and wishes to be rewarded to comply with God's order to kill the Americans and plunder their money wherever and whenever they find it (World Islamic Front, 1998).

Terrorism is, in fact, the first of three stages of violent change by an organization not in political power. As strength grows, guerrilla war wages against the ruling government's military. Potentially, the strength of the challenging organization grows so that its army can fight and defeat the ruling group's military forces. Such was the strategy pursued by Mao Tse-tung in China and Ho Chi Minh in Vietnam, in their respective "wars of national liberation" (Pike, 1966; Tse-Tung, 1966). Closer to home were the patriots of the American Revolution who utilized similar strategy to win the freedom from the British colonies in America. The British would have considered the U.S.' "Founding Fathers" to be the equivalent of terrorists by today's definition.

Asymmetric Warfare

Although terrorists can do significant damage and inflict large numbers of casualties, and potentially mass casualties in the tens of thousands, terrorism is actually a reflection of weakness, not strength. A more recent concept to clarify this is called "asymmetric warfare," in which the weak attack the strong and use the enemies' strengths against them or simply make their strengths irrelevant to the battle (Cordesman, 2001; Gray, 2002; Mack, 1975; McKensie, 2002; Metz & Johnson, 2001; Staten, 1999). For example, on September 11th, the terrorists used American transportation infrastructure against the United States. Commercial jets turned into very powerful bombs by a handful of men armed only with box cutters and the desire to die for their cause.

The anthrax attack in the fall of 2001 used well-developed American infrastructure to distribute a biological agent. The U.S. Postal Service provided the delivery mechanisms to attack United States leaders and citizens. Both the 9/11 and the anthrax attacks illustrate successful uses of asymmetric warfare. Blair, Keel, Nix and Vlosich (2003) describe asymmetric warfare in some detail.

The Bioterrorism Formula

A full understanding of bioterrorism preparedness and response cannot be achieved effectively by considering these phenomena in a vacuum or informed only by the bioterrorism literature in its current form. To articulate the overall conception we, as editors, pursued in the development of this volume, have developed what we call the "bioterrorism formula."

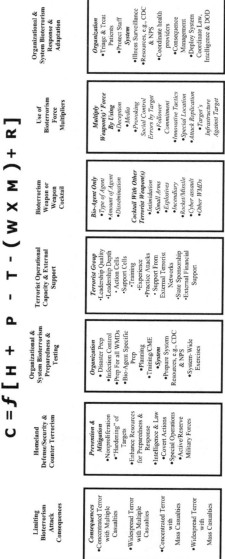

$$c = f[H + P - T - (w \times M) + R]$$

Limiting Bioterrorism Attack Consequences	Homeland Defense/Security & Counter Terrorism	Organizational & System Bioterrorism Preparedness & Testing	Terrorist Operational Capacity & External Support	Bioterrorism Weapon or Weapon Cocktail	Use of Bioterrorism Force Multipliers	Organizational & System Bioterrorism Response & Adaptation
Consequences •Concentrated Terror with Multiple Casualties •Widespread Terror with Multiple Casualties •Concentrated Terror with Mass Casualties •Widespread Terror with Mass Casualties	*Prevention & Mitigation* •Nonproliferation •"Hardening" of Targets •Enhance Resources for Preparedness & Response •Intelligence & Law •Covert Actions •Special Operations •Active/Reserve Military Forces	*Organization* • Disaster Prep •Infection Control •Prep For all WMDs •Bio-Agent Specific Prep •Planning •Training/CME *System* •Prepare System Resources, e.g., CDC & NPS •System-Wide Exercises	*Terrorist Group* •Leadership Quality •Leadership Depth • Action Cells •Support Cells •Training •Experience •Practice Attacks • Support From External Terrorist Networks •State Sponsorship •External Financial Support	*Bio-Agent Only* •Type of Agent •Amount of Agent •Dissemination *Cocktail With Other Terrorist Weapon(s)* •Intimidation • Small Arms •Explosives •Incendiary •Rocket/Missile •Cyber assault •Other WMDs	*Multiply Weapon(s)' Force By Using* •Deception •Media •Provoking Social Control Errors by Target •Follower Commitment •Innovative Tactics •Special Location •Attack Replication •Target's Infrastructure Against Target	*Organization* •Triage & Treat Patients •Protect Staff *System* •Illness Surveillance •Resources, e.g., CDC & NPS •Coordinate health providers •Consequence Management •Deploy System •Coordinate Law, Intelligence & DOD

Fig. 1. The Bioterrorism Formula: Healthcare Preparedness and Response in Their Security and Terrorist Contexts.

In a summary way, this formula brings together the myriad of activities by a wide range of organizations that are involved in preventing, preparing for and responding to a bioterrorist attack. However, the formula also includes the perspectives we believe terrorist organizations have on how best to achieve their goals, i.e. powerful attacks with significant consequences.

Our bioterrorism formula is $C = f[H + P - T - (W \times M) + R]$.

We formulated this "simplistic" formula to concisely portray the many classes of variables involved in limiting the consequences of bioterrorism at the same time as variables associated with terrorist capability and actions are attempting to maximize those consequences. Its solution can be operationalized only conceptually. The pluses and minuses suggest a force field of limiting and maximizing variables.

This symbolic use of a formula represents the following:

- Limiting the **Consequences** of Bioterrorism is a
- **function** of
- **Homeland** Security/Defense & Counter Terrorism *plus*
- Organizational & System **Preparedness** & Testing *minus*
- **Terrorist** Operational Capacity & External Support *minus*
- *the multiplicative effect* of the Use of a Bioterrorism **Weapon** or Weapon Cocktail *times* the Use of Bioterrorism Force **Multipliers** *plus*
- Organizational & System Bioterrorism **Response** & Adaptation.

The key formula variables display in Fig. 1. Under the formula variables are some illustrative details to clarify what we mean by each concept.

This formula serves as our heuristic for the key factors that make up the essential pieces of bioterrorism preparedness, attack and response that we have identified. This formula may appear a gross simplification of the variables (or clusters of variables) involved, but we have provided this to try and put what is normally thought of bioterrorism preparedness issues in their equally important security and terrorist contexts. As specialized scholars or focused organizational practitioners, we must focus on "our" part of the formula. Here, we have sought to bring experts' insights on as many of these factors as possible into one volume. The above figure illustrates where each of the articles presented fits into the bioterrorist equation (Fig. 2).

BIOTERRORISM PREPAREDNESS, ATTACK AND RESPONSE: THE BOOK

In the volume, the assembled authors together to address some dimensions of all the factors in the formula. These include homeland security, homeland defense,

Limiting Bioterrorism Attack Consequences	• *All Papers Discuss the Consequences of Attack*
Homeland Defense/Security & Counter Terrorism	• *The International Threat of Biological Weapons: Legal and Regulatory Perspectives* • *Integration or Disintegration? An Examination of the Core Organization and Management Challenges at the Department of Homeland Security* • *The Role of the Reserve Forces in Defending the Homeland*
Organizational & System Bioterrorism Preparedness & Testing	• *The Environmental Jolt of Likely Bioterrorism* • *Multiprovider Systems as First Line Responders to Bioterrorism Events: Challenges and Strategies* • *Bioterrorism Visits the Physician's Office*
Terrorist Operational Capacity & External Support Bioterrorism Weapon or Weapon Cocktail Use of Bioterrorism Force Multipliers	• *Modeling the Environmental Jolt of Terrorist Attacks: Configurations of Asymmetrical Warfare* • *Cocktails, Deceptions, and Force Multipliers in Bioterrorism* • *The Bioterrorism Formula: Facing the Certainty of the Uncertain Future*
Organizational & System Bioterrorism Response & Adaptation	• *Chaos and Complexity in a Bioterrorism Future* • *Changing Organizations for Their Likely Mass-Casualties Future* • *Civil-Military Relations in an Era of Bioterrorism: Crime and War in the Making of Modern Civil-Military Relations* • *Bioterrorism Preparedness and Response: A Resource Guide for Health Care Managers* • *Responding to Bioterrorism: A Lesson in Humility for Management Scholars*

Fig. 2. The Bioterrorism Formula and this Volume: Where the Articles Fit into this Formula.

related to the nature of bioterrorism attacks, including examining terrorists' use of weapons and force multipliers, together with the many key issues related to preparing capably and responding effectively. Since many of the readers of this volume will come from a healthcare leadership background and serve in organizational or medical roles, we felt it imperative to provide not only as many new ideas about what they normally think about but also new insights into the broader context around bioterrorism preparedness and response.

We start the book with several premises:

• That the research on bioterrorism prevention and response has not approached the bioterrorism problem from a sophisticated *management and organizational* perspective, consistent with the development of the fields of management, more generally, and health care management, more specifically;

- That health care management scholars could bridge the gap between our management and organizational theories and the likely challenges of bioterrorism prevention and response facing health care organizations and their leaders;
- That homeland defense and security perspectives have not been fully incorporated into the existing research on bioterrorism preparedness; and
- That the perspectives of the terrorists themselves and the options they have to attack have not been addressed fully in the bioterrorism literature.

To help address the limitations discussed in our premises, we have commissioned a bevy of well-known scholars in the field of healthcare management, thoughtful analysts on counter-terrorism, homeland security and defense as well as some well-known in other fields like military sociology, and more, junior colleagues and practitioners to address issues of bioterrorism preparedness, attack, and response.

Thematically, this volume arranges in four parts.

- Several papers addressing "biological agents and terrorist agents" comprise Part 1.
- Part 2 focuses on "chaos, complexity and change."
- Part 3 concentrates on whether and how "organizations respond – or not" to bioterrorism.
- Finally, Part 4 addresses macro issues involved in "changes and challenges in defending the homeland."

Throughout this volume, the authors have viewed each topic through the lens of one or more management theory. Among the perspectives represented here are organizational change theory, organization structure theories, chaos theory, complexity theory, leadership theory, self-organization theory, theory relating to development of learning culture, alliance theory, stakeholder management theory, and network theory.

Biological Agents and Terrorist Agents

This is the first paper in this part. The second paper in this part is "The International Threat of Biological Weapons: Legal and Regulatory Perspectives" by Paul Westney and James Hoffman. It examines the international threat of biological weapons from a legal and regulatory perspective. The paper begins with an overview of the threat of biological weapons of mass destruction and their unique challenges and political role. A brief history of biological weapons control is followed by proposed revisions in the international legal structure. The paper concludes with a synthesized approach to biological weapons control.

"Cocktails, Deceptions, and Force Multipliers in Bioterrorism" is by Blair and Vlosich (2003). It examines the different ways in which terrorists plot an attack to increase their effectiveness on a target population. The paper shows how a terrorist to enhance an attack can use various force multipliers, and lays a foundation for which a reader can deduce the possible outcomes of an attack. By integrating various scenarios, the paper explains how a bioterrorist attack relates to the health care environment.

The fourth and final paper is by John Blair and his colleagues "Modeling the Environmental Jolt of Bioterrorism Attacks Modeling the Environmental Jolt of Terrorist Attacks: Configurations of Asymmetrical Warfare" develop various attack scenarios under which biological weapons might be delivered through the use of asymmetrical warfare. The paper compares real life terrorist attacks on the United States to the optimal outcomes a terrorist wishes to occur. The paper gives the reader a better understanding of the various configurations a terrorist will use to bring about asymmetrical warfare on a target population.

Chaos, Complexity and Change

Chaos and complexity theory offers a theoretical perspective for viewing the bioterrorism threat, which is quite different from the more traditional planning perspective. Rather than plan for likely scenarios, chaos and complexity theory focuses on creating adaptable organizations "on the ground" which are more responsive in identifying and responding to actual bioterrorism events.

In the first paper, McDaniel (2003) provides us with an essay concerning how health care leaders might use some ideas from chaos and complexity theory to think more creatively about the meaning of bioterrorism for their organization. The paper is titled "Chaos and Complexity in a Bioterrorism Future." After discussing the basic concepts of complexity science, McDaniel briefly suggests ways leaders might engage in effective action to be more successful in responding to bioterrorism. A case study of how St. Vincent's Health System responded to 9/11 is provided to illustrate the salient points. The essay concludes with some research questions for the future.

The second paper by Friedman and Marghella (2003) is "The Environmental Jolt of Likely Bioterrorism." In this paper, the authors build on the existing theoretical concepts of organization change and environmental jolts to study bioterrorism events. It begins with a possible bioterrorism scenario and then reviews of some of the theory and practice associated with managing change in health care organizations. The authors then consider how health care organizations react and respond to specific environmental jolts outside the leaders frame of reference. They conclude with a series of suggestions and recommendations to assist health care leaders in preparing for a bioterrorism event.

In the third paper entitled "Changing Organizations for Their Likely Mass-Casualties Future," Begun and Jiang (2003) focus their discussion on the management of "surprises" of all types, including a possible bioterrorism event. They investigate how organizations can change in order to be better prepared for unforeseen catastrophic events. The paper goes on to point out that conventional organizational change strategies go only so far in preparing for a bioterrorism event. Concepts from complexity science complement those traditional approaches. The paper goes on to outline guidelines for preparing health care organizations for bioterrorism events based on both traditional organizational change and newer complexity science perspectives. It concludes with a research agenda on the topic of how health care organizations can effectively implement changes to better prepare for a future bioterrorism event.

Organizations Respond – Or Not

The question addressed in this section of our book is how health care organizations have and might prepare for and respond to a bioterrorism event. The papers in this section outline and discuss the various authors' perspectives concerning bioterrorism preparation and response. These papers focus on where we are at present in terms of preparation and potential response, the "gaps" which must be filled, and how such gaps might be addressed.

The first paper in this section is "Multiprovider Systems as First Line Responders to Bioterrorism Events: Challenges and Strategies" and is written by Fottler, Scharoun and Oetjen. They explore the possibility of multiprovider systems as first-line responders to a bioterrorism event. While there is no certainty regarding any terrorist act, the authors believe a bioterrorist event is most likely to occur in a large metropolitan area and the most-likely "first-line responders" (after the problem is identified) will be urban multiprovider systems or networks. The authors point out that the lack of empirical research or first hand case study research on the topic necessitates a theoretical approach to outlining the potential benefits of multiprovider systems or networks in addressing the bioterrorism issue. The paper begins with an outline of the nature and challenges of bioterrorism and the current "state-of-the-art" in preparation on the part of four categories of health care organizations. It concludes with recommendations for structuring multiprovider systems to enhance their ability to plan for and respond to bioterrorism events.

"In Bioterrorism Visits the Physician's Office," Wolper, Ganz, and Peterson address the role of the physician office in the event of a bioterrorism event. The authors believe little attention has been devoted to primary care physicians and specialists to whom patients may be referred. Their focus is on the need for physician offices to be able to clinically, operationally, and managerially respond to virulent new organisms. Office-based physicians could be the first providers of care in the event of a bioterrorism event. The paper reviews the manner in which physician offices operate with regard to policies, procedures, responsibility, and authority; provides an overview of the bioagents and related symptoms about which physician offices should be alert; and recommends a range of operational and procedural changes that will enhance their ability to respond to patient exposure to bioagents.

The third paper is by Malvey, Fottler, Buck and Fry and is called "Responding to Bioterrorism: A Lesson in Humility for Management Scholars." This paper uses stakeholder management theory to help healthcare executives identify and manage their key stakeholders in the event of a bioterrorism event. The authors discuss the nature of bioterrorism and the challenge it represents for healthcare organizations, consider the adequacy of traditional planning responses, propose integrative response planning among a network of organizations, compares management of a bioterrorism event with management of day-to-day operations, and propose key stakeholder identification and management as the most visible approach to the problem. The paper concludes with additional recommendations for future research.

Holubik and Tomlinson (2003) provide valuable guidance to the complex bioterrorism topic addressed in this part of the book. Their paper is entitled "Bioterrorism Preparedness and Response: A Resource Guide for Health Care Managers."

Changes and Challenges in Defending the Homeland

Section four addresses homeland security issues at the macro level (i.e. above the level of the individual health care organization). These papers will help health care management scholars and practitioners see the "big picture" in preparing for and responding to a bioterrorism event.

In their paper "The Role of the Reserve Forces in Defending the Homeland" Zapanta, Wightman and Eder (2003) address the emerging issues in homeland defense within the context of the "total force." The U.S. Military's total force is made up of both active as well as the reserve components. The reserve component of the military includes the Army National Guard, the Air National Guard, the Army Reserve, the Air Force Reserve, the Navy Reserve, the Coast Guard Reserve, and the Marine Reserve. The authors provide insights into the homeland defense, which is the responsibility of the Department of Defense. Although their respective roles are still evolving, homeland defense complements homeland security.

The second paper by Hunter, Kelty, Kestnbaum and Segal (2003) addresses civil-military relations in an era of potential bioterrorism. In "Civil-Military Relations in an Era of Bioterrorism: Crime and War in the Making of Modern Civil-Military Relations," the key issue is how to achieve balance in the relationship between civilian and military authorities. The authors begin with a historical review of civilian-military relationships prior to and after 9/11 and then consider various models for such relationships during national security crises. They go on to discuss bioterrorism as an act of war rather than merely a criminal act and conclude that the future of civilian-military relations depends largely on factors that have yet to play themselves out. They note that we are in the midst of a change in these relations, the permanence of which will coincide with the longevity of the national security crisis.

In the third paper, Dodge, Whitehead and Gerber (2003) consider the newly created Department of Homeland Security (DHS) and some of the salient organizational and management issues, which could facilitate or impede its effectiveness. In "Integration or Disintegration? An Examination of the Core Organization and Management Challenges at the Department of Homeland Security, "the authors focus particularly on the successful integration of the eclectic 22 different agencies, which comprise DHS to facilitate their successful execution of tasks associated with countering terrorism and bioterrorism. The paper concludes with a recommendation that DHS adopt a differentiated network structure to address the leadership, management, and organizational challenges discussed earlier.

A review of all of the papers in this volume provides a wide variety of theoretical perspectives, managerial strategies, managerial tactics, and future research topics that shall be of interest to health care management scholars, health care executives, and disaster planning in government and the private sector. All of the approaches discussed here should be viewed as complementary perspectives for viewing the bioterrorism event.

FACING THE CERTAIN UNCERTAIN FUTURE

Therefore, should the population of the United States still worry about adequate preparedness? A Gallup poll announced just prior to the 9/11 of 2003, says people still are concerned. The collective conclusions of the authors of this book are that they should be.

Help prepare not only your organization but also work with others in your community. You all need to be ready to "defend in depth" your patients, co-workers and your own loved ones. The authors have shared some truths about the issues. Truth of the new reality of bioterrorism in the United States can in fact, kill. Unfortunately, it would have already killed your sister and nieces if they had experienced what we discussed, above, hypothetically.

The readers of this book are likely to be managers of healthcare organizations that will encounter bioterrorism in the future. For such organizations, there are three related issues: bioterrorism preparedness, response, and change or adaptation in response to a direct attack or based on experience from other attacks.

Bioterrorism attacks can vary dramatically based on type of agent used, the availability of the agent, how it is disseminated, the target selected, its vulnerability to biological attack, and the likely success of the attack. Although no one can foretell the specifics of these, the healthcare systems must have response plans in place for any number of biological, chemical and radiological events, as each type of WMD has a set of characteristics, dangers, need for specialized equipment and very specialized treatment means and procedures. Bioterrorism preparedness should be bio-agent specific and involve significant levels of planning.

The preparation for a bioterrorist attack by a healthcare facility is by no means simple or inexpensive, in time or in money. Outlined and practiced procedures, instructional manuals, the development of communication networks and training of personnel on bioagents and terrorist tactics are vital to any facility, healthcare or not, to be ready for a potential attack.

This paper serves only as an introduction to how intricate and at times, extremely difficult, preparing for an event of such magnitude can be for the administrators of a healthcare facility. The other authors in this volume have set out to help us all in the continual struggle to look anew at the reality of what it means to achieve dynamic and strategic preparedness, significant attack prevention, and mitigation and rapid and effective response. Such new perspectives must continue to challenge the fundamentally unworkable *status quo*.

Political leaders, military leaders, business leaders, healthcare leaders, and healthcare professionals must be fully committed to the large variety of profound organizational and system changes to improve effectiveness and to save lives and reduce the extent of terror. Ongoing problems and limitations should be identified, resources changed, terrorist organizations and leadership undermined, system and organizational plans revised and new international, national and local exercises and "war games" planned and implemented.

Perhaps the most difficult will be to really accept a future that is certain to be uncertain. The uncertainty will come from several sources:

• Today's terrorist, whom we are learning to understand, is likely not to be tomorrow's terrorist.

- Today's terrorist weapons and tactics that we are learning to, prevent, mitigate, prepare for and respond to are likely to not be tomorrow's terrorist weapons and tactics.
- Today's growing belief that we achieving homeland security and have successfully stopped terrorist attacks, which is increasingly the situation we are in two years after 9/11, will be shattered by the next attack tomorrow.
- Today's and tomorrow's terrorists will always be able to attack and open society like that in the United States, with its myriad of soft targets, and tomorrow we need to have gotten preparedness and response right – at both the organizational and system levels.
- Today's societal and organizational leaders will need to have understood all the variables in the formula and then challenged themselves and their organizations to have acted on the variables in the bioterrorism formula that they can change. This is not negotiable if we are to be ready for the certainty of our collective uncertain future – because today's and tomorrow's terrorists will certainly have focused on their variables – their asymmetric organizations, strategies, tactics, weapons/cocktails and force multipliers.

REFERENCES

Ahiquist, G., & Burns, H. (2002). *Bioterrorism: Improving preparedness and response*. Miami: Booz, Allen & Hamilton.

Alibek, K. (1999). *Biohazard:* The chilling true story of the largest covert biological weapons program in the world, told from the inside by the man who ran it. New York: Random House.

Anderson, P. (2003). *Threat-vulnerability integration: A methodology for risk assessment*. Washington, DC: Center for Strategic & International Studies.

Anser Institute for Homeland Security (2002). Silent vector. *Briefing slides* (October 17–18).

Arquilla, J., & Ronfeldt, D. (2001). *Networks and netwars*. Santa Monica, CA: RAND.

Bardi, J. (1999). Aftermath of a hypothetical smallpox disaster. *Journal of Emerging Infectious Diseases*, 5(4), 547–551.

Bartlett, J. (1999). Applying lessons learned from anthrax case history to other scenarios. *Emerging Infectious Diseases*, 5(4), 561–563.

Baxter, J., & Downing, M. (Eds) (2002). *The BBC reports on America, its allies and enemies, and the counterattack on terrorism*. Woodstock, NY: Overlook Press.

Beelman, M. (2002). The dangers of disinformation in the war on terrorism. *International Consortium of Investigative Journalists* (February 2).

Begun, J., & Jiang, J. (2003). Changing organizations for their likely mass-casualties future. In: J. Blair, M. Fottler & A. Zapanta (Eds), *Bioterrorism Preparedness, Attack and Response, Volume 4, Advances in Health Care Management*. London: JAI Press/Elsevier.

Benjamin, D., & Simon, S. (2002). *The age of sacred terror*. New York: Random House.

Berkowitz, B. (2003). *The new face of war: How war will be fought in the 21st century*. New York: Free Press.

Blair, J., Holubik, C., Keel, R., Roberson, A., & Tomlinson, S. (2003). Bioterrorism preparedness. In: L. Wolper (Ed.), *Health Care Administration* (4th ed.). New York: Jones and Barlett.

Blair, J., Keel, R., Nix, T., & Vlosich, W. (2003). Modeling the environmental jolt of bioterrorism attacks. In: J. Blair, M. Fottler & A. Zapanta (Eds), *Bioterrorism Preparedness, Attack and Response, Volume 4, Advances in Health Care Management*. London: JAI Press/Elsevier.

Blair, J., & Vlosich, W. (2003). Cocktails, deceptions and force multipliers in bioterrorism. In: J. Blair, M. Fottler & A. Zapanta (Eds), *Bioterrorism Preparedness, Attack and Response, Volume 4, Advances in Health Care Management*. London: JAI Press/Elsevier.

Bolkcom, C., & Elias, B. (2003). Homeland security: Protecting airliners from terrorist missiles. *Report for Congress* (February 12).

Boot, M. (2002). *The savage wars of peace: Small wars and the rise of American power*. New York: Basic Books.

Boyd, A., & Sullivan, M. (1997). *Emergency preparedness for transit terrorism*. Washington, DC: National Academy Press.

Bozzette, S. et al. (2003). A model for a Smallpox-vaccination policy. *The New England Journal of Medicine*, 348(5), 416–425.

Byman, D. et al. (2001). *Trends in outside support for insurgent movements*. New York: RAND.

Carr, C. (2002). *The lessons of terror*. New York: Random House.

CDI (2003). Pascal's new wager: The dirty bomb threat heightens (February 4).

Center for Disease Control (1994). Addressing emerging infectious disease threats: A prevention strategy for the United States executive summary. Summary of notifiable diseases, 1945–1994. Morbidity and Mortality Weekly Report 43, 70–78.

Central Intelligence Agency (2003). Terrorist CBRN: Materials and effects (U). Directorate of Intelligence, CTC 2003–40058. Washington, DC: U.S. Government Printing Office.

Cieslak, T. J., & Eitzen, E. M. (2000). Bioterrorism: Agents of concern. *Journal of Public Health Management Practice*, 6(4), 19–29.

Clancy, T. (1994). *Debt of honor*. New York: G. P. Putnam's Sons.

Clancy, T. (1996). *Executive orders*. New York: Berkley Books.

Cordesman, A. (2001). Defending America: Asymmetric and terrorist attacks with biological weapons. *Center for Strategic and International Studies* (February 12).

Cosgrove, M. (2002). Terrorist strategy and global economic implications. *International Business & Economics Research Conference*. Las Vegas, NV.

DeCarlo, A. (2002). *Business continuity: Embracing the unexpected – or else*. White Paper, AT&T.

Dodge, N., Whitehead, C., & Gerber, B. (2003). Integration or disintegration? An examination of the core organization and management challenges at the Department of Homeland Security. In: J. Blair, M. Fottler & A. Zapanta (Eds), *Bioterrorism Preparedness, Attack and Response, Volume 4, Advances in Health Care Management*. London: JAI Press/Elsevier.

Dolnik, A. (2003). Die and let die: Exploring links between suicide terrorism and terrorist use of chemical, biological, radiological, and nuclear weapons. *Studies in Conflict & Terrorism, 26*, 17–35.

Emerson, S. (2002). *American Jihad: The terrorists living among us*. New York: Free Press.

Epstein, J. M. (2002). Modeling civil violence: An agent-based computational model. Center on Social and Economic Dynamics. Washington, DC: Brookings Institution.

Fauci, A. (2002). Bioterrorism: A clear and present danger. *The Haskins Lectureship in Science Policy* (November 15).

Fielding, N., & Fouda, Y. (2003). *Masterminds of terror: The truth behind the most devastating terrorist attack the world has ever seen*. New York: Arcade Publishing.

Fricker, R., Jr., Jacobson, J., & Davis, L. (2002). *Measuring and evaluating local preparedness for a chemical or biological attack.* Santa Monica, CA: RAND.

Friedman, G. (2003). Two years of war. *The Stratfor Weekly* (September 9).

Friedman, L., & Marghella, P. (2003). The environmental jolt of likely bioterrorism. In: J. Blair, M. Fottler & A. Zapanta (Eds), *Bioterrorism Preparedness, Attack and Response, Volume 4, Advances in Health Care Management.* London: JAI Press/Elsevier.

Gamson, W. (1975). *The strategy of social protest.* Homewood, IL: Dorsey Press.

Grow, R. W., & Rubinson, L. (2003). The challenge of hospital infection control during a response to bioterrorist attacks. *Biosecurity and Bioterrorism: Biodefense Strategy, Practice, and Science, 1*(3).

Gunaratna, R. (2002). *Inside Al Qaeda: Global network of terror.* New York: Berkley Books.

Gurr, N., & Cole, B. (2000). *The new face of terrorism.* London: I. B. Tauris.

Heinrich, J. (2001). Bioterrorism: Public health and medical preparedness. Testimony before the Subcommittee on Public Health, Committee on Health, Education, Labor and Pensions, United States Senate. GAO-02–141T.

Henderson, D. (1998). Bioterrorism as a public health threat. *Emerging Infectious Diseases, 4*(3), 488–492.

Henderson, D. A. (2002). Public health preparedness: Science and technology in a vulnerable world. *Science and Technology in a Vulnerable World*, 33–40.

Hoffman, B. (1998). *Inside terrorism.* New York: Columbia University Press.

Hoge, J. F., Jr., & Rose, G. (2001). *How did this happen?* New York: Public Affairs.

Holubik, C., & Tomlinson, S. (2003). Bioterrorism preparedness and response: A resource guide for health care managers. In: J. Blair, M. Fottler & A. Zapanta (Eds), *Bioterrorism Preparedness, Attack and Response, Volume 4, Advances in Health Care Management.* London: JAI Press/Elsevier.

Hudson, R. (1999). The sociology and psychology of terrorism: Who becomes a terrorist and why? *Report Prepared by the Federal Research Division, Library of Congress* (September).

Hunter, E., Kelty, R., Kestnbaum, M., & Segal, D. (2003). Civil-military relations in an era of bioterrorism: Crime and war in the making of modern civil-military relations. In: J. Blair, M. Fottler & A. Zapanta (Eds), *Bioterrorism Preparedness, Attack and Response, Volume 4, Advances in Health Care Management.* London: JAI Press/Elsevier.

Huntington, S. (1996). *The clash of civilizations and the remaking of the world order.* New York: Simon & Schuster.

Inglesby, T. (1999). Anthrax: A possible case history. *Emerging Infectious Diseases, 5*(4), 556–560.

Inglesby, T., Grossman, R., & O'Toole, T. (2001). A plague on your city: Observations from TOPOFF. *Confronting Biological Weapons* (February 1).

Inglesby, T. V., Henderson, D. A., & Barlett, J. G. (1999). Anthrax as a biological weapon: Medical and public health management. *Journal of the American Medical Association, 281*, 1735–1745.

Institute of Medicine National Research Council (2002). *Countering bioterrorism.* Washington, DC: National Academic Press.

Jackson, B. et al. (2002). *Protecting emergency responders: Lessons learned from terrorist attacks.* Santa Monica, CA: RAND.

Karasik, T. (2002). *Toxic warfare.* Santa Monica, CA: RAND.

Kean, T. H., & Hamilton, L. H. (2003). *First interim report of the national commission on terrorist attacks upon the United States.* Washington, DC: 9/11 Commission.

Krebs, V. E. (2002). Mapping networks of terrorist cells. *Connections, 24*(3), 43–52.

Kuong, J. (2002). *Protecting your enterprise from the new global threats of terrorism; cyber terrorism and infrastructure attacks – a plan of action for survival and business continuity in the new global threat environment.* Wellesley Hills, MA: Management Advisory Publications.

bin Laden, U. (1996). Declaration of war. In: B. Rubin & J. Rubin (Eds) (2002), *Anti-American Terrorism and the Middle East: A Documentary Reader* (pp. 137–142). Oxford: Oxford University Press.

Laqueur, W. (1999). *The new terrorism: Fanaticism and the arms of mass destruction.* New York: Oxford University Press.

Laqueur, W. (2001). *A history of terrorism.* New Brunswick, NJ: Transaction Publishers.

Laqueur, W. (2003). *No end to war: Terrorism in the 21st century.* New York: Continuum.

Leavitt, J. (2003). Public resistance or cooperation? A tale of smallpox in two cities. *Biosecurity and Bioterrorism: Biodefense Strategy, Practice, and Science, 1*(3).

Lugar, R. (1999). The threat of weapons of mass destruction: A U.S. response. *Nonproliferation Review* (Spring-Summer).

Macintyre, A. G., Christopher, G. W., Eitzen, E., Gum, R., Weir, S., DeAtley, C., Tonat, K., & Barbera, J. A. (2000). Weapons of mass destruction events with contaminated casualties: Effective planning for health care facilities. *Journal of the American Medical Association* (January 12), *283*(2), 242–249.

Macintyre, B., & Deatley, C. (2001). *Ambulances to nowhere: America's critical shortfall in medical preparedness for catastrophic terrorism.* BCSIA Discussion Paper 2001–15, ESDP Discussion Paper ESDP 2001–07, John F. Kennedy School of Government, Harvard University.

Mack, A. J. R. (1975). Why big nations lose small wars: The politics of asymmetric conflict. *World Politics, 27*(2), 175–200.

Malvey, D., Fottler, M., Buck, G., Jr., & Fry, R. (2003). Responding to bioterrorism: A lesson in humility for management scholars. In: J. Blair, M. Fottler & A. Zapanta (Eds), *Bioterrorism Preparedness, Attack and Response, Volume 4, Advances in Health Care Management.* London: JAI Press/Elsevier.

McDaniel, R. (2003). Chaos and complexity in a bioterrorism future. In: J. Blair, M. Fottler & A. Zapanta (Eds), *Bioterrorism Preparedness, Attack and Response, Volume 4, Advances in Health Care Management.* London: JAI Press/Elsevier.

Metz, S., & Johnson, D. V., II (2001). *Asymmetry and U.S. military strategy: Definition, background and strategic concepts.* Carlisle, PA: Strategic Studies Institute, U.S. Army War College.

Miller, B. (2002). Study urges focus on terrorism with high fatalities, cost. *Washington Post* (April 29), A03.

Miller, J., Engelberg, S., & Broad, W. (2001). *Germs: Biological weapons and America's secret war.* New York: Simon & Schuster.

Miller, J., & Stone, M. (2002). *The cell: Inside the 9/11 plot, and why the FBI and CIA failed to stop it.* New York: Hyperion.

Moore, D. M. (2003). Public little concerned about Patriot Act: Wants civil liberties respected, but feels Bush administration has not gone "too far" in restricting liberties. *September 9: Gallup News Service.* Washington, DC: Gallup Organization.

Moore, D. M. (2003). Worry about terrorism increases: Fifty-four percent of Americans expect new acts of terrorism in the United States in the next several weeks. *September 8: Gallup News Service.* Washington, DC: Gallup Organization.

Moore, R. (2003). *The hunt for Bin Laden: Task force dagger.* New York: Random House.

National Strategy For Homeland Security, Office Of Homeland Security (July, 2002).

National Strategy to Secure Cyberspace (February 2003).

O'Toole, Mair, & Inglesby (2002). Shining light on "dark winter". Center For Civilian Biodefense Strategies, Johns Hopkins University, Baltimore, Maryland.

Office of Homeland Security (2002). *State and local actions for homeland security.* Washington, DC: U.S. Government Printing Office.

Perl, R. (1997). Terrorism, the media, and the government: Perspectives, trends, and options for policymakers. *CRS Issue Brief* (October 22).

Peters, K. (2003). As bioterror threat grows, federal capacity to respond shrinks. *Government Executive Magazine* (July 8).

Pike, D. (1966). *Viet Cong: The organization and techniques of the national liberation front of South Vietnam.* Cambridge, MA: MIT Press.

Pillar, P. R. (2001). *Terrorism and U.S. foreign policy.* Washington, DC: Brookings Institution Press.

Posner, G. (2003). *Why America slept: The failure to prevent 9/11.* New York: Random House.

Preston, R. (2002). *The demon in the freezer.* New York: Random House.

Preston, R. (1997). *The cobra event.* New York: Ballantine Books.

Rappaport, D. C. (1988). Theories of terrorism: Instrumental and organizational approaches. In: *Inside Terrorist Organizations.* New York: Columbia University Press.

Richardson, R. (2003). *Computer Security Institute/FBI computer crime and security survey.* Washington, DC: U.S. Government Printing Office.

Ridge, T. (2003a). TOPOFF 2: National combating terrorism exercise. *U.S. Secretary of Home land Defense.*

Ridge, T. (2003b). *America two years later.* Speech given on September 11, 2003.

Salvucci, A., Jr. (2001). *Biological terrorism, responding to the threat: A personal safety manual.* Carpinteria, CA: Public Safety Medical.

Scheitzer, G. (2002). *A faceless enemy: The origins of modern terrorism.* Cambridge, MA: Perseus Publishing.

Segarra, A. (2002). Agroterrorism: Options in congress. *Report for Congress* (July 17).

Smith, D. (2002). *Business continuity management: Good practice guidelines.* Manual, Business Continuity Institute.

Staten, C. L. (1999). Asymmetric warfare, the evolution and devolution of terrorism; The coming challenge for emergency and national security forces. *Journal of Counterterrorism and Security International,* 5(4), 8–11.

Stern, J. (1999). *The ultimate terrorists.* Cambridge, MA: Harvard University Press.

Thompson, R. (1989). *Make for the hills: Memoirs of far Eastern wars.* London: Leo Cooper.

Tse-Tung, M. (1966). *Selected military writings of Mao Tse-Tung.* Peking, China: Foreign Languages Press.

United States Department of State (2003). *Patterns of global terrorism: 2002.* Washington, DC: Center for Strategic and International Studies.

United States General Accounting Office (2003). Hospital preparedness: Most urban hospitals have emergency plans but lack certain capacities for bioterrorism response. *Report to Congressional Committees* (August).

U.S. National Strategy for Combating Terrorism (2002). September 20.

Vatis, M. A. (2001). *Cyber attacks during the war on terrorism: A predictive analysis.* Hanover, NH: Institute for Security Technology Studies at Dartmouth College.

Wheelis, M. (2002). Biological warfare at the 1346 siege of Caffa. *Emerging Infectious Diseases,* 8(9), 971.

Williams, P. (2002). *Al Qaeda: Brotherhood of terror.* New York: Alpha.

Wolper, L., Gans, D., & Peterson, T. (2003). Bioterrorism visits the physician's office. In: J. Blair,
     M. Fottler & A. Zapanta (Eds), *Bioterrorism Preparedness, Attack and Response, Volume 4,
     Advances in Health Care Management*. London: JAI Press/Elsevier.
World Islamic Front (1998). Statement: Jihad against Jews and Crusaders (February 23). In: B. Rubin
     & J. Rubin (Eds) (2002), *Anti-American Terrorism and the Middle East: A Documentary
     Reader* (pp. 149–151). Oxford: Oxford University Press.
Yahoo News (2003). International exercise launched to simulate smallpox bioterror attack (Monday,
     September 8).
Zapanta, A., Wightman, R., Jr., & Eder, M. (2003). The role of the reserve forces in defending the
     homeland. In: J. Blair, M. Fottler & A. Zapanta (Eds), *Bioterrorism Preparedness, Attack and
     Response, Volume 4, Advances in Health Care Management*. London: JAI Press/Elsevier.

THE INTERNATIONAL THREAT OF BIOLOGICAL WEAPONS: LEGAL AND REGULATORY PERSPECTIVES

Paul S. Westney and James J. Hoffman

ABSTRACT

The following article looks at the biological weapons problem from different perspectives to evaluate the international threat of biological weapons from both a legal perspective and a regulatory perspective. Biological weapons fall into a category all their own with unique characteristics as weapons of mass destruction in which suggestions for new directions should be explored with respect to historical failures. Biological weapons regulation is currently predicated on a certain legal framework, and through that a presentation is shown by a synthesized approach to biological weapons control.

In spite of nuclear disarmament and a heightened sense of cooperation between the world's military superpowers, weapons of mass destruction (WMD) have persisted as a central security concern in the post-Cold War years. The root of the threat, however, has shifted considerably in the direction of unconventional weaponry, which includes both chemical and biological warfare agents. While there are in fact treaty regimes in place to check these threats, the effectiveness of international regulation, based on a brief history of state noncompliance, cannot be guaranteed. Additionally, there is compelling evidence that suggests that violent, non-state-aligned actors have sought unconventional weapons to

Bioterrorism, Preparedness, Attack and Response
Advances in Health Care Management, Volume 4, 25–49
Copyright © 2004 by Elsevier Ltd.
ISSN: 1474-8231/doi:10.1016/S1474-8231(04)04002-9

significantly increase their destructive potential, and therefore their political leverage.

The most contemporary example of such a group is the Al-Qaeda organization, which is headed by the Saudi Arabian-born billionaire Usama bin Laden. This group has, in recent months, been rooted out of its operative bases in Afghanistan, and the subsequent investigation of Al-Qaeda's military documents yielded conclusive evidence that this radical, and extremely dangerous, Islamist group had been pursuing biological and chemical weapons capabilities. There were, in fact, vials of chemical agents (which were reportedly obtained from China) and extensive plans, as well as a minimal, although fully functional, lab in which these agents could be cultivated. Although it is impossible to determine whether this group has in fact deployed chemical or biological agents in combat or terrorist scenarios, this proof is telling of a threat that has become residual among extremist groups (Rhode, 2001). The possibility of weapons of mass destruction falling into the hands of non-state actors is a frightening prospect that merits close attention. The global population would undoubtedly be deeply, perhaps irreparably, scarred by an effective, large-scale unconventional weapons attack, and under no circumstances should the international community leave the opportunity for such an attack to occur.

The purpose of this article is to evaluate the biological weapons problem from several perspectives. In order to fully evaluate the international threat of biological weapons and evaluate this threat from both a legal perspective and a regulatory perspective, this article first outlines the particularities of the biological weapons problem and highlights these weapons' unique characteristics as weapons of mass destruction. Next, an analysis of the Biological and Toxin Weapons Convention (formerly the BWC) is given along with suggestions for new directions it should explore with respect to its historical failures. The newly implemented advances in the Chemical Weapons Conference will be employed as a yardstick for prospective changes in the BTWC. Biological weapons regulation is currently, and should continue to be, predicated on this legal framework, and thus this section of the analysis will establish a foundation for rest of this article. The article concludes by presenting a synthesized approach to biological weapons control.

THE THREAT OF BIOLOGICAL WEAPONS OF MASS DESTRUCTION: AN OVERVIEW

There is little military history and hard evidence to prove that biological weapons might be as potentially destructive as advanced conventional weapons of mass destruction. It may be premature to equate them with nuclear weapons or certain forms of chemical weapons. However, the particularities of biological weapons in terms of practicality, efficiency, and potential virulence have encouraged both states and non-state-aligned actors to pursue these weapons as weapons of choice. Thus, biological weapons must be considered weapons of mass destruction for reasons unlike those concerning nuclear, chemical, and other advanced conventional weapons of mass destruction. The dimensions of biological weapons are outlined below as a means to trace the difficulties they present in both scientific and political terms.

Biological Weapons: A Unique Problem

A primary difference between biological weapons and conventional weapons of mass destruction is that biological weapons, in terms of attainment and deployment, are potentially far more covert than other forms of WMD. These invisible agents, as proved by the anthrax attacks on the United States, are virtually impossible to trace without a claimant actor or in the absence of warlike circumstances. Deployment of biological agents does not require military facilities, and therefore easily falls outside the scope of conventional interstate war. It is also more likely that biological attacks will target civilians, rather than military installations, and therefore circumvent the traditional protocol of war in the state system. As such, a biological attack will not necessarily bring about interstate warfare. In this manner, biological weapons, if properly deployed, may significantly weaken states' existing military strength.

Predictably, covert action has become attractive as a means of deploying biological agents, as their effectiveness can essentially be compounded by resulting social unrest or paralysis in the public health sector. It is important to recall that biological agents are in fact living creatures that, when employed, will temporarily live within a target community. A biological attack may have the capacity to pervade even the basest elements of daily life, yet even a selective usage of these agents could have social implications similar to a widespread epidemic. This fact is perhaps most evident in the anthrax attacks against the United States, a scenario in which postal deliveries, a quotidian and politically inconsequential element of American life, became a potentially deadly apparatus of bioterrorism. The ambiguity surrounding this attack has created enormous difficulties for American military planners and the general public alike.

The production of biological weapons can also be covert. The little history there is to draw from regarding biological warfare proves that specialized, and easily identifiable, military facilities are not a necessary prerequisite to the production of these agents. In fact, pharmaceutical production plants are fully capable of making biological weapons, and accordingly have been suspected of doing so in certain instances (Scharf, 1999). Even in the case of large-scale, state-promoted biological weapons productions, which was seen most clearly in the Soviet arms production program, international regulatory bodies found it difficult to discern these facilities from other, accepted military installments (Alibek, 1999). And at the other extreme, militant groups and individuals have at their disposal sufficient means to create small, although functional, in-house labs, which are perhaps the most shockingly untraceable of all modes of producing biological weapons (Tucker, 2000). Such labs are capable of fueling terrorist action and will continue to challenge the security of the international community.

In general, weapons of mass destruction are rooted in pure science. Biological weapons are consistent with this trend. However, in contrast to other weapons of mass destruction, especially nuclear weapons, biological weapons require surprisingly little expertise. A dilettante's knowledge of microbiology, when combined with the massive amount of published material on the subject, will suffice to develop the hardier strains of biological weapons agents, such as anthrax and tularemia. Further, many biological agents with potential as weapons occur naturally and can be harvested in nature for subsequent cultivation in the laboratory. Michael T. Osterholm, a leading biological weapons expert, maintains that for every $1500 spent on nuclear warfare technology, one cent is required to produce an equally destructive biological weapons arsenal (Osterholm, 1999). Therefore the production of these agents cannot be easily monopolized by the state and, at the state level, cannot be concentrated in the hands of the military and economic elite. Biological weapons have the capacity to cause a relative breakdown in the state-centered global power hierarchy.

Biological weapons have become attractive to individuals, non-state-aligned groups, and less developed states for the very reasons that make these weapons easily produced without prohibitively advanced or expensive human resources and technology. The production process of these weapons requires a fraction of the expenses necessary even for an equally destructive chemical weapons production program. Therefore such programs do not necessarily rely on extensive military budgets or state backing. Although the weaponization of these agents, which is to say the insertion of biological agents into conventional warheads and related delivery systems, may significantly increase the scope of a biological attack. This process is by no means a necessary precondition for such action since biological weapons may in reality be far easier to employ anonymously. This is especially true when dealing with contagious agents, which essentially transform humans into mobile delivery systems. Weaponization has been preferred by states interested in biological weapons, but weaponized biological agents have thus far not been openly used in combat scenarios. Non-weaponized biological agents can be an effective way for states, groups, and individuals to sidestep the prohibitive features of the traditional military construct.

Defense by states against the trend towards devolution of power, however, is quite difficult. Civilian defense capabilities are predicated on a thorough understanding, and perhaps stockpiling, of organisms that have been suspected of being cultivated for biological attacks. Accordingly, a proper defense shield depends upon military and public health facilities that may be similar or identical to those used for offensive purposes. This ambiguity presents a particularly difficult scenario for the international community in that states' intentions must be made clear, although, as history as shown, are virtually impossible to quantify or verify.

There is a fundamental problem with biological weapons prevention at the state level. Research undertaken for public health purposes may in fact result in a biological arms race, given the proper circumstances (Falk, 1990). In short, it is difficult to discern the difference between offensive and defense research by looking at a state's facilities and scientific endeavors. But it may be a critical mistake for states to forego research on biological agents for the sake of international political harmony. Smallpox is a disturbing example of a biological weapon that has become exponentially stronger as a result of an absence of proper research and vaccine production. Although declared eradicated in the late seventies, smallpox exists in states' type culture collections and was weaponized most extensively by the former Soviet Union. In light of this development, the global vaccine supply for the smallpox virus is now insufficient to provide adequate civilian defense in the event of an outbreak. In fact, it is primarily for this reason that the Soviet Union found smallpox to be an attractive biological agent for weaponization. Although it is improbable that smallpox will be used for offensive purposes, states must be equipped for even the remotest scenarios involving biological warfare, and public health officials can make no assumptions concerning the status of a particular agent.

Biological weapons use, even in wartime scenarios, has been deemed repugnant to the human conscience, and as such is a military action that has over time become morally loaded. Military institutions in developed states have therefore distanced themselves from biological weapons and have typically preferred conventional weapons. Biological weapons implicitly rely on the manipulation of living organisms for military purposes, and in an abstract sense are a means of pitting life against life. But perhaps the most important consideration in this regard is the effect biological agents have on the health of target populations. The international community has renounced advanced conventional and nuclear weapons as well, but on slightly different grounds. In contrast to these weapons, whose footprint is left almost immediately, biological weapons extend their victims' deaths and accordingly magnify their suffering. Particularly virulent strains, such as the filovirus Marburg, induce unspeakable levels of pain and may take well over a month to kill infected citizens. The deployment of such strains is almost universally considered to be a violation against humanity, regardless of the political or military circumstances. Biological weapons, including those that are mild by comparison, have essentially become taboo.

There is also a crucial ambiguity concerning the long-term effects of releasing biological agents into the natural environment. Biological weapons are highly imprecise in this regard. This fact has been a critical factor in discouraging the use of biological weapons by states capable of producing effective conventional weaponry; given the proper weather conditions, a state employing biological weapons may infect its own troops instead of or in addition to the target group. And in the case of a contagious biological agent such as smallpox, the destructive effects, when coupled with inadequate containment facilities in the public health sector, spread indiscriminately. The destructive potential of such weapons, therefore, may be chilling, and more importantly, well beyond the control of the aggressor party. Biological weapons thus create the possibility of human actions resulting in an artificially induced plague. Such a scenario may indeed be a remote possibility, but in the case of biological weapons remote possibilities may result in global catastrophe. There can be little doubt that biological weapons are in fact weapons of mass destruction and should be treated accordingly by the international community.

The Political Implications of Biological Weapons of Mass Destruction

Biological weapons bear a crucial similarity to conventional weapons of mass destruction in that they have been sought and developed for political purposes. While it has been established that a considerable portion of biological weapons development and use will most likely be covert, there may also be a significant amount of overt biological weapons production by states in unfavorable political situations. Such states may rely on biological weapons as a staple of their military arsenal, and as is true with all weapons on the state level, have a political weight that cannot be discounted. After the declared destruction of biological arsenals in both the United States and the former Soviet Union, the political implications of biological weapons have been for the most part regionalized to areas with particularly sensitive political balances, such as the Middle East and Korea. Although biological weapons have nominally left the hands of the military superpowers, the effects of this shift are quite serious.

The defining features of biological weapons, as outlined above, have afforded less developed states the opportunity of increasing their political and military leverage vis-à-vis the nuclear superpowers. Global military strength during the Cold War era revolved around an arms race that caused military elitism among the most technologically advanced states. Military power was thus concentrated in the hands of few disproportionately prominent actors. As a result, states lacking sufficient human, financial, and technological resources became largely inconsequential. The development of biological weapons, however, has opened a window through which weaker and poorer states can assert their power on a military level. The limits of biological weapons arsenals prevent these states from being placed on equal footing with the developed superpowers, but developing such weapons is a comparatively simple and cheap mechanism with which to increase global political presence and influence. There is considerably less military exclusivity in the post-Cold War era, and it is fair to assume that the monopoly of military strength once held by few has eroded to a certain degree.

The relations between North Korea and the United States exemplify this trend. North Korea is a state that is virtually devoid of human and natural resources, has no trace of a developed market, and is economically negligible on a global scale. It stands as one of the last closed states and maintains little interaction with the international community. But there is a strong suspicion that the North Korean biological and chemical development program is one of the most diverse in existence (Moodie, 1999). North Korea has therefore commanded disproportionate attention from the American intelligence community and has been given acute attention in interstate relations. The obvious concern is to shield the American public from an enigmatic, and potentially aggressive, North Korean regime.

Biological weapons have also been influential in regional balance of power issues. While this is certainly true in the North Korean case, it is perhaps most saliently evident in the case of the Middle East, where there is a sharp imbalance between the military capabilities of Israel and its Arab neighbors. Israel possesses both nuclear and biochemical arsenals and, because the political relations of the region are highly contentious, has inspired Syria to proclaim itself as a military counterweight to Israel's dominance (Tucker, 2000). The preferred, and perhaps the only, pursuit for Syria, given its limited military resources, has been to develop a biochemical arsenal. In the Middle East, however, the problem of biological weapons extends well beyond the Israeli-Syrian conflict. Libya has sought biochemical weapons to disrupt the power imbalance on a broader, although still localized, scale, and Iraq has in large part compensated for inability to develop nuclear weapons by pursuing a similar arsenal. The relatively simple production process of biological weapons coupled with Middle Eastern states' military capabilities has made the pursuit of these weapons attractive throughout the region.

Simple balance of power logic indicates why these weapons have been so appealing in certain circumstances. As in the Syrian case, the presence of biological weapons is in large part a pursuit towards deterring Israel from using these weapons against Syrian civilians. Thus, Syria has equipped itself, to the best of its abilities, to stage an in-kind retaliation to a potential Israeli attack. In this regard its development of a biochemical arsenal can be considered to be a critical step in the defense of Syrian civilians. The Israeli-Syrian relationship illustrates how biological weapons have become deeply rooted in international politics. This is an issue that will almost inevitably endure, provided that the current mechanism for regulation of these agents has not effectively halted production and research. Effective regulation must incorporate avenues for resolving conflict at the political level, as well as eliminate the option of biological weapons production.

The International Regime for Biological Weapons Control: A Brief History

International concern surrounding biochemical weapons was crystallized in 1925 with the drafting of the Geneva Protocol and has continued to be a primary focus of the international community. The limitations of the protocol quickly became obvious, and the aim of the second major treaty concerning biological weapons, the Biological Weapons Convention, was drafted to improve upon the Geneva Protocol, which in practical terms was quite weak. International negotiations have continued well after the entry into force of the BWC in order to develop the standing treaty regime into a working preventive mechanism.

While the Geneva Protocol was indeed the first major international agreement that directly assessed the usage of biological weapons, biological agents were included only tangentially. The protocol was, in fact, geared primarily towards preventing the use of chemical weapons in wartime scenarios mainly as a response to the experience of World War I. In reality, the protocol has most likely reduced the visible history of chemical warfare to a minimum, and the same may in face be true of biological warfare. But as an agreement, its scope is far too narrow to adequately address the contemporary concerns surrounding biological warfare agents, which have developed considerably since the Protocol was drafted.

A primary limitation of the Protocol was that it restricted the use of biological and chemical agents only in wartime scenarios. The provisions of the Protocol were such that there was considerable room for interpretation, or perhaps misinterpretation, of the definition of war. Additionally, states that were either in violation of or not party to the Protocol fell outside of its scope. States were also afforded the independence to research and develop such weapons, and the Protocol quite significantly failed to established whether the impermissible agents were to be both lethal and nonlethal or for use against humans, animals, and plants. Additionally, scientific progress since the Protocol was drafted has further weakened its provisions. In 1925, certain potentially weaponized agents were not yet known, such as viruses, and the wording pertinent to biological weapons remained unclear (der Haar, 1991). It was not long before talks between the states party to the Protocol resumed in the late 1960s to update the parameters of the regulatory literature established in 1925.

The resulting Biological Weapons Convention was therefore established primarily for the purpose of building upon the preventive framework outlined by the Geneva Protocol. The language of the BWC is considerably stronger and more decisive concerning the use of biological weapons under any circumstances. It in fact prohibits their deployment without regard for the particularities of states' individual military scenarios, in-kind retaliation notwithstanding (der Haar, 1991). In this regard, the scope of the BWC is far more extensive than the Geneva Protocol.

The BWC also has major shortcomings. Perhaps the most significant limitation of the BWC is its failure to address the issue of independent research, production, and stockpiling of biological agents, which it claims are permissible for defensive purposes. It must be recalled that defensive production programs may in fact stoke the production of biological arms unilaterally. Such a provision might therefore increase the difficulty of eliminating the threat of a biological attack by both state and non-state actors. The BWC also failed to update the provision of the Geneva Protocol in terms of what may be considered an appropriate target of a biological attack. The issue of such attacks aimed at plants and animals has not been addressed.

The most politically relevant failure of both the Biological Weapons Convention and the Geneva Protocol, however, is the absence of an effective enforcement structure. Both agreements do not create avenues for international confidence-building measures, which leaves open the possibility of a biological arms race, and are almost fully dependent on voluntary compliance. Further, the degree of international compliance to the existing treaty provisions cannot be effectively investigated. Similarly, there are no codified regulations in place to deal with treaty violations, and appropriate punitive or retaliatory measures for a biological attack have not been established. The BWC and the Geneva Protocol are accordingly little more than international declarations against the use of biological weapons. It is therefore difficult to rely on such provisions as means to alleviate national security concerns.

There have been major efforts to overcome these deficiencies. Since 1972, there have been three Review Conferences geared towards reinforcing and refining the provisions of the BWC. These conferences have sought to: (1) ensure that the effectiveness of the treaty will not decrease as a result of scientific developments; (2) create a consultative committee consisting of party representative in order to gather information, including on-site visits, and effectively field complaints without referring to the United Nations Security Council; (3) establish a universally accepted interpretation of the wording of treaty provisions; (4) encourage the exchange of information among states party to the Convention; and (5) make the treaty a legally binding document, and, in a broad sense, strengthen the BWC. There is little doubt that these measures are necessary. However, the BWC may in fact be irreparably flawed. Without an empowered regulatory body capable of investigating the levels and locations of the production of biological weapons there can be few reassurances, and the BWC remains woefully dependent on states' intentions and international compliance.

The United Nations Special Committee (UNSCOM) has perhaps been the most promising prospect for sealing the gaps in the BWC's fundamental structure. UNSCOM's history, however, is unimpressive. There are two primary cases of unsuccessful on-site inspections that cast doubt upon the future success of UNSCOM investigations. The first of these, which was conducted in Soviet production facilities in the 1980s, yielded little in the way of results. Soviet scientists, who had been alerted to the inspections well beforehand, were able to effectively disguise their facilities in such a way that it was extremely difficult for UNSCOM to declare without reasonable doubt that the Soviet Union was developing biological agents for use as offensive weapons (Alibek, 1999). This instance is particularly shocking because the United Nations' weapons inspectors had been given extensive tours of Soviet facilities, and were as such at the very locus of production, and remained unable to determine the actual intent of the Soviet facilities. The Soviet Union continued to produce staggering amounts of biological agents at an advanced level and with remarkable efficiency. The Soviet government consistently denied the existence of this program, and the Russian government continued to do so after the fall of the Soviet Empire. It has only been through the detailed account of a defector from the Soviet Union, who was a prominent biological arms expert, that the West has learned of the extent and sophistication of its biological weapons development program (Alibek, 1999).

UNSCOM's inspections of Iraqi facilities were equally disappointing. While the international community has little doubt that Saddam Hussein's administration has been actively pursuing biological and chemical arms, on-site inspections were unable to collect enough evidence to make a convincing claim against the Iraqi government. There is a strong suspicion that Iraq, during the Gulf War of the early 1990s, positioned several weaponized biochemical agents, which had been inserted into the warheads of long-range missiles, along its western border (Tucker, 2000). Such suspicions must be carefully investigated. The presence of weaponized biological warfare agents in Iraq's arsenal could have extensive policy ramifications in the international community. International representatives party to the BWC and Geneva Protocol must learn from, and improve upon, the frustrating history of UNSCOM.

Even though there is a threat of terrorists using biological weapons, the available avenues for official international action concerning biological weapons control have been considered to be "working but not succeeding" (Haass, 1999). International regulatory bodies are therefore in need of thorough revision, or perhaps fundamental reconstruction. It is clear that biological weapons present a grave concern to the international community. The existing biological weapons control regime must focus its attention on empowering regulatory structures to act as is necessary to eliminate the problem of biological weapons. In order to more fully address this issue, the next part of this article examines the international legal structure and makes suggestions for new directions the BTWC should explore with respect to its historical failures.

NECESSARY REVISIONS IN THE INTERNATIONAL LEGAL STRUCTURE

The international legal regime surrounding the use and spread of biological weapons is constructed solely by treaty agreements. This construct was spawned in 1972, but has been subject to a considerable amount of revisions and renegotiations, and the process of refining the original draft of the Biological and Toxin Weapons Convention remains open. There is little doubt in the international community that the BTWC will not provide a comprehensive solution to the problems of proliferation, deterrence, and non-compliance regarding biological weapons. Rather than take the comprehensive approach, the BTWC has given rise to an Ad Hoc Group of experts to deal with biological weapons in a streamlined fashion. The understood function of this Group is to "increase transparency, enhance confidence in compliance and to help deter non-compliance," (Pearson & Dando, 1999) which may effectively curtial the proliferation of biological weapons. Under no circumstances will such a regulatory structure erase biological weapons from the list of modern security concerns. However, there remains a crucial role for an international legal structure intended to arrest the possibility of the use of such weapons.

The primary emphasis of the legal framework of biological weapons control must necessarily be placed on an independent verification regime. The standing treaty order, which includes a significant number of States parties and addresses the primary concerns of the biological weapons prevention community, is in practice unable to effectively execute the stated goals of the text. States parties therefore remain quite skeptical of the degree of biological weapons control in an international order with few, if any, concrete confidence-building measures. The aim of the contemporary revisions to the BTWC should accordingly be to translate the written goals into a codified and effective procedural framework, which in turn would serve to reduce the ambiguities currently surrounding major issues, such as stockpiling, research, and covert production. In the modern security era, it needs to be recognized that determined responses to non-compliance are necessary to underpin the other elements of the web of deterrence. There is little point in establishing strong rules if these can be flouted with impunity (Pearson, 1998).

Because this trend has in fact held true in recent history, the creation of concrete measures for responding to non-compliance is becoming progressively more relevant. A revised treaty regime slated to include an independent verification Protocol would without question bolster participation on all sides in the activities of the BTWC.

But the legal content of the BTWC has the potential to stretch even further. The BTWC quite noticeably lacks provisions for dealing with states who release biological agents on civilian populations and ground troops. In this regard, the standing BTWC addresses only one half of the problem of biological weapons. There is also room for improvement in terms of international norms against the use of biological weapons. The BTWC has the potential to implement programs to involve the scientific communities of the states parties and encourage further involvement in the Convention. Additionally, the administrative framework of the BTWC is such that is has the potential to enforce harmonization of domestic policy with the provision of the Convention. Clearly there are possibilities to improve the current treaty framework for biological weapons. Some of the more promiment opportunities and programs are outlined below, with an emphasis placed on the implementation of a verification regime.

The Parameters of a Verification Regime

BTWC officials have had the benefit of the entry into force of a similar verification regime – that of the Chemical Weapons Convention (CWC) – to draw upon for guidance in the incipient stages of expanding the scope of the BTWC. The verification regime of the CWC relies rather heavily on on-site inspections and declarations by States parties of sites of production as well as statements of intended purpose. Additionally, the CWC has established a list of relevant chemicals and scaled the inspections procedure according to potential virulence and industrial use. The CWC recognizes that a significant number of potentially harmful chemicals are in fact industrial staples and will inevitably occur in States parties' industrial systems. These chemicals receive little attention from the inspection regime. By contrast, certain chemicals, such as nerve agents (i.e. VX gas) and their predecessor chemicals dominate the focus of the regime. Together with the Organization for the Prevention of Chemical Weapons (OPCW), the CWC regime has constructed a three-tiered scaling system for organizing the attention of weapons inspectors to militarily relevant production sites.

Biological weapons, however, present a unique threat to weapons inspectors. While the collective regime of the CWC and the OPCW may provide an excellent model for a parallel regime for biological weapons, the formula for prevention and inspection is not quite as straightforward as that of chemical weapons. It is once again the unique characteristics of biological warfare agents that demand special attention in terms of inspection and regulation of production. Whereas a militarily significant chemical weapons production plant will almost inevitably have one of several indications of unacceptable intent, biological weapons do not necessarily follow this pattern.

Perhaps the most immediate concern in this regard is the discrepancy between size and quantity of the two agents. Unlike chemical weapons, a sophisticated biological weapons arsenal, depending on the agent, may require a base specimen "orders of magnitude smaller" than that of an identically destructive chemical weapons arsenal. This trend is most apparent when dealing with contagious biological agents, such as smallpox, which have the potential to trigger an epidemic with the release of only grams of the base agent (Pearson & Dando, 1999). This inverse relationship between virulence and difficulty of prevention indeed makes the job of biological weapons inspectors a difficult one. Further, a significant number of potentially weaponized or otherwise harmful biological agents occurs naturally. These organisms may be culled from natural sources and subsequently cultivated in the laboratory indefinitely. With regard to such organisms – anthrax is perhaps the most significant example – the biological weapons control regime may face significant difficulties in eliminating states' ability to procure the necessary base agents for creating a biological weapons arsenal. Similarly, the cultivation process of biological agents does not require specific base substances, as in the case of chemical weapons, which makes the procurement patterns of potentially aggressive states extremely difficult to track. Without the benefit of instrusive and thorough on-site inspections, therefore, the biological weapons prevention regime may have limited practical application.

It is also significant that biological weapons inspectors will in large part not be able to determine which biological agents are being produced in a given facility without conducting the necessary scientific research. This point takes on a particularly strong relevance in terms of the inspections themselves in that it may be difficult, if not impossible, for inspectors to discern the true function of a plant without testing each individual culture in stock for banned agents. It is indeed unfair to demand that weapons inspectors be able to determine each organism in every culture collection they encounter. Rather, these inspectors should be well-trained in spotting other, non-biological indicators that may prove a country or party's guilt, or at the very least call for further, scientific investigation. The case of the first UNSCOM visit to Soviet production plants as well as subsequent UNSCOM action in Iraq provide clear examples of investigations whose results came primarily from hardware and testing facilities and not from stockpiled biological agents. The inspection team charged with analyzing the Soviet biological plants was largely unable to elicit any genuine information from their Soviet guides. The inspectors were kept out of research laboratories that contained sensitive materials and given false answers to probing questions regarding the status of certain facilities. It became apparent, however, that the Soviets had been pursuing biological weapons for non-peaceful purposes after a room was discovered in which test explosions – the Soviet scientists had produced small bomblets to deliver biological agents – had left dents in the laboratory wall. UNSCOM therefore was able to attain conclusive evidence of military research in Soviet facilities not through finding their extensive stockpiles of biological agent, but rather through insight into military development procedures and investigating the Soviet facilities accordingly.

The Iraqi case produced similar results through similar procedures. Like the Soviets, the Iraqi scientists vehemently denied any wrongdoing when interviewed by UNSCOM. The relevant production plants had been well prepared for UNSCOM's visit and were predictably empty of any contentious chemical or biological products. Headed by Rolf Ekeus, the inspection team, through a rigorous series of interviews and on-site inspections, was able to find a significant number of discrepancies between the accounts of the Iraqi scientists. Some gave inconsistent answers in consecutive interviews. The team persisted through what was initially a frustrating and fruitless investigation and managed to distinguish the actual function of several fermenting vats that had been declared as agricultural products. These vats had in fact been used to produce shocking amounts of chemical weapons. The international community therefore had effectively attained enough information to appropriately reprimand the Iraqi administration.

Thus, the general guidelines for a biological weapons inspection regime are clear, although they must be followed under a widespread understanding of the implicit limitations of the investigative approach to biological weapons control. Even with the capacity for on-site inspection, the BTWC does not fully remedy its flawed basis of multilateral cooperation. This implies that there may still be room for covert production of biological weapons agents even with a reinforced treaty structure and an inspection regime. The following framework has been proposed by several biological weapons experts and should reduce the amount of covert production of biological agents to a minimum.

Inspection Procedures

The point of departure for the biological weapons inspection regime must necessarily be identical to that of the chemical weapons regime. The international scientific community is obligated to outline in detail which biological agents should be of paramount importance to weapons inspectors based both on potential virulence and applicability to prophylactic research. However, as established above, the likelihood of discovering particular agents during an on-site inspection is discouraging. The purpose of biological weapons inspectors in terms of fieldwork should therefore be limited in scope and geared towards both clarifying states parties' scientific intent in having the production facilities as well as verifying the declarations made in accordance with the BTWC.

The structure of such a regime will be complex and multilayered. It has been asserted that "it is vital that the regime is a three-pillar regime" composed of agencies in charge of fielding and analyzing initial declarations from the States parties, conducting routine on-site inspections, and conducting challenge inspections, as requested by states parties (Pearson & Dando, 1999). This structural setup allows for the regime to direct its resources to issues regarding biological weapons appropriately and without bureaucratic waste, as well as in a timely and efficient manner. Although there will inevitably be, to a certain degree, organizational and bureaucratic challenges to such a regime, its general function is such that serious infractions can be addressed independent of cumbersome international meetings. The specifications and requirements of each of the pillars of the proposed Protocol regime to the BTWC are outlined below, with particular attention being paid to similar provisions of the CWC.

Mandatory Declarations

The first pillar will be without question the most bureaucratic and least directly involved pillar of the three. The purpose of this arm of the Protocol regime will be to coordinate the declarations, as required by the revised BTCW, of states parties' sites relevant to the provisions of the Convention. Each member is required to submit a detailed list of the production patterns of each relevant site, which will be visible to each state party. It is important to note that the list will have its limitations. Indeed, "verification, confidence, and trust does not come about in the

CWC regime because all relevant items are monitored, and the same will be the case under the BTWC Protocol regime" (Pearson & Dando, 1999). In other words, the actual list of the CWC and the projected one for the BTWC operate within realistic confines. Their purpose is to guide the states parties' intelligence units and the Protocol inspection regime while affording these actors a significant degree of flexibility within which to draw conclusions. The desired effect of the process of declarations and the subsequent visibility among states parties is to increase transparency between states and thus build confidence in the efficacy of the BTWC Protocol regime. This process, if implemented fully and correctly, should reduce the perceived necessity of maintaining biological weapons stockpiles.

Unlike the CWC, the BTWC does not require States parties to destroy scientific equipment relevant to the Convention. The degree of dual-purpose usage of biological research equipment is simply too broad to realistically expect states parties to forgo their right to conduct peaceful and prophylactic research for the sake of the success of the revised Protocol regime. Instead, members are only asked to declare whatever relevant equipment they expect to maintain. It will be the function of the other two pillars of the Protocol regime to inspect and monitor the declared facilities to ensure compliance. In a similar vain, the BTWC Protocol must respect the interests of states parties' commercial sectors and national security interests. Thus, there is a proposed devolution of control of inspections and visits to the member states such that sensitive information will not become public under the guidelines of the BTWC. Pharmaceutical research employs facilities similar to those relevant to the production of biological weapons agents, but violating commercial interests would certainly jeopardize the long-term survival of the BTWC Protocol regime.

The first pillar will also be responsible for creating a schedule of inspections and coordinating investigations around an agreed-upon severity of each potentially weaponized biological agent. This process should parallel that of the CWC, which allots more visits to sites involved in the production of Schedule 1 chemicals and does not set a time limit for each inspection. (Schedule 1 chemicals are those determined to be the most dangerous. Schedule 2 chemicals are less dangerous, and Schedule 3 chemicals are common chemicals, which may be harmful, that are often found in industrial processes.) It is also significant that Schedule 1 sites are not given advance warning of CWC Protocol regime visits, although Schedules 2 and 3 sites are given 48 and 120 hours, respectively. This arm of the Protocol regime will therefore standardize inspection procedures and scale the attention of the regime such that the most threatening biological agents will be monitored most closely. Accordingly, its purpose is to ensure that the verification regime's actions are spent such that they bear the greatest humanitarian results for the international community.

Thus, the first pillar is expected to provide avenues for increased transparency and guidance for biological weapons inspectors with strong consideration paid to states parties' individual interests. This summary procedure, which will be implemented on an annual basis, is intended to frame an information-based foundation upon which the other two pillars will operate as well as further cooperation between states under the aegis of the BTWC. While the construction of the first pillar will not remedy some of the fundamental flaws of the original BWC, specifically its strong dependence on states' participation and declaration of intent, it is a critical point of departure for the implementation of an independent Protocol regime designed to enforce the BTWC to the greatest degree possible.

Routine On-Site Inspections

The second pillar of the Protocol regime should be designed to routinely inspect the sites listed in the annual mandatory declarations for validity and verification of non-aggressive research. The inspection teams should be composed of ad hoc groups of experts from the scientific communities of the states parties. According to the parameters of inspections, as set out by the first pillar of the Protocol regime, the second pillar is to help the process of putting the principles set out in the text of the BTWC into practice. The execution of the terms and conditions of the BTWC will rely most heavily on the periodic information-gathering visits by each team.

The scope of these visits, however, will by necessity be rather limited. Using the CWC as an example, it is clear that there will inevitably be close cooperation between industrial figures and the technical secretariat to constrain the degree to which the visits will be intrusive. Chemical-producing plants have been demarcated into areas in which contentious chemicals are produced and areas in which they are not, and the inspectors are allowed access "solely to those particular areas . . .; the facility agreements preclude access to other areas" (Pearson & Dando, 1999). The purpose of such limitations, which certainly seem ill-fitting to a regime whose sole purpose is to be intrusive as a means to verify the Chemical Weapons Convention, is to avoid potential commercial or national security information violations. The provisions of the Convention are in place to serve a particular, and rather narrow, function, and the international community must take care to prevent its practice from spilling over into other areas of interest for the states parties.

Routine inspections will be enforceable through the legislative procedure of the Convention itself. States parties who deny weapons inspectors access to production sites are to be assumed to be in violation of the Protocol and treated accordingly.

In terms of the Convention, such states will not be granted votes during the periodic meetings and will be given no power in BTWC negotiations. Suspicions of serious violations should be deferred to the United Nations and its weapons inspectors.

Challenge Inspections

The provision of the CWC that grants states parties the privilege to call for challenge inspections on suspected aggressor states is perhaps the most vital in terms of the practical effectiveness of the Protocol regime. This dimension "represents the CWC regime at its most intrusive," as the range of sites available to inspectors performing challenge visits exceeds that of the routine inspections (Pearson & Dando, 1999). Further, the nature of challenge inspections offers states an opportunity to apply the provisions of the CWC to individual scenarios and is the most versatile application of the Protocol regime. The same will necessarily be true of the emerging BTCW Protocol regime. Because "the relevant amounts of biological weapons agents are much smaller and they are to be found primarily in the laboratory rather than the factory" (Pearson & Dando, 1999) the provision of challenge inspections will be even more crucial to the survival of the BTCW Protocol regime than it already is to the CWC regime. Challenge inspections are for several reasons more necessary for the elimination of biological weapons than they are for chemical weapons.

In order to successfully verify compliance with the BTWC, inspectors will have to focus on research rather than outright production in their on-site visits. This fact "requires the monitoring of all activities involving biological agents and the possession of any amount of such agents" (Pearson & Dando, 1999). The routine inspections and mandatory declarations cannot realistically provide such a broad base of information. Biological challenge inspections will therefore require a significant amount of time sorting through biological stocks, as well as subsequent labwork, to test the types and individual strains of biological agents[1] in accordance with relevant equipment and vaccination research. These visits will therefore be used sparingly and only in the event of notable and troubling absences of information. Challenge inspections will therefore carry out the duties which would be unreasonable on a broad scale in order to ensure full compliance with the Convention.

The effectiveness of the biological challenge inspections will rely quite critically on the degree to which these inspections are coordinated by and called for through the annual routine inspections. There is a triangular relationship between the functions of the three pillars of the proposed BTCW Protocol regime that involves states' intelligence communities and BTCW administrators equally in the process of eliminating the threat of biological weapons research and storage. The backbone of this relationship, the first pillar organized around information and intelligence, necessarily begets the annual procedure of routine, organized on-site visits according to the information gathered voluntarily from states parties. The resulting wave of information coming from these visits is to be visible by the states parties, who synthesize their intelligence communities with the verification regime by processing the given information and call for challenge inspections when deemed necessary. Further information gathered from these sessions is to be incorporated into the function of the first pillar, which in turn will widen the scope of the routine inspections and force states parties to refine their declared information if necessary. It is the hope of the international community that this mutually-reinforcing Protocol regime structure will, through increased involvement by the states parties, reduce the likelihood of covert biological weapons production through perceived necessity by degrees.

Other Applications of the BTWC

Extension of the Biological and Toxin Weapons Convention should go well beyond the establishment of a verification regime to solidify its stated goals as a regulatory legal framework. While the implementation of the verification regime remains the primary concern of BTCW planners and will dominate international discourse until finalized, there are additional measures that will strengthen the BTWC and encourage states parties to uphold its principles. These measures are not as critically important as the verification regime for the immediate future, but are equally vital to the long-term success of the BTWC.

Appropriate Retaliation

As is, the BTWC deals almost exclusively with biological weapons prevention and counterproliferation and leaves the hypothetical issue of a post-biological attack scenario unaddressed. Biological weapons become particularly difficult to address in terms of retaliation. Because they are by nature a morally loaded military tool, the logic of in-kind retaliation does not necessarily apply on a global scale.[2] Currently, the United States is in a unique position of formulating appropriate military responses to modern security threats and has not ruled out the possibility of nuclear retaliation. However, if American retaliation were deemed inappropriate by the international community, it may face serious consequences in both political and social terms.

It is therefore the necessary for the BTWC to establish exactly how state parties should effectively punish aggressor states, parties, or individuals after the use of biological weapons has been broached. Specific military options are well outside the scope of this analysis, and it will accordingly not attempt to reach a practical solution to the problem. It must be recognized, however, that there is no recognizable military order to the biological weapons control regime, which may become a serious issue in the event of a biological attack. The Biological and Toxin Weapons Convention must ensure that states parties do not exacerbate an already unsavory military scenario through overcompensation and potential use of equally chilling weapons of mass destruction as a preferred retaliatory tool.

Strengthening the Norm Against Biological Weapons

In addition to the creation of a verification regime, which will by nature increase the degree of confidence between the states parties, the BTWC could expand to other areas which would produce the same effect. This critical step once again involves information collection, but the nature of this activity in terms of strengthening the norm against biological weapons is such that it extends to the scientific communities of the states parties. It has been proposed that this contribution should come through investigations of unusual outbreaks in states parties' territories. While this appears to be fairly routine in character, its impact is actually quite significant. Creating a regime for scientific investigation within the scope of the international community for biological weapons prevention will in turn create a professional path for scientists and epidemiologists through which to use their skills positively and for the benefit of the international community. Such a regime will also reduce the likelihood that some gifted scientists will be recruited by potentially aggressive states or organizations for violent purposes. On-site scientific inspections should reinforce the gravity with which the international community considers biological weapons prevention as well as its willingness to take the necessary measures to appropriately address the problem. The purpose of the scientific community in the larger community of biological weapons prevention will thus be to prevent covert biological weapons development and testing in light of the revised Biological and Toxin Weapons Convention.

Creating a scientific regime within the broader community should begin with a comprehensive internship program. Such a program would be set in place to offer young scientists the opportunity to dedicate their scientific careers to the purpose of eliminating the spread of biological agents for illicit uses. Again, the internship program would also encourage unilateral involvement by the states parties, which would reinforce the provisions of the Convention. The ostensible goal of such a program would be to essentially professionalize the field of biological weapons prevention and thus develop a crop of dedicated scientists to the pursuit of eliminating biological weapons.

Domestic Legislation

BTWC administrators are obligated to force states parties to craft their domestic laws to fit the principles laid out in the Convention. The means of enforcing this provision will necessarily be identical to those that should be in place for routine and challenge inspections. The administrative arm of the Convention will therefore be responsible for ensuring that international laws regarding biological weapons extend to the domestic sphere.

Domestic implementation of international treaty agreements "is required for almost all treaties" and the necessity of extending such a provision to the Convention is rather straightforward (Isaacs, 1990). The logic is simply to prevent individuals within the purview of states parties' respective legal systems from conducting research and development that has been outlawed by the BTWC. This provision is particularly relevant in light of the extensive pharmaceutical and private research sector, which could potentially harbor terrorists or disgruntled individuals willing and able to release biological agents against civilian populations or hand off the knowledge or materials to terrorist organizations. The United States, in fact, has been slow in implementing domestic legislation against illicit biological research, and even in recent years a private research assistant working for a biotechnology company or an independent scientist engaged in experiments in new biological weapons technologies would not necessarily face sanctions under U.S. law (Isaacs, 1990).

Within the United States, the noticeable lack of domestic policy can be attributed in large part to internecine bureaucratic problems. Implementation legislation has been in the works in Congress since shortly after the drafting of the 1972 Biological Weapons Convention,[3] but has lagged behind its European counerparts in terms of enacting policy concordant with the revised BTWC. Such action is unquestionably a vital support system to the Convention.

Enforced domestic action will also strengthen confidence between states parties and help further the norm against biological weapons in all scenarios. At the international level, the effectiveness of the Biological and Toxin Weapons Convention is contingent on the strict translation of international principles to the domestic sphere insofar as international law is unable to monitor clandestine action within states' borders. Like the process of inspection, effective implementation of the

Protocol regime will rely heavily on close coordination with national extensions of the BTWC administration.

From the preceding analysis it is clear that there is in fact a major role for international law in the field of biological weapons control. International law must necessarily be the backbone for an effectively biological weapons prevention regime and must therefore be comprehensive in scope and effective in practice. The standing legal framework has several fundamental flaws, the correction of which must be the primary focus of BTWC administrators in the coming years. There are, however, inherent limitations in the abilities of any final legal product resulting from the international negotiations that have been taking place periodically since the drafting of the Biological Weapons Convention in 1972.

The treaty should focus its attention to areas in which its provisions can realistically be implemented. Generally speaking, this implies that the treaty's primary target should be the international state system and state-funded militaries and research facilities. The proposed verification regime will involve a high degree of state participation and will therefore operate most significantly at the state level. While there will necessarily be, to a certain degree, devolution of control to the states parties in terms of domestic law and enforcement, the international forum must limit itself so as to remain effective as a standing legal document.

Thus, the question of international law regarding biological weapons control begets equally problematic questions concerning the issue of complete prevention at all levels. This is at present one of the most pressing issues to the international community as well as one of the most difficult. However, there is room for optimism. It is the authors' belief that comprehensive biological weapons control measures can be successfully implemented, given the proper treatment by international planners. Several of these control measures are briefly identified below and then the article concludes by presenting a synthesized approach to biological weapons control.

CONTROL MEASURES AND A SYNTHESIZED APPROACH TO BIOLOGICAL WEAPONS CONTROL

There are several measures that can be utilized to help solve the problem of biological weapons control. These measures include: active protection measures such as intelligence sharing along with preemptive and preventive strikes; passive protective measures such as vaccines, detection along with training and preparation; and international pathogen control through collection and regulation. There is little debate surrounding the necessity of such measures, and it seems apparent that the future of state security will be in large part contingent on how well states adapt to the shifting threat of weapons of mass destruction. An international approach to these problems, however, will require significantly more coordination at the interstate level and will be considerably more complex than the national approach.

To date, it remains unanswered how the international community will most effectively implement an integrated approach geared towards biological weapons prevention. The proposed advances are multisectoral, transnational, and involve both public and private actors. Currently, independent actors involved in biological weapons prevention do so voluntarily. It is clear that such a diffuse framework for prevention is not a sufficient adaptation to the problems presented by biological weapons. There is therefore an acute need for fundamental change in the international preventive structure.

Such change is a distinct possibility in the contemporary political atmosphere for several reasons. After the breakdown of the Cold War political order, and especially after the terrorist attacks of September 11, 2001, the international community has placed a strong emphasis on adapting to new security threats. This fact has held true primarily because state security threats have become less clearly-defined, and there can be no assurances as to when or from where security threats will arise. The changing political order has also eased tensions between major states capable of pioneering new international legislation for sweeping issues, such as the biological weapons problem, and the prospects for cooperation are indeed promising.

While tensions between major states have declined to a significant degree, the threat these states are facing from aggressive non-state actors has sharply increased. International crime has essentially supplanted the political tension that arose from the bipolar political order that existed in recent history. The international community must take definitive measures to encourage cooperation against non-state actors. Additionally, it must recognize that such action must happen soon since delaying the implementation of biological weapons defense measures only extends potential aggressors' windows of opportunity and places the global population at risk of a biological attack. The current political environment is indeed favorable for such broad measures. States should therefore exploit this political climate before any unforeseen changes occur.

Conclusions

At first glance, this analysis of biological weapons control may offer little more than grim prospects for the future of state security and global health. The development and cultivation of biological weapons agents over the past few decades represent the human spirit and the capabilities of science at their absolute worst. Recombinant DNA technologies and other advanced modern scientific pursuits, such as genome-mapping, present opportunities for biological weapons to become uncontrollable weapons of mass destruction capable of sidestepping all current biological defense measures to inflict staggering levels of terror on the global population. We, as humans, are indeed lucky that there has thus far not been an efficient, large-scale biological weapons attack and can only hope that the hypothetical situation of such an attack will never become a reality.

However, there is cause for optimism in spite of the bleak truths surrounding biological weapons. While biological weapons are perhaps the most difficult weapons of mass destruction to approach, and are therefore a significant threat, their presence is also a challenge that must be met by the international community. It is the belief of this analysis that the multinational actors involved in biological weapons prevention will elevate themselves to this challenge and come to the defense of the global population as a whole. The positive prospects for biological weapons control do exist, and these prospects must be explored and expanded upon to maintain state and public security in the modern era.

Biological weapons may, in certain circumstances, be the ultimate killers, but they will always be at odds with the strength of the human conscience. It is a testament to humans' resilience that weapons such as smallpox have remained in the laboratory and have yet to be unleashed on defenseless populations as a means of progressing a political agenda or personal cause. Although concrete measures for controlling these weapons internationally are unquestionably important, there is also an embedded prevention mechanism in the human spirit that may be by degrees more effective than international regulation efforts. In fact, the history of repugnance by and attempted prevention of biological and chemical weapons is far more extensive than the actual use of these weapons.

The threat presented by biological weapons is perhaps the most serious security threat in terms of weapons of mass destruction, but the probability of their use for the purpose of causing mass casualties, in combat situations or otherwise, is rather slim. Regardless, the pursuit of biological weapons regulation is a serious one that should dominate the international community's collective security agenda in the near future. A continued pursuit in this field, coupled with the intrinsic preventive measures of biological weapons, should reduce the likelihood of a biological attack even further. Barring unforeseen scientific advances, this effort will be of paramount importance to the general health of both current and future generations.

NOTES

1. It is important to note that certain strains of particular biological agents may be engineered to be more or less virulent to human beings or to be strictly for the purpose of

vaccination. Thorough lab research may therefore be a true indicator of a state's intentions for keeping biological agents on hand.

2. In spite of the moral dimension of biological weapons, in-kind retaliation has been mentioned in certain regional scenarios, particularly that of the Middle East. Almost all Middle Eastern states, with some exceptions, have developed significant biological weapons arsenals and have voiced their willingness to use them in the event of a biological attack.

3. Implementation legislation was in fact drafted by the U.S. Congress as soon as 1973, but was never acted upon.

REFERENCES

Alibek, K. (1999). *Biohazard*. New York: Random House.

Falk, R. (1990). Inhibiting reliance on biological weaponry. In: S. Wright (Ed.), *Preventing a Biological Arms Race* (p. 242).

Haass, R. (1999). Strategies for enhanced deterrence. The chemical weapons threat. In: S. Drell, A. D. Sofaer & G. D. Wilson (Eds), *The New Terror: Facing the Threat of Biological and Chemical Weapons* (p. 404). Stanford: Hoover Institution Press.

Isaacs, J. (1990). Legislative needs. In: S. Wright (Ed.), *Preventing a Biological Arms Race* (p. 291). Cambridge: MIT Press.

Moodie, M. (1999). The chemical weapons threat. In: S. Drell, A. D. Sofaer & G. D. Wilson (Eds), *The New Terror: Facing the Threat of Biological and Chemical Weapons* (p. 28). Stanford: Hoover Institution Press.

Osterholm, M. T. (1999). Living terrors: What America needs to know to survive the coming bioterrorist catastrophe. In: S. Drell, A. D. Sofaer & G. D. Wilson (Eds), *The New Terror: Facing the Threat of Biological and Chemical Weapons*. Stanford: Hoover Institution Press.

Pearson, G. (1998). Countering biological warfare: An overview. In: A. Kelle, M. R. Dando & K. Nixdorff (Eds), *The Role of Biotechnology in Countering BTW Agents* (p. 26). Dordrecht: Kluwer Academic.

Pearson, G., & Dando, M. R. (1999). The emerging protocol: An integrated, reliable, and effective regime. *In Strengthening the Biological Weapons Convention*, Briefing Paper Number 25. Bradford: Department of Peace Studies.

Rhode, D. (2001). At a terrorist training camp, arms, manuals, and a noose. *The New York Times* (December 5).

Scharf, M. (1999). Enforcement through sanctions, force, and criminalization. In: S. Drell et al. (Eds), *The New Terror: Facing the Threat of Biological and Chemical Weapons*. Stanford (p. 458). Hoover Institution Press.

Tucker, J. (2000). Toxic terror: Assessing the use of chemical and biological weapons. Cambridge: MIT Press.

COCKTAILS, DECEPTIONS, AND FORCE MULTIPLIERS IN BIOTERRORISM

John D. Blair and K. Wade Vlosich

ABSTRACT

Terrorists' threats pose a grave danger to the health care environment in which we live. In the following paper, we look at how bioterrorist plots can effect a given population and show ways to dissect terrorist actions. We look at variables that use various cause and effect relationships, and lead the reader down a path of being able to use information presented in a real life or fictitious bioterrorist attacks. We seek to inform the reader reasons why preparedness is essential in dealing with the likelihood of the following scenarios.

At 9:15 p.m. eastern standard time on January 29, 2002, in the United States Capitol building in Washington, D.C., the President of the United States, George W. Bush, Jr., during his State of the Union speech, refers to North Korea as being in an axis of evil, drawing a parallel to the axis of evil during World War II. What the president does not realize is those few words are a catalyst for the following scenario. In the summer of 2003, fearing possible invasion as in Iraq, North Korea now seeks out other means to ensure the survival of their country.

After the fall of the Soviet Union many top Russian scientists, around 10,500, became a proliferation risk and took their knowledge of weaponized smallpox to various countries including North Korea. The famous speech by Bush places

Bioterrorism, Preparedness, Attack and Response
Advances in Health Care Management, Volume 4, 51–77
© 2004 Published by Elsevier Ltd.
ISSN: 1474-8231/doi:10.1016/S1474-8231(04)04003-0

North Korea in a bind; like any country, they only want to survive the passage of time. In order to prevent an attack on their homeland, North Korea must now use unconventional warfare against the United States. The leaders of this Asian country realize that, in order to prevent their own downfall, they must attack the United States homelands.

Prior to this event, the head of Special Operations for North Korea had been in contact with members of Al Qaeda. Al Qaeda wanted to form a strategic alliance with North Korea in order to attack the "Great Satan" known as the United States. Al Qaeda sought the biological, chemical and nuclear proliferation of weapons from North Korea for use against the U.S.

After seeing the events unfold in Afghanistan, North Korea realizes aligning itself with Al Qaeda at this time would advance the United States' case against itself. Kim Il-chol, the Special Operations leader for North Korea, decides that in order to alleviate the pressure placed on his country he must use some form of attack to divert the United States away from Korea. After a meeting in January 2002 with Abdullah Ahmed Abdullah, a member of Al Qaeda, Kim decides that a Deception Operation and biological attack through unconventional warfare by way of force multipliers is the perfect way to fix the current situation.

Kim makes sure that Abdullah Ahmed Abdullah is his direct contact with Osama bin Laden, and no one else knows about their secret meetings. In an agreement with Al Qaeda, a plan by Kim Il-chol is set in motion containing three stages. The first stage is a Deception Operation by way of disinformation. The second stage is a biological attack on American citizens in Mexico. The third and final stage is a biological attack on Mexican citizens living on the U.S./Mexico border.

> By way of deception, thou shalt do war.
>
> Author Unknown

The first stage of the attack, the deception event, occurs in December of 2003 and involves Al Qaeda releasing a tape to the Arab news station Al Jazeera. The tape states the following: "For the actions taken against Iraq and its children, the hand of Saddam will strike the children of the United States with a deadly plague." The group lead by Abdullah Ahmed Abdullah poses as members of Saddam's regime in order to implicate Iraq in the second and third stages of the attack. This group has direct contact with no one other than Osama bin Laden. The deception course of action implicates Iraq in the biological attack on the United States. The desired perception is that the fallen Iraqi army used weapons of mass destruction. The Americans are vilified in the public eye for all the months of touting that Saddam could use weapons of mass destruction against the U.S. Kim Il-chol, being the studious observer of American culture, realizes that the people of the United States believe in the grandiose of Saddam Hussein being the wizard of Oz, rather than the reality that there is simply a man behind the curtain pulling all the strings.

North Korea knows that the United States will take into account that Saddam's regime released a tape stating that they would strike against the children of the United States. With the release of smallpox, the United States assume Saddam initiated the attack. The use of deception acts as a force multiplier by inflaming Americans to believe the wrong person was behind the attack. The real terror in the situation is that no one will be the wiser that North Korea and Al Qaeda made the attack. Al Qaeda does not care about putting more pressure on Iraq, since Osama does not like the secular leader of Iraq, Saddam Hussein.

North Korea deduces that the United States media will play a crucial role in swaying public opinion towards the attack occurring by Iraq. Many of the political commentators, whose opinion is used to sway public beliefs, will only taut the slogan, "I told you so." The media thrives on any source that it digs up; since by the time stages two and three are complete none of the agents are still in Mexico. Although, the media's podium reinforces President Bush's case for war and may allow him victory in the upcoming election of 2004, North Korea's plan will meet their own needs and those of Al Qaeda.

And now I saw a pale horse, and its rider's name was Death. And there followed after him another horse whose rider's name was Hell. They were given control of one fourth the earth, to kill with war and famine and disease . . .

Paul, *Revelation*

The second stage of the attack involves North Korean Special Forces infiltrating Mexico with weaponized smallpox, which Korea received precursors from Russia during the early 1990s. The Koreans use four cells consisting of two people located in each of the six popular spring break cities of Mexico. The spring break cities where the attack occurs are Cancun, Tijuana, Cozumel, Mazatlan, Acapulco, and Matamoras. The target areas of interest are clubs in which the travelers visit. The targets are United States students traveling to the Mexican cities during the spring break week occurring between the dates of March 13, 2004 to March 20, 2004. The target concentration provides a tactical advantage to the Koreans by allowing the Operation Forces to remain relatively small.

According to numbers provided by the United States Department of State, Korea knows that 100,000 students visit Cancun during this March spring break week. The terrorist groups in Cancun consist of four cells with two members placed in clubs, which frequented visits occur during spring break. The cells each contain one man and one woman. Each cell rotates between a different clubs every night, so as not to look suspicious by going to the same club on repetitive nights. The first group, Cihuacaoty, positions itself at Carlos n Charlie's. The second group,

Mictlantcuhtle, positions itself at Coco Bongo. The third group, Quetzalcoatl, situates itself at Dady'O. The final group, Mictlan, locates itself at La Boom. They each carry weaponized smallpox in mace-like containers into the club and release it upon the spring breakers. These clubs are excellent spreading grounds for the disease since the victims are in tight quarters and the atmosphere is hot. In addition, the virus spreads by the use of fans and fog machines within the club.

According to Spring Break Travel, a travel agency for college students within the U.S., North Korea estimates over 25,000 students visit Tijuana during this spring break weekend. The terrorist group in Tijuana consists of four cells of two placed at the spring break hot spots. The cells consist of a mixture of men and women. Each cell rotates to a different club every night, starting on Saturday March 13, 2004. The first group, Anguta, positions itself at Baby Rock. The second group, Raven, situates itself at Iguanas Ranas. The third group, Nesaru, positions itself at Tilley's Fifth Ave. The final group, Michabo, locates itself at Zoo'll Bar-Galeria. They each carry weaponized smallpox into the club and release it upon the spring breakers that frequented the clubs.

North Korea also calculates that over 20,000 students visit Cozumel during this spring break weekend. The group in Cozumel has four cells of two placed at clubs that are frequented during spring break. The cells consist of men only. Each cell rotates to a different club every night, starting on Saturday March 13, 2004. The first group, Loki, positions itself at Carlos n Charlie's. The second group, Ragnarok, situates itself at Neptuno Dance Club. The third group, Hel, positions itself at Fat Tuesday. The final group, Fenrir, locates itself at Hard Rock Café. They carry weaponized smallpox into the club and release it upon the spring breakers.

Over 40,000 students visit Mazatlan during this spring break weekend. The group in Mazatlan consists of four cells of two positioned at clubs that are frequented during this week of spring break. The cells in this town consist of women only. Each group rotates to a different club every night in order to go unnoticed. The first group, Seth, positions itself at Senor Frogs. The second group, Osiris, positions itself at Mango's. The third group, Thoth, locates itself at Bora Bora. The final group, Isis, locates itself at Valentino's. They carry weaponized smallpox into the club and release it upon the spring breakers that frequented the clubs.

North Korea deduces that 50,000 students visit Acapulco during this spring break weekend. The group in Acapulco has four cells of two placed at clubs that are frequented during spring break. The cells consist of men only. Each cell rotates to a different club every night, starting on Saturday March 13, 2004. The first group, Mithra, positions at Baby'O. The second group, Mihr, locates itself at Zucca. The third group, Apam Nepat, positions at Enigma. The final group, Zarathustra, locates at Hard Rock Café. They carry weaponized smallpox into the club and release it upon the spring breakers that frequented the clubs.

Finally, over 30,000 students visit Matamoras during this spring break weekend. The group in Matamoras has four cells of two placed at clubs that are frequented during spring break. The cells consist of men and women. Each cell rotates to a different club every night, starting on Saturday March 13, 2004. The first group, Araziel, positions at Crazy Lazy's. The second group, Liber, positions at Chaparral. The third group, Hermes, positions itself at Los Dos Republicos. The final group, Hades, locates at Utopia. They carry weaponized smallpox into the club and release it upon the spring breakers that frequented the clubs.

The scheme of maneuver, a description of how arrayed forces that accomplish the commander's intent, consists of spraying the weaponized smallpox on the spring break travelers at the previous clubs in order to unconventionally attack the U.S. Based upon Kim's plans the Special Operations forces travel virtually unnoticed with all of the other students. Smallpox is the weapon of choice, since it is easily aerosolized and carries a fatality rate is 30%. The incubation period is between 7 and 17 days, and it follows along the early symptoms of smallpox.

After receiving the smallpox, students travel back to the United States; with this, the terrorists of North Korea never have to enter the United States to attack the U.S. citizens. The students travel back on airplanes to their colleges. The strategic plan involves the students going to the return destinations such as Atlanta, Baltimore, Boston, Buffalo, Charlotte, Chicago, Cincinnati, Columbus, Cleveland, Dallas, Denver, Detroit, Hartford, Houston, Indianapolis, Kansas City, Los Angeles, Miami, Nashville, New Orleans, New York, Minneapolis, Orlando, Philadelphia, Phoenix, Pittsburgh, Portland, Raleigh, Richmond, San Francisco, Seattle, and Syracuse and spreading the disease. The targets, thus, travel to the major cities without the students ever knowing that they have the smallpox virus. The strategic advantage in this stage occurs in the fact that North Korea never has to enter into the U.S. and can leave the scene of the crime before anyone is the wiser.

The North Koreans realize that using the students to carry the smallpox virus will utilize the United States own infrastructure against itself. The infrastructure includes both the student citizens and American airplanes. The students become the means to disseminate the virus throughout the entire country by way of air, allowing the second stage of the onslaught to take full swing. The United States will not know until long after the students start arriving home that the symptoms are the smallpox virus.

North Korea realizes this cause's extensive damage to the U.S. economic system, especially the travel industry. With the knowledge that smallpox infected the nation's youth, the media outlets will relay the information that smallpox can live in the airplanes seat coverings. This causes a reduction in air travel, like that which occurred after 9/11. The major difference being that the airlines will stop flights to decontaminate all areas of the plane to ensure passengers that they will not receive the disease from traveling on their airlines. With the current economic situation of the airline industry, some will file for bankruptcy. This will also effect business activities and reduce other aspects of the U.S. economy.

This will cause greater effects on global travel, resulting in a downfall of the global economy. The majority of international airlines, to be on the safe side, will also have to decontaminate planes. They will also have to decontaminate the airports throughout the U.S. that connect to the rest of the world. Restriction of air travel will hinder the transportation of goods and massive delays will take place.

North Korea statistically determines that if smallpox infects only one-third of the students in each of the six spring break cities in Mexico, then 87,450 victims carry the disease. This denotes the possibility of spreading the disease throughout the planes that they were traveling back on, in the airports, other means of transportation, or any close contact with the infected. Non-spring breakers infected lead to numbers that are unable to estimate. When students' start arriving home, some show signs of symptoms, since they normally begin occurring 7–17 days after being exposed. Given that the death rate is 30%, then 26,235 spring breakers will die from the disease. Based upon statistics found from the Dark Winter Scenario played out by the U.S. government, each initial case of smallpox could infect an average of 13.3 people per one person infected. Kim realizes that these numbers occur on the small side since the Dark Winter scenario used a small, localized population, not a population that was traveling throughout the U.S. Because the spread is non-localized, the numbers would dramatically increase. The human casualties could increase since students might pass the disease to over 13.3 people based upon travel back to their respective universities and the close proximity of university life. They also pass the disease on through countless cities in which they are traveling through to return home.

The disease blankets the entire U.S. in less than 2–3 days. Kim knows through media outlets that if cases of smallpox are discovered in a major city, then federal, state, and local responders draw a ring around those cases, focusing first on people the victims had contact with over the previous two weeks. Those infected receive vaccination with the possibility of quarantine, thus causing voluntary mass inoculations nationwide. This causes a great burden on the health care system with a panic of people running to their health care providers in the United States.

Students on these planes easily transmit the virus since most are coughing and throwing up, and it goes unnoticed because most think they are just hung over from spring break. People infected with smallpox exhale small droplets that carry the virus to the nose or mouth of close contacts. The greatest risk comes from prolonged close contact exposure (within seven feet) to an infected person. The longer somebody is in close contact with an infected person, the greater the chance of transmission. Indirect contact is less likely to transmit the virus, but infection still can occur via fine-particle aerosols or inanimate objects carrying the virus. For example, contaminated clothing or bed linen could spread the virus. This means that the planes carrying the passengers could become a place spread the virus, along with layovers and hotels.

This stage has strong implications for the U.S. stockpile of vaccines for biological weapons. About 1,000 out of every 1,000,000 vaccinated people experience reactions to the smallpox vaccine that were serious, but not life threatening. Most of these reactions involved spread of vaccine virus elsewhere on the body. In the past, between 14 and 52 people out of 1,000,000 vaccinated for the first time experienced potentially life-threatening reactions. Infection rates would also rise considering the fact that American health care systems place the worried well in the same areas as the infected. The health care system is likely to see a collapse, since many workers will fear contracting the disease and not come to work.

In this stage, the North Koreans would have accomplished spreading the disease across the entire United States, while never even entering the country. The implications of this would be far reaching since the United States would then remember the message from stage one and believe that Saddam Hussein was behind the attack. This proves ample for the United States in their current situation; because they believe the attack was actually from Iraq thus proving they were right all along about the weapons of mass destruction stockpile. Although, they would be none the wiser that it was in fact not Iraq.

> We are the carriers oh health and disease – either the divine health of courage and nobility or the demonic diseases of hate and anxiety.
>
> Joshua Loth Liebman

The third stage involves a force multiplier by Special Operations units located into the towns of Juarez and Del Rio. They spray the Mexican populations visiting the markets with the weaponized smallpox. The population does not have to be widely infected in order to accomplish the final means. With this action, North Korea will cause a shut down of the U.S./Mexico border. The United States will fear anyone coming in from Mexico will be carrying the smallpox disease. The media, through the media platforms of protégées of Billy O'Reilly and Pat Buchanan who are for restricting access across the border, further damage the United States economically and aid in the shutdown of the border.

After the completion of all the attack components, the United States puts North Korea on the backburner in order to deal with smallpox epidemic. This is a basic market test by Korea to see if it will be able to sell weapons of mass destruction to other entities. North Korea will then be able to continue to make chemical, biological, and nuclear weapons without the United States interfering in their operations. They sell the weapons to Al Qaeda since much of the attention will focus on finding Saddam Hussein and the people that caused the epidemic. Thus, Al Qaeda ship weapons received from Korea to their strong hold in Sudan, where they will use nuclear weapons as bargaining tool against the United States. North Korea profits from the weapons sales and from the less imposing restrictions placed by the United States. The United States' knowledge of who to challenge, punish, or even attack will be diminished since the actual terrorist will be hiding in virtual bunkers.

INTRODUCTION

The previous hypothetical scenario shows how a smallpox attack can have devastating results for the United States, with dire implications to the health care systems. The scenario shows how terrorist can disseminate a biological agent into a target population with ease and with the aid of the target. Health care providers must plan for circumstances, such as these, which are likely to occur in the future.

The following article presents itself to show the reader a model by which terrorist may follow when attacking. It shows the stages that present what weapons terrorists use, what factors aid in causing terror, what the various mixes of attacks produce, and the ways that they can effect the target population.

Terrorist attacks against the United States on its own soil are a reality that has become all too clear. The likelihood of another attack like that of September 11th is inescapable, and preparations should be made by all heath care workers to prepare their environment for this inevitable future. Terrorism, according the Department of Defense Dictionary of Military Terms (2003), is the calculated use of unlawful violence or threat of violence to produce fear that intends to coerce or intimidate governments or societies in the pursuit of goals that are political, religious, or ideological in nature. In the health care arena, most systems would not be prepared from the onslaught of a bioterrorist attack, like that of the smallpox scenario shown previously.

Currently, scholars are making preparations that would counter these doomsday scenarios, but even such scenarios could cause the U.S economy to suffer as much as $1 trillion in losses (Miller, 2002). The exact loss estimates for health care are not calculable and would likely take up the majority of the costs. Although, out of all attacks, biological attacks cause the greatest increase for health care costs. For example, a "few kilograms of anthrax can kill as many people as a Hiroshima-size nuclear weapon" (Siegrist, 1999). By using force multipliers, terrorists can cause unprecedented amounts of damage and in some instances go undetected.

TERRORIST CHOICE OF WEAPONS

Terrorist are coming up with many innovative ways in which they can strike selected targets, ranging from the use of weapons of mass destruction to conventional weapons. Weapons of mass destruction are biological, chemical, radiological, or nuclear weapons that are capable of killing or injuring a large number of people. These weapons can often be found in small quantities, but have more extreme consequences than conventional weapons. The vast majority of these weapons found are in the United States and Russia. Conventional weapons are manufactured or improvised firearms, which include firearms, cyber assaults, and explosives (Table 1).

Chemical weapons are chemical agents modified by laboratories to produce injurious to deadly effects in targets. They include incendiaries, poison gases, herbicides and other types of chemical substances that can kill, harm or temporarily incapacitate the recipient. Chemical weapons release vapors, aerosols, gases, or liquids that assault a human's skin, blood, nerves, or lungs. According to the Mayo Clinic (2003), chemical weapons production is easy, but the use is not. The agent's effectiveness is limited in many ways; age, purity, dispersal, quantity, and release conditions. Another limiting factor to chemical weapons occurs with the danger they represent to the user. Some common chemical agents are Sarin gas, Mustard gas, VX, and hydrogen cyanide.

Chemical weapons pose a problem for the United States in the area of disposing chemical agents, and how to properly defend the civilian population from a terrorist chemical attack (Mauroni, 2000). Chemical agents pose a problem for health care providers for the simple fact that in the event an attack occurs the mortality rate is high, and the mass panic of the target population floods emergency responders.

Table 1. Potential Terrorist Weapons.

Weapons of mass destruction
Chemical, *e.g. Sarin Gas*
Biological, *e.g. Anthrax, Smallpox*
Radiological, *e.g. Dirty Bomb from Medical Waste*
Nuclear, *e.g. Suitcase Bomb*
Conventional weapons
Firearms, *e.g. 0.50 Caliber Sniper Rifle, M-16, AK-47*
Cyber assault, *e.g. Virus, Worm*
Military grade explosives, *e.g. C4*
Self constructed explosives, *e.g. Fertilizer Bomb*
Rocket launchers, *e.g. Rocket Propelled Grenades*
Shoulder-fired heat seeking anti-aircraft missiles, *e.g. US Stinger*

Currently, assessments into the preparedness of a chemical attack show that the responders would not be ready for a chemical attack (Karasik, 2002), so health care providers need to establish plans in the event that an attack would occur.

Biological weapons, also referred to as bioterrorism, use organisms to spread various agents in order to inflict harm or kill others. These are the most likely choice of weapons by terrorists since they are easy to conceal and disperse among a population. Biological weapons take the form of viruses, bacteria, or other microorganisms that assault a human in a variety of different ways (Salvucci, Jr., 2001). Most common forms of biological agents seek to incapacitate or kill the intended target. These agents spread by way of aerosol, through contact with objects, or from infected people. Common biological weapons are anthrax, bubonic plague, and smallpox. As seen in the scenario, the aerosolized smallpox was a biological agent that infected a mass population with little to no attention.

Biological weapons do not differentiate between any individuals. The use of such agents could spread through a target population rapidly, thus causing implications to the health care system. A response to such a terrorist attack will involve weeks to months of constant work in order to combat the spread of the biological agent. Plans to effectively survive such an attack include providing proper garments for protection (Jackson et al., 2002), setting up areas to separate the worried well from the infected, and keeping stockpiles of vaccines.

Radiological weapons spread radioactive material by contaminating individuals, equipment, buildings, and terrain (Laqueur, 1999). The radioactive agent acts in a toxic manner from which exposure leads to harmful or deadly consequences. These weapons are often difficult for a terrorist to acquire, but they have far-reaching effects like a nuclear bomb. "Dirty bombs" are a common expression used to represent these radiological agents (CDI, 2003). These weapons affect the body by means of contamination through direct contact or ingestion into the body. Radiological weapons spread either through a device set off near radiological material or as a weapon. Types of radiation used by terrorists are gamma, alpha, and beta radiation.

Nuclear weapons are a deadly weapon, which cause mass casualties and ravages the environment. The weapons are bombs that detonated by means of atom fission, commonly referred to as hydrogen or atom bombs. These weapons can diminish the human body to ashes on contact, and if not in direct contact can cause severe radiation damage. Scientists build these weapons, and many find their way to black markets for terrorists to purchase.

It has been established for many years that nuclear weapons are not only an immediate danger to the target population, but radiation from such an attack can produce ailments in the target population's offspring (Teller & Latter, 1958). This effect could cause future medical environments to change focus on existing a diseases and have to prepare for future generations of diseases that may not show up in offspring for many years to come. Health care providers are currently not prepared to treat massive patient influxes that have radiation damage or direct nuclear weapon contact.

Terrorist most often use *conventional weapons* as a way to attack targets, but in an uncommon way compared to traditional military forces. Firearms are a broad category of weapons that include handguns to rocket launchers. Most firearms and bombs assembled by terrorists improvise on standard models. Bomb or explosives fall into terrorist hands by way of military or commercial complexes. Often time terrorists will improvise and make bombs out of household fertilizer. A variation of a conventional weapon is a cyber assault. Cyber attacks occur in the manner of a terrorist using a hacker to gather information on an entity or inflicting damage through use of a virus or worm (Mitliaga, 2001).

Terrorist uses of rocket launchers cause damage to the target, and they are an easy weapon to smuggle into various countries. This proves potentially harmful to the United States with the porous borders to both the north and the south. Shoulder-fired missiles are another potential weapon that needs more measures to counter. Currently, the Homeland Security Office is evaluating measures to counter missile attacks against airlines by considering the installation of infrared countermeasures on aircraft and improving airport and regional security (Bolkcom & Elias, 2003).

TERRORIST USE OF FORCE MULTIPLIERS

Terrorist have many weapon choices to select from when inflicting harm on a given target population. The choices range from single use to a variety of additional weapons that produce a "cocktail." The cocktail mixture of weapons proves the most viable and allows the terrorist to create exponentially greater damage to the target. In order to seek optimal damage, the terrorist organization will use force multipliers for a given situation. According the Department of Defense (2003), "force multipliers are a capability that when added to and employed by combat force, significantly increases the combat potential of that force and thus enhances the probability of successful mission accomplishment."

A cocktail assortment of attacks has been used to describe many different aspects of military actions or plans. The most famous of the cocktail assortment derives from the Russian term Molotov Cocktails, which classically is a bottle filled with gasoline with a rag sticking out of the neck (Methvin, 1970). Cocktail assortments are common among terrorist groups because of ease of use, and these assortments are highly effective.

As a crow bar helps to leverage an object, so does a terrorist use a force multi-plier to enhance the capabilities over the target. The scenario above uses deception, use of innovative tactics, and use of target infrastructure as few examples to show force multiplication. There are varying amounts of force multipliers used by any terrorist unit to exaggerate the threat and impose heightened terror on given target. A few force multipliers used by terrorist organizations include; use of decep-tion, use of media, use of target infrastructure, use of location, use of replication, use of innovative tactics, follower commitment, and provoke social control errors (Table 2).

From the following table, the information relays different ways in which terror-ists drastically inflict harm upon the target population. Special Operations forces use the various tactics to enhance the strategic plan over an enemy to obtain an advantage. Through the strategies, the terrorist can see their goals extracted and placed in the laps of the target. Many of the force multipliers listed occur through-out countless ages to achieve an upper hand in any battle.

Use of Deception

The use of deception, as seen in the first stage of the smallpox scenario, is to cause misperception in the enemy by means of distortion, manipulation, or falsification of evidence to produce a desired reaction (Gerwehr & Glenn, 2000). False-flagging, as shown by North Korea, is that act of causing blame on one entity when they are not the deceiving force. The deception action for the scenario takes all the related events together to form the deception operation. This is a tactical advantage to any force since the intelligence capabilities of the enemy will spend time focusing on the perceived target rather than the deceiver. The level of sophistication in a deception operation is counteractive with higher levels of sophistication, which yield less effective measures (Gerwehr & Glenn, 2003). Thus meaning if North Korea made plans that are more intricate, then the operation is likely to fail.

The use of deception can take others forms to add force multipliers. Sensory saturation is another way in which information deludes intelligence sources. These operations occur by providing enemy intelligence with to much information for it to discern, causing false leads. A way to induce this deception is by way of psychological operations. PSYOP groups can release pamphlets or media adver-tisements to confuse opposing intelligence on which is the proper lead to take. One problem that these forces face occurs with the PSYOPS having to compete with highly specialized media outlets that may act against the goals of the psychological operation (DOD, 2000).

Table 2. Terrorist Use of Force Multipliers.

Use of deception
  False Flagging
  Disinformation
  Diversion
  Spoofing
  Sensory saturation
  Concealment

Use of media
  Gain attention
  Sway public attention
  Legitimacy
  Cause damage to enemy
  Distraction
  Create false terror

Provoking social control errors by target
  Excessive quarantine
  Closing borders
  Attack Wrong Country/Groups
  Forced Vaccination
  Excessive use of force
  Violations of civil rights

Follower commitment
  Willing violence against non-combatants
  Willing tactical suicide
  Persistence in the face of known retaliation

Use of innovative tactics
  Situation specific opportunities *e.g. Terrorist use of weather to disseminate biological agent*
  Exposed/contaminated people as the primary disseminators of biological weapons
  Airplanes as bombs

Use of location
  Spatial differences
  Urban/congested areas
  Rural areas
  Multiple cities
  Symbolism of location

Use of replication
  Multiple attacks *e.g. Terrorist attacks Orlando and Las Vegas*
  Multiple targets *e.g. Civilian, military, animals, or plants*

Use of target's infrastructure against target
  Transportation systems
  Energy
  Communication systems, including the internet
  Finance
  Government

Sensory saturation potentially causes a fatal blow to physicians and responders. With too much information, physicians might not be able to see the underlying ebb and flows of other diseases that might surface. For instance by focusing on primarily smallpox, another terrorist cell could release anthrax agents among the population. The infected go to primary care physicians and are misdiagnosed due to the unseen cases of smallpox. These diagnoses might mean the difference between treating lethal and non-lethal case correctly.

Use of Media

Terrorists often use the media to amplify the effect of terror on a target population. The media culture in a time of bioterrorism often make reports that occur in haste, rumor, or rushes to judgment (Clark, 2001). This can only enhance the terrorists' mechanisms. Terrorist, governments, and the media often have opposing views when covering terrorist events. Perspectives of the media sometimes result in "tactical and strategic gains to the terrorist operation and the overall terrorist cause" (Perl, 1997, p. 1). As in the case of North Korea, it wants to see nothing more than the U.S. economy dwindle by the media over exaggerating the smallpox ordeal.

In order to enhance fear, terrorists realize they must publicize an event in order convey their specific message. The media portraying the terrorist actions place the terrorist cause in front of the average person who rarely glimpses other ends of the world. The media occasionally gives legitimacy to the terrorist in the eyes of the viewers often skewing the actual reality of the situation. In some instances, terrorist use the media as a means of conveying a message from one cell to another. This also poses a problem for reporters who can not differentiate between actual or fake information from the terrorist cell (Beelman, 2002).

Media studies portray repeatedly that mass collectives can be swayed by outside sources to fall in line with a certain topic area (Mutz, 1998). Terrorists use perceptions to influence people on their various campaigns. The U.S. population bases much of what they believe to be true by way of media programming, so when the media portrays a terrorist plight in a certain way the terrorists may seem the righteous. Although the media may not want to show a terrorist organization, they will often times show violence to increase their viewer numbers based upon viewer preferences to these acts of violence (Crigler, 1996).

By adding the simple multiplier of media, the terrorists increase the weight that the health care system carries in a time of crisis by causing panic to the population. The media's force amplification of the smallpox epidemic sends the sick to physician offices, but also increases the number of worried well. This act is detrimental in creating terror, especially in the case of North Korea by having the sick and worried patients flood hospitals and physician practices, thus inundating the U.S. health care system. Health care providers must also be aware that with extreme circumstances the information received from media outlets may be false. Governments may actually use the media to disseminate a ruse that would help to neutralize the immediate threat posed by terrorists (Perl, 1997).

Provoke Social Control Errors

Terrorist like to see their enemies provoke social control measures. These measures usually have a negative effect on the target population. In the case of smallpox, many people may not go and receive vaccination for fear of the negative side effects. Thus, forced vaccination may be imposed. The health care field will have to deal with unwanted mass vaccinations, but this social control measure necessitates with a smallpox epidemic. Another problem that will need to be resolved with vaccinations occurs with how fast additional production of vaccines can reach a target population in the time needed (Henderson, 1998).

Excessive quarantine is another measure that terrorist wish to occur among the target population. No person wants mass quarantine, which cause loss of revenues to the household. Mass quarantine helps to amplify the use of media in targets because they will watch more television that will undoubtedly increase terror. People that resist mass quarantine may face excessive use of force to keep them indoors, which is another social control measure that brings into play a whole new set of variables involving civil liberties and law enforcement issues (Ahiquist & Burns, 2002). The use of smallpox from the scenario causes infected to be placed in a ring that through uncertainty may circle entire local populations.

The effects of the U.S. government closing down the borders would have a negative impact. A large percentage of the U.S. economy flows between the north and southern borders. In addition, much of the border region is an integrated economic area with a unique cultural identity (Spenner & Staudt, 1998). By provoking this social control error, the impacts to the economies and relations between nations would cause negative implications to occur throughout the entire time of the border closing.

Use of Target's Infrastructure Against Target

The use of target's own infrastructure is a common way in which a terrorist enhance terror. By taking out energy infrastructure, a special operation force cause widespread panic and chaos. This occurred even with out a terrorist strike, as shown in the northeastern blackouts of 2003. Energy is a vital resource to any hospital or building in order to serve the patients.

Terrorists also target transportation systems as means of hurting the target population economically (Kuong, 2002). If the enemy cannot move and sell goods, then it does not make money. These systems aid the terrorist in the transfer of bioterrorist agents, as seen in the scenario. If a shutdown of transportation systems occurs, then patients may never receive medical due to lack of transportation or vital vaccines may not be shipped.

Communication systems are an area that if disrupted cause mass panic and multiply the terror effect. In the age of information, people want to know what is going on, without this information panic begins to ensue. This damages health care systems for the simple fact that if proper information is not disseminated to the population the masses will exodus to the nearest hospital.

Use of Innovative Tactics

Innovative tactics often produce devastating effects to the enemy's population. Bioterrorism experts, when planning for preparedness, often over look the opponents' unseen tactical strategies. For instance, planes were used as bombs during September 11th, but no one strategically planned to counter such a situation. These innovative tactics caused shockwaves to run across the airline industry. They also make other types of attacks more feasible for other terrorist groups.

In the scenario, using the people as the actual weapons not only reduced a valuable target resource, but it spread the disease to the far off regions of the United States. Another way terrorist use innovative tactics occurs when the agents, like anthrax, spread throughout an enemy population by the wind or through a mail system. This actually occurred in 2001 when a terrorist laced mail with anthrax spores killing five people in the U.S. Innovative tactics pose a threat to the health care environment. By using people as biological weapons, the physicians and staff face mass casualties and a high number of the worried well.

Weather is often seen as a force multiplier that advance terrorist plans. The use of weather can influence the spreading of biological agents, like anthrax, by providing a system that easily disseminates. The future tactics of weather modification might fall into the hands of terrorists, which could prove potentially harmful. It is now projected that United States in the future will be able to disrupt enemies via small-scale tailoring of natural weather patterns to defeat or coerce an adversary (House et al., 1996).

Use of Location and Replication

The use of location and replication furthers the terrorist agenda by adding two force multipliers. The use of location in a congested area allows for easy spread of disease with agents such as smallpox or anthrax. The uses of replication in multiple cities on different target populations add to the terror felt. This will leave people with the thought that no place is safe from terrorist retribution. Multiple attacks can have a negative effect on the health care environment. Vaccines supplies become confused if too many populations need the vaccination at the same time. This causes the diseased patients to stay in populist areas, i.e. hospital waiting rooms, which aid in spreading the disease.

Follower Commitment

Commitment of the follower to the terrorist organization boosts the effectiveness of the force multiplier. Outside influences cause a different attitude towards violence to emerge among terrorist populations where they only wish to maximize aggression toward the perceived enemy (Hudson, 1999). Violence against non-combatants induces a greater sense of terror since it leaves the enemy population with the thought that nowhere or no one is safe. These kinds of violence occur in the scenario as seen by the terrorists attacking and using innocent civilians to carry the biological agent.

Suicide bombers also aid the force of terror. They show a population that they are willing to go to any lengths for their cause. Thus, arousing the feeling in the target population that this battle is not worth effort it, and possibly giving into the terrorist demands. Health care systems in these situations must prepare for casualties caused by conventional and non-conventional weapons, while realizing that these situations cause panic among the worried well.

TYPES OF TERRORIST ATTACKS

If a terrorist uses biological weapons with force multipliers, then the effects devastate the enemy. The use of force multipliers can range from low to high on a scale in which the terrorist wishes to cause harm to the enemy. When combined with the type of terrorist weapon used, varying degrees of attacks form into place. This attack matrix is one that references the strategy in which the terrorist will lay out a plot. The attack can either be solitary or a cocktail, a mixture of terrorist weapons to create a concoction that induces significant amount of terror within the target area (Fig. 1).

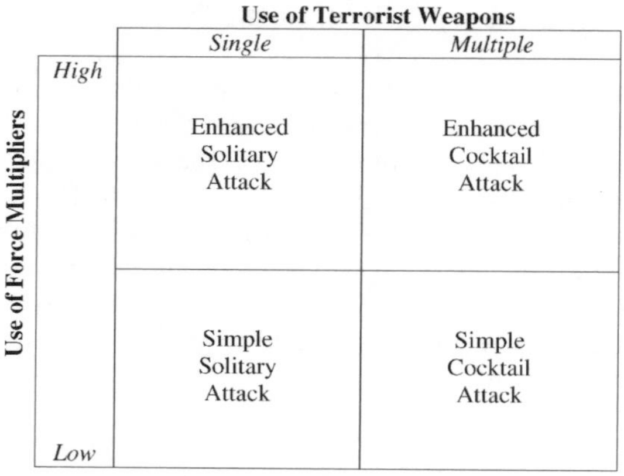

Use of Terrorist Weapons

	Single	*Multiple*
High	Enhanced Solitary Attack	Enhanced Cocktail Attack
Low	Simple Solitary Attack	Simple Cocktail Attack

(Row label at left, vertical: Use of Force Multipliers)

Fig. 1.   Types of Terrorist Attacks.

The *simple solitary attack* uses few force multipliers and a single weapon. An example of this occurs in a scenario written by Inglesby in 1999 involving anthrax. This scenario used anthrax spores and disseminated the biological agent over a football stadium. At the end, 4,000 citizens died with the first 10 days of the anthrax attack with others dying later from refusal to take antibiotics. The scenario shows that proper preparedness in the case of antibiotics and vaccinations for infected populations needs improvement. In addition, it portrays a situation in which location spreads biological agents more readily. Inglesby (1999) states that with modest preparation the outcome of the scenario could change, but experts need to collaborate to achieve positive futures.

The *enhanced solitary attack* occurs with the use of many force multipliers, but only a single weapon. The anthrax attack of 2001 is a prime example of how a single weapon with added force multipliers inflicts terror. The attacker used anthrax and disseminated it through the mail. By sending the anthrax to media personalities, the anthrax attack heightened terror among the United States population. Key infrastructure by way of the mail system was used to release the anthrax spores across the nation. The problem that the medical community must face in this type of attack is diagnosing the illness as the systems receive patients (Cordesman, 2001). Public health officials must prepare for possible local vaccinations if the situation arises, and learn how to diagnose accurately anthrax symptoms.

The *simple cocktail attack* occurs when multiple weapons are in use with a low amount of force multipliers. The effects of a cocktail attack often confuse the target population in ways of organizational preparedness. An example of this occurs with the Blair bioterrorism scenario (2003) in which terrorist cells combine an anthrax attack with a cyber assault. This action disrupted health care lines of communication and prevented proper diagnosis of the agent. Health officials must plan for alternate ways of retrieving information given the instance that multiple attacks may occur. They must also set up plans that involve the local, state, and federal government, if health care systems want to be accurately prepared for a biological attack (Bartlett, 1999).

An *enhanced cocktail attack* uses a high amount of force multipliers and multiple weapons. In the North Korean attack on American spring break students in Mexico and border town markets, terrorist use multiple amounts aerosolized smallpox in different locations. By the using replication, use of location, use of media, and use of innovative tactics, the effects of terror exponential rise. Combining all these factors create a problem for the health care environment by spreading the terror nationwide and causing panic. In a mass casualty situation, local and regional hospitals flood with the injured and the worried well.

TYPES OF CONSEQUENCES FROM TERRORIST ATTACK

By combining the choice of weapons, the potential force multipliers, and the types of attacks, effects of the terrorist plan can mete out actual casualties and the extent of terror. Based upon the types of attack a mixture of casualties, multiple to mass, and the extent of terror, local to international, can drastically cause a meltdown of the health care systems if proper preparedness implementation does not take place. The effects can be concentrated to widespread based upon the extent of terror produced in the target population. The effects based upon the number of casualties range from limited to mass. When discussing the extent of terror in this paper we are referring to only geographical location, not how deep the terror runs within an individuals mind. Table 3 illustrates the various factors that inflict terror in the minds of the target population.

Table 3. Psychological Factors that Enhance Terror.

Anxiety
Physical health
Social well being
Sleep
Proximity to terrorist act
The number of casualties
Population bias towards the terrorist attack
Personalization bias (exaggerate terror occurring to them based upon previous events)

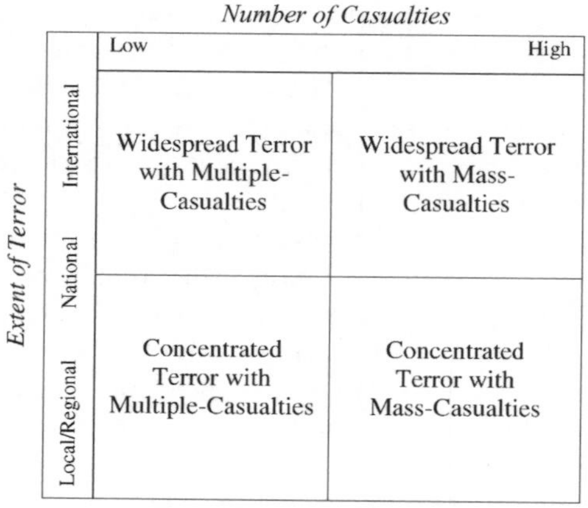

Number of Casualties

		Low	High
Extent of Terror	International	Widespread Terror with Multiple-Casualties	Widespread Terror with Mass-Casualties
	National		
	Local/Regional	Concentrated Terror with Multiple-Casualties	Concentrated Terror with Mass-Casualties

Fig. 2.   Types of Consequences from Terrorist Attacks.

Terror exists when a person has a fear of the unknown (Rachman, 1978). The fears that people contrive stem from not being able to control the environment around them. This loss of control leads to helplessness, avoidance, or phobias (Denny, 1991). The extent in which one perceives terror involves many different areas, such proximity to the terrorist event, frequency of terrorist acts, and the severity of terrorist acts. This gives rise to the areas in which terror can be felt (Fig. 2).

Concentrated Terror with Multiple Casualties

The use of concentrated terror refers to the extent of terror that is limited to a local or regional area. In a given outcome such as this, the casualties remain in the lower numbers when compared to a national epidemic. This type of consequence effect largely only local responses. In order to combat a bioterrorist attack public health agencies must set up networks with hospitals, law enforcement, and the local community. This strengthens the response mechanism for an occurrence of a bioterrorist attack (Mahmoud, 2002).

The Inglesby scenario in which anthrax was released is a type that involves concentrated terror with multiple casualties. The aerosolized anthrax released upon the target population in the football stadium caused a few number of casualties when compared to numbers on a mass scale, such as the U.S. population as a whole. The consequences of such an action dealt heavy blows to the local responders and possibly increased terror throughout the region.

Communication of local information is crucial for this type of consequence. Cities should set up clear channels of communication, and alternate channels if those relays become disrupted as in the Blair Scenario. Clear communication is essential in order to alleviate panic and restrict the worried well from flowing into public health agencies.

Another effective way to combat a concentrated event is to make sure that physicians know how to properly diagnosis biological agent symptoms. Most physicians would not accurately diagnosis a smallpox case if it presents itself. Leaders in the local heath care arena should set up symposiums in which health care providers remain updated on how to properly diagnosis a bioterrorist attack.

Widespread Terror with Multiple Casualties

The consequence of widespread terror with multiple casualties refers to a large geographical location feeling the impact of bioterrorist attack. This is type of attack occurred with the 2001 anthrax attack. If anything can be taken away from this attack, it is that the United States was not prepared for this event.

The widespread terror from this action occurred through the terrorist sending anthrax ridden messages to the media. The media programs were flooded with the incident causing people throughout the U.S. to feel that nowhere was safe. The use of innovative tactics was another force multiplier, by way of the terrorists using the mail system to disseminate anthrax.

Widespread terror occurs with the use of force multipliers even though the target population was concentrated. This will cause health care agencies to flood with the worried well who are not in the infected area. The United States will have to increase its laboratory capabilities in order to properly diagnosis the event of a bioterrorist attack. Most of the common problems that will occur on this level are disease detection and surveillance. Accurate diagnoses are important so that proper response can be divvied out to the population (Mahmoud, 2002).

Concentrated Terror with Mass-Casualties

The consequence of concentrated terror with mass-casualties refers to a local or regional geographic location receiving a bioterrorist attack. The mass-casualties occur when a biological agent infects a large population. U.S. intelligence is imperative to counter this type of attack. Intelligence agencies need to carefully analyze any likelihood of this kind of consequence.

The Blair Scenario (2003) uses a concentrated attack on one city by spraying anthrax over a target population for one aspect of the scenario. This dissemination of this causes mass casualties to occur among the target population. A problem that health care providers will face in this type of situation is mass panic to hospitals and other service areas. Providers must come up with clear response plans for such an event. They must also use alternate ways of control in order to separate the infected population from the worried well.

Widespread Terror with Mass-Casualties

Widespread terror means that the terror spreads to a national or international level with a high amount of mass-casualties. This consequence would have dire effects on the health care system by giving mass vaccinations, having to mass quarantine, and restricting trade with other countries. Not only do these consequences occur, but also they affect the entire national political structure.

This kind of terror occurred from the North Korea scenario. Terror disseminated throughout the U.S. to nearly every major university location. This only causes the terror to amplify throughout the world seeing that they may have come in contact with the infected. A complete response plan must be carefully thought out in order to not cause downfalls of various structures within the U.S. Responders must have a clear understanding of the what, where, and when social control measures should be implemented. Although the population may resist the control aspects, these measures will essential for keeping up a response system that not cause a collapse of the health care system.

ATTACK MAP

Figure 3 illustrates how the various components of a bioterrorist attack can be combined together to form a map of the likely consequences of a bioterrorist attack. The figure combines the number of weapons used, force multipliers, the types of terrorist attacks, and the consequences from the terrorist attacks. Through the process of creating the map, the authors understand that many numerous factors go into producing a terrorist attack. For this specific figure, we show only the components illustrated within the text.

SCENARIO SUMMARY TABLE

Table 4 is a comparative summary of the various scenarios used throughout the article. The table illustrates how the scenarios relate to one another and the basics

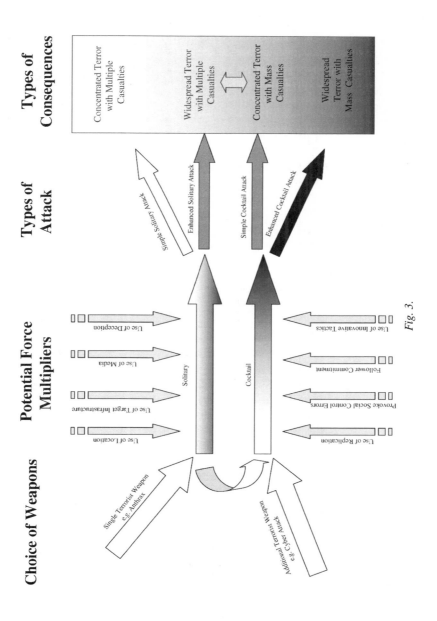

Fig. 3.

Table 4. Scenario Summary Table.

	Vlosich Scenario	Blair Scenario	2001 Anthrax Attack	Dark Winter Scenario	Inglesby Scenario	Topoff Scenario
Weapons used	Smallpox	Anthrax Cyber assault	Anthrax	Smallpox	Anthrax	Plague
Force multipliers used	Use of deception Use of media Use of innovative tactics Provoke social control errors	Use of media Use of replication	Use of media Use of replication	Use of innovative tactics Use of media Use of location	Use of innovative tactics	Use of innovative tactics Provoke social control errors Use of location
Type of terrorist attack	Enhanced cocktail attack	Simple cocktail attack	Enhanced solitary attack	Enhanced solitary attack	Simple solitary attack	Enhanced solitary attack
Types of consequences from terrorist attacks	Widespread terror with mass-casualties	Concentrated terror with mass-casualties	Widespread terror with multiple-casualties	Widespread terror with multiple-casualties	Concentrated terror with multiple-casualties	Widespread terror with multiple-casualties

of how they correspond with each other. It should be noted here that the enhanced solitary attack and the simple cocktail attack could have interchangeable consequences in the area of widespread terror with multiple casualties and concentrated terror with mass casualties.

CONCLUSIONS

In short, our analysis of potential bioterrorist attacks and their consequences, which are selective a few involving a variety of different categories, can help to illustrate how attacks might formulate. With a growing number of new attack weapons and weapon cocktails that can be used upon a target population, the degrees to which a bioterrorist attack analysis accurately predicts a consequence could drastically change. The implications of any bioterrorist attack could be staggering and produce huge casualty and economic losses, or they could be limited, like the 2001 anthrax attacks that produced minimal casualties with enhanced terror.

Health care providers must always prepare for a biological catastrophe in which the hypothetical scenarios, like the one written in this article, can bring about. Preparedness and response needs constant updates within the health care field in order to meet the ever-changing demands placed upon it. If proper preparation and response are met, then health care providers will be able to adequately meet the needs of their constituents.

Future research into bioterrorist attacks should include the new types of weapons, weapon cocktails, force multipliers, the preparedness of the United States, and how well response mechanisms are put into place. Preparedness and response cannot be effectively researched without a full awareness of weapons and force multiplication factors. This suggests the need for multi-disciplinary teams involving counter-terrorist experts as well as health care ones. This more complete research will help the U.S. as a whole, and health care organizations specifically face the certainty of our uncertain future.

REFERENCES

Ahiquist, G., & Burns, H. (2002). *Bioterrorism: Improving preparedness and response*. Miami: Booz, Allen & Hamilton.

Bartlett, J. (1999). Applying lessons learned from anthrax case history to other scenarios. *Emerging Infectious Diseases, 5*(4), 561–563.

Beelman, M. (2002). The dangers of disinformation in the war on terrorism. *International Consortium of Investigative Journalists* (February 2).

Bolkcom, C., & Elias, B. (2003). Homeland security: Protecting airliners from terrorist missiles. *Report for Congress* (February 12).

CDI (2003). *Pascal's new wager: The dirty bomb threat heightens* (February 4).

Clark, R. (2001). Bioterrorism and media understanding. *Poynter Online* (May 11). Available at: http://www.poynter.org/content/content_print.asp?id=3569&custom=.

Cordesman, A. (2001). Defending America: Asymmetric and terrorist attacks with biological weapons. *Center for Strategic and International Studies* (February 12).

Crigler, A. (Ed.) (1996). *The psychology of political communications.* Ann Arbor, MI: University of Michigan Press.

Denny, M. R. (Ed.) (1991). *Fear, avoidance, and phobias: A fundamental analysis.* Hillside, NJ: Lawrence Erlbaum.

Department of Defense (2000). The creation of dissemination of all forms of information in support of psychological operations (PSYOP) in time of military conflict. *Report of the Defense Science Board Task Force* (May).

Department of Defense (2003). Dictionary of military terms (June 5).

Gerwehr, S., & Glenn, R. (2000). *The art of darkness: Deception and urban operations.* Santa Monica, CA: RAND.

Gerwehr, S., & Glenn, R. (2003). *Unweaving the web: Deception and adaptation in future urban operations.* Santa Monica, CA: RAND.

Henderson, D. (1998). Bioterrorism as a public health threat. *Emerging Infectious Diseases, 4*(3), 488–492.

House, C. T. et al. (1996). Weather as a force multiplier: Owning the weather in 2025. Research Paper Presented to United States Air Force, Department of Defense in August.

Hudson, R. (1999). The sociology and psychology of terrorism: Who becomes a terrorist and why? *Report Prepared by the Federal Research Division, Library of Congress* (September).

Inglesby, T. (1999). Anthrax: A possible case history. *Emerging Infectious Diseases, 5*(4), 556–560.

Jackson, B. et al. (2002). *Protecting emergency responders: Lessons learned from terrorist attacks.* Santa Monica, CA: RAND.

Karasik, T. (2002). *Toxic warfare.* Santa Moncia, CA: RAND.

Kuong, J. (2002). *Protecting your enterprise from the new global threats of terrorism; cyber terrorism and infrastructure attacks A – plan of action for survival and business continuity in the new global threat environment.* Wellesley Hills, MA: Management Advisory Publications.

Laqueur, W. (1999). *The new terrorism: Fanaticism and the arms of mass destruction.* New York: Oxford University Press.

Mauroni, A. (2000). *America's struggle with chemical-biological warfare.* Westport, CT: Praeger Publishers.

Mayo Clinic (2003). Biological, chemical weapons: Arm yourself with information (February 6). Available at: http://www.mayoclinic.com/invoke.cfm?id=MH00027.

Methvin, E. (1970). *The riot makers; the technology of social demolition.* New Rochelle, NY: Arlington House.

Miller, B. (2002). Study urges focus on terrorism with high fatalities, cost. *Washington Post* (April 29), A03.

Mitliaga, V. (2001). Cyber-terrorism: A call for governmental action? 16th BILETA Annual Conference (April 9–10).

Mutz, D. (1998). *Impersonal influence.* New York: Cambridge University Press.

Perl, R. (1997). Terrorism, the media, and the government: Perspectives, trends, and options for policymakers. *CRS Issue Brief* (October 22).

Rachman, S. (1978). *Fear and courage.* San Francisco: W. H. Freeman and Company.
Salvucci, A., Jr. (2001). *Biological terrorism, responding to the threat: A personal safety manual.* Carpinteria, CA: Public Safety Medical.
Siegrist, D. (1999). The threat of biological attack: Why concern now? *Emerging Infectious Diseases,* 5(4), 508.
Spenner, D., & Staudt, K. (Eds) (1998). *The U.S.-Mexico border: Transcending divisions, contesting identities.* Boulder, CO: Lynne Rienner Publishers.
Teller, E., & Latter, A. (1958). *Our nuclear future: Facts, dangers, and opportunities.* New York: Criteron Books.

MODELING THE ENVIRONMENTAL JOLTS FROM TERRORIST ATTACKS: CONFIGURATIONS OF ASYMMETRIC WARFARE

John D. Blair, Robert K. Keel, Timothy W. Nix and K. Wade Vlosich

ABSTRACT

When modeling environmental jolts from terrorist attacks, various aspects should be analyzed in order to properly present an accurate configuration. The following article discusses how asymmetrical warfare has an impact on the outcomes of a terrorist attack. The several dimensions of terrorist attacks can be extracted to deduce the ways that asymmetrical warfare can damage the health care system. The article will relate real life terrorist attacks and hypothetical scenarios to better inform the reader about the weak attacking the strong, and then explain how this relates to health care providers.

Terrorist attacks are potentially significant environmental changes, which are likely to "jolt" the healthcare organizations, other organizational responders, the community, or regions attacked as well as broader national or international stakeholders. The consequence of these attacks can vary both in the number of casualties (dead, ill, or not yet ill) and the extent of "terror" created among key stakeholders in the area of the attack and elsewhere.

Bioterrorism, Preparedness, Attack and Response
Advances in Health Care Management, Volume 4, 79–116
© 2004 Published by Elsevier Ltd.
ISSN: 1474-8231/doi:10.1016/S1474-8231(04)04004-2

It is critical to recognize that a terrorist attack is not a tornado or an earthquake, nor is a bioterrorist attack a plane crash at the airport, in a country field, or even inside a city. As a society, we have been able to prepare to effectively respond to these events. Even though these events are dangerous, which will create a level of fear and potentially many casualties, they are a knowable phenomena and, to some extent, predictable, i.e. they are known environmental jolts that have been routinized in preparedness and response.

Terrorism, however, is a very different kind of phenomenon. Terrorism is officially defined as "any premeditated, unlawful act dangerous to human life or public welfare that is intended to intimidate or coerce civilian populations or governments. This description covers kidnappings, hijackings, shootings, and conventional bombings. It also includes attacks involving chemical, biological, radiological, or nuclear weapons. Terrorists can be United States citizens or foreigners, acting in concert with others, on their own, or on behalf of a hostile state (Metz & Johnson, 2001)."

The fear created by the anticipation of an attack may paralyze civilian populations, causing more actual economic damage than the attack itself. Terrorists attempt to destabilize targeted societies and cause such fear that the citizens of the targeted countries demand that their government accommodate the terrorists' demands.

Terrorist using weapons of mass destruction (chemical, biological, radiological, or nuclear) are most likely to create what organizational theorists see as a true environmental jolt. For example, bioterrorist attacks create an ongoing fear and danger not only to those initially under direct attack, but also to their families and friends. In a bioterrorist attack, most who are exposed to such an attack will not even be aware of the timing or location of the attack nor that they are at risk. More uniquely, the patients may very likely be potentially dangerous to the healthcare professionals who will take care of them.

Bioterrorism defines as "the unlawful use, or threatened use of microorganisms or toxins derived from living organisms to produce death or disease in humans, animals or plants. The act is intended to create fear and/or intimidate governments or societies in the pursuit of political, religious or ideological goals" (Metz & Johnson, 2001). Due to terrorist goals, biological weapons are appealing to terrorists because of the silence of their attack and the great number of potential victims.

Terrorism and bioterrorism are more likely to create a major environmental jolt than war itself, unless it is on one's own national territory. The 2003 War in Iraq (Iraqi Freedom) has had more political conflict in the United States, which the conflict itself jolts organizations or even the public. The same, of course, is not true of the Iraqi people and particularly not the case for the members of the former Sadaam Hussein regime.

How then, can terrorists successfully deliver major environmental jolts to civilian populations and the organizations designed to protect them and their health? The answer is to be found in the nature of "asymmetric war," a complex phenomenon that has many dimensions.

As Blair, Fottler and Zapanta (2003) point out, although terrorists can do significant damage and inflict large numbers of casualties, and potentially mass casualties in the tens of thousands, terrorism is actually a reflection of weakness, not strength. A more recent concept to clarify this is called asymmetric warfare, in which the weak attack the strong and use the enemies' strengths against them or simply make their strengths irrelevant to the battle. For example, on September 11th and again in the Anthrax attacks, the terrorists used American transportation and mail delivery infrastructure against the United States.

It is the purpose of this paper to delineate what we see to be the key dimensions of asymmetrical warfare and to provide a method to map the different potential configurations that can appear as a result of different terrorist attacks. In doing so, we believe we will have also modeled the nature of the differing environmental jolts facing organizations and the nation. How organizations can adapt to such jolts are addressed in the next section.

THE CONCEPT OF ASYMMETRIC WAR

The "bioterrorism formula" presented by Blair, Fottler and Zapanta (2003) in their first chapter of this book illustrates a number of key dimensions of the threat of bioterrorism and preparing for it effectively. Their formula provides a context for terrorist attacks and points to the importance of examining the specific form of terrorist weapons used in a specific attack. Insights into these weapons and the many combinations of weapons and ways to enhance their impact is found in the piece by Blair and Vlosich (2003) on weapons, weapon cocktails and force multipliers.

Terror organizations are becoming less state-controlled and more entrepreneurial. They have the ability to act without direction from above and they will become even less centralized as U.S. counter-terror efforts become more effective. Eventually the lack of outside support will make them less effective but also harder to track down. A goal of terrorists is to increase their power and freedom of action while they minimize the capabilities and increase the constraints on targeted governments.

Terrorist organizations are able to tie down large amounts of government resources while expending little relative effort. Terrorists have the initiative and can often strike when and where they want – despite extremely large investments in homeland security, homeland defense, and counter terrorism.

In conventional terms, asymmetry illustrates when one side of a business competition or a military conflict is greatly stronger than its opponent. The weaker side does not confront the stronger opponent directly, but must employ an "indirect approach," a term developed by military theoretician B. H. Liddell-Hart. "The indirect approach is as fundamental to the realm of politics as to the realm of sex. In commerce, the suggestion that there is a bargain to be secured is far more potent than any direct appeal to buy. And in any sphere it is proverbial that the surest way of gaining a superior's acceptance is to persuade him that it is his idea! As in war, the aim is to weaken resistance before attempting to overcome it; and the effect is best attained by drawing the other party out of his defenses" (Liddell-Hart, 1967, p. 18).

According to a current military definition, asymmetrical warfare occurs when "adversaries are likely to attempt to circumvent or undermine U.S. strengths while exploiting its weaknesses, using methods that differ significantly from the usual mode of U.S. operations" (McKenzie, 2000). The indirect approach avoids a direct attack on enemy strength. It is an attack on the targets weaknesses with the terrorist's strengths. Although, the definition of asymmetry cannot place itself in one specific category in regards to what actions are asymmetric since one country's asymmetric threat is another's standard of fighting (Gray, 2002).

Asymmetric warfare enlists many aspects of unconventional weapons, such as chemical, biological, nuclear, or radiological weapons. These weapons give the aggressor an upper hand to an opponent who could defeat the aggressor on a conventional battlefield. Terrorists use asymmetry along with force multipliers by producing a deadly cocktail of attacks to wreck havoc on the target country. Transformation of policies, strategy, and forces are a necessity for the United States to be able to stay on equal ground with a terrorist who uses asymmetrical warfare (Department of Defense, 2002).

Positive Asymmetry

When conventional capabilities far exceed those of the competitor, opponent, or enemy positive asymmetry occurs. The objective of attaining positive asymmetry is defined as "using differences to gain an advantage" (Metz, 2001, p. 6). An example Metz gives is the attempt by the U.S. military to obtain technology and training advantages over any potential enemies such as the old Soviet Union, which relied upon numerical superiority in an attempt to gain positive asymmetry.

Positive asymmetries can only be used to advantage when both sides are competing with similar types of capabilities. In the example given above, the Soviet Union had positive asymmetry with respect to overall amount of military hardware and clearly had superiority on that capability (Cohen, 1993). If that were the only capability that was important in military conflict, then the Soviet Union would have had an undeniable advantage. However, the United States countered that by creating positive asymmetry with respect to military technology, intelligence, and training, and they had superiority on those capabilities (Vandenbroucke, 1993).

Sometimes investors use asymmetrical information to make a profit in the stock market. Meaning that the profit attained is from information not available to other investors, this often occurs from insider information (Levine & Hoffer, 1991). In addition, investors buy stock below what the stock would sell for if other investors had complete information. The investor with the additional information knows more about the real situation than others. As a result, he can take advantage of the fact that he knows what a stock is really worth, rather than what people without the information believe it is worth. There is informational asymmetry between this trader and the rival investors. In an effort to make the market competitive, governments have tried to minimize asymmetries in stock trading by outlawing insider information.

Positive resource asymmetry in business firms occurs when firms have an advantage in comparison to their competitors on a capability, such as access to less expensive raw materials, more qualified personnel, or access to information unknown to competitors. Gnyawali and Madhaven (2001) state that network position, or number and quality of network memberships, create resource asymmetries.

Dissimilar firms are often the targets of an attack because the attacker has a positive resource asymmetry. Chen (1996) states that firms, which have similar resource capabilities, retaliate against rivals when the firm is attacked. Therefore, firms that are alike often will not confront one another. Likewise, nations that posses relatively equal military ability are not likely to attack each other, even if those abilities are based on different types of capabilities.

Negative Asymmetry

Negative asymmetry occurs when ones capabilities are significantly weaker than the opponent's capabilities. Under this circumstance, direct competition will lead to certain defeat. However, in negative asymmetry, competition derives from the inability of the opponent to take advantage of strong capabilities. In this case, the asymmetry becomes "a difference that an opponent might use to take advantage of one's weaknesses or vulnerabilities. It is, in other words, a form of threat" (Metz, 2001, p. 6). McKenzie (2000, p. 2) refers to negative asymmetry as "leveraging inferior tactical or operational strength against American vulnerabilities to achieve disproportionate effect with the aim of undermining American will in order to achieve the asymmetric actor's strategic objectives."

Terrorism as Negative Asymmetry

Terrorists will not attack the United States in a conventional manner with tanks and aircraft because our superior technology and our highly trained service personnel would defeat the terrorists. Terrorists like to attack undefended targets, such as the World Trade Center. They will not attack armed soldiers who can defend themselves, but the terrorist's groups attack civilians and unprepared military personnel who feel that they are in a safe area (Hoffman, 1998).

Because of the conventional forces required to attack the United States, terrorist must use an asymmetric approach by the utilization of surprise attacks on undefended targets. The United States cannot defend all possible targets from terrorist attack, thus the millions of soldiers required to defend every potential target in the United States and overseas stretches the strategic resources thin (Greenfield, 2002). The terrorists only have to be stronger than the United States at the time and place of an attack. These well-publicized attacks often have a strategic impact beyond the number of deaths caused by the attack. One possible outcome of these attacks is that they keep a stronger power such as the United States from becoming involved in a conflict or they cause the stronger country to withdraw from a country or region (McKenzie, 2000).

Negative asymmetry gives rise to various forms of terrorist attacks, which entail detailing plans to be able to effectively counter the likelihood of an attack, such as a terrorist plot to use shoulder-fired missiles on a target country's airlines in order to produce casualties (Bolkcom & Elias, 2003). Because the terrorist is not able to strike the target country on equal ground, it must use negative asymmetry to its advantage. Another popular terrorist tactic is suicide bombing. Terrorist often resort to these tactics because they possess inferior conventional weapons as compared to the target. The use of suicide bombing gives the weaker terrorist organization a strong tactical advantage over other forms of delivery (Dolnik, 2003). Terrorist will also use their negative asymmetry to attack the target in financial areas. It is estimated that an agro-terrorist attack to livestock could cost the U.S. economy between $10 billion and $30 billion (Segarra, 2002).

Relevance for this Paper

The concept of asymmetry helps to explain why terror groups attempt biological attacks on the United States. It also explains why rogue nations provide biological weapons to terrorist organizations. These rogue nations, without the luxury of defeating the United States on the conventional battlefield, could strike United States cities with biological weapons. The difficultly in proving who gave the biological agent to the terrorist group allows the rogue nation to escape retaliation from the United States. Unlike the soldiers on the battlefield who are prepared for biological and chemical attacks, the civilians living in large cities are not prepared for a biological attack. A biological attack on civilians is an effective means of attacking the will of a government.

This paper shows how terrorist actions in a cocktail weapon attack or single use can inflict harm on the United States and the entire health care system. The use force multipliers are a concept that relatively few health care providers understand when trying to combat terrorist implications that occur upon their organization. The paper will also show how the terrorist's organizational capacity and external support enhances the extent of terror felt within a population along with the actual number of casualties. The organizational and system preparedness and response must reflect adequately the implications of terrorist attack and its effects on the health care environment.

MODELING A TERRORIST ATTACK

Necessary Elements for Successful Negative Asymmetry

In order for negative asymmetry to be effective against a strong government such as the U.S., certain key elements need placement. These elements expose the vulnerability of the stronger government. McKenzie discusses five features of asymmetry that describe this type of vulnerability: (1) disparity of interest; (2) the will of the stronger opponent as the ultimate target; (3) the seeking of strategic effect; (4) the need for effectiveness; and (5) the dynamic process of threat and response. In addition, terrorists also use another element to their advantage; (6) the element of time.

Disparity of Interest
If there is a disparity of interest between the weaker and stronger nation with the weaker nation having a strong interest in the conflict and the strong nation having a weak interest in the conflict, then the weaker nation or group initiates the use asymmetrical methods. The weaker nation attempts to make the costs exceed the benefits to the stronger nation. However, the weaker nation must inflict enough damage to weaken the will of its enemy without inflicting so much damage that the stronger country becomes resolved to avenge its losses. McKenzie uses the Japanese attack on Pearl Harbor as an example.

The Japanese attack only intended to destroy the offensive capability of the United States' Pacific Fleet so that the United States could not oppose the Japanese conquest of Asia and the Western Pacific. Instead, the attacks made the United States wage total war on Japan and demand nothing less than unconditional surrender. The disparity of interest translates in basic terms to disparity of commitment (Myers, 2002). Examples of this occur by both the U.S. war with Vietnam and Russia's war with Afghanistan. The commitment of the followers' leads to the disparity, which is also a force multiplier as referred to earlier in the book.

Attacking the Will
Each party in a conflict needs the will to engage in the conflict. The ultimate goal of asymmetrical warfare is to attack the will of the stronger opponent. The weaker power knows that it is unable to destroy the military power of its opponent, so it attacks the will of the government and its citizens to continue the conflict. If costs incurred by the dominant party outweigh the value of winning the conflict, then the weaker party prevails.

According to Kreighbaum (1997), if a terrorist subdues an enemy's moral force, then it can cause three times the potential advantage as defeating the target's physical forces. A terrorist realizes that the psychological impact on the will of a nation changes the course of battle in which they engage. The extent of terror imposed on the will, as explained by Blair and Vlosich (2003), only adds to the psychological ramifications that a country faces during a terrorist attack. Thus, this causes the will of a nation stretched to the breaking point.

Strategic Effect
The ultimate goal of every asymmetric action, even at the tactical and operational level, is to produce a strategic effect. Every small terror attack that kills only a few civilians intends to have political and psychological ramifications at the strategic level. A terrorist attack is an attempt to intimidate the citizens of the targeted country and to make them think that the government cannot protect them. The effects of an asymmetric attack go beyond the number of injured and killed. The casualties that occurred to the U.S. with the attack on the World Trade Center were low, but the strategic effect of the attack by Al Qaeda led to cumulative costs to the global economy, which produced many political ramifications (Cosgrove, 2002).

The strategic effect of many terrorist actions lend themselves to casualties and their affect on the organizations or systems preparedness and response, but most often health care providers do not realize the long term implications of isolation, political turmoil, and policy changes within the government. By having the target provoke social control errors, the terrorist achieves a strategic effect of going far beyond casualties and affecting the actual infrastructure and the will of the nation

(Prins, 2002). Health care providers must realize that terrorist attacks have a wide range of implications, which could cause drastic policy changes in the fields that they work in.

Effectiveness

The asymmetric attacks insist on effectiveness. Publicity is paramount for an attack, and the terrorists must convince the individual citizen of the targeted country that they are a credible threat. The publicity by the media leads to many strategic gains for the terrorist group by enhancing an attack in the minds of the target population (Perl, 1997). This action is a force multiplier that only strengthens a terrorist's attack.

The attacks must achieve disproportionate results, such as the September 11 attacks where the lives of the hijackers were traded for the lives of thousands of Americans, in order to produce a believable threat in the minds of the target population. In addition, people are likely to join the side that is obviously winning the war. People are less likely to join a side that is losing or appears ineffective. The weaker country cannot afford a war of attrition with a stronger nation. Lovelace, Jr. (1997) urges that in order to combat the effectiveness of terrorist attack techniques and equipment, we must develop methods for decontamination, treatment of mass casualties, and disaster relief.

The Dynamic Process of Threat and Response

Each side in a conflict reacts to the actions of the other side in an unending process. "Our own actions and strategic choices will drive the nature of the asymmetric threat. As we refine operational practices, potential adversaries will look for ways to counter. This process of action-reaction is inescapable" (McKenzie, 2000, p. 11). This produces an ending effect that leads to a stalemate in action if resolution has no conclusion.

The public health field plays a crucial role in the response to a terrorist attack. First, health care providers need to accurately detect an attack, followed closely by a diagnosis of its nature. Then, it is necessary to evaluate the scope of the exposure, by identifying and implementing the appropriate control measures (Smith & Thomas, 2001). Health care providers' ability to respond greatly improves if these actions occur, if not then it could cause negative outcomes from the terrorist attack.

The Element of Time

Terrorist organizations enjoy a time component of asymmetry. They take advantage of the American desire for quick, decisive results and stretch the conflict out until Americans are dissatisfied with a war that appears to have no end, with apparently endless casualties. Sir Robert Thompson, advisor to President Nixon, coined a term describing the North Vietnamese strategy in the Vietnam War, the "cost formula." "In strategic terms, if an insurgent movement at a cost which is indefinitely acceptable can impose costs on the government which are not indefinitely acceptable, then, although it may be losing every battle, it is winning the war. The will of the government will inevitably crack" (Thompson, 1989).

A MODEL OF THE DIMENSIONS OF TERRORIST ATTACKS

The elements of negative asymmetry as described above are obviously intertwined. For example, if there is a greater disparity of interest, then it is more likely that the will of the dominant party is broken. Unlimited time makes the terrorist more effective while reducing the will of the dominant party. In addition, if the strategic effects of the terrorists are high, then it will unduly affect the response mechanism and the timeliness of the retaliation.

The four major dimensions of terrorist attacks, shown in Fig. 1, separate into two halves or spheres. These two spheres are the terrorist threat, on the left, and the terrorist impact, on the right. As the dimensions described below increase (towards the outside of the circle), the more effective the attacks will become, and the greater the presence of the six elements listed above, making the terrorist more likely to succeed.

As indicated at the bottom of the figure, the left half of the model represents the terrorist threat. It is composed of two major dimensions; terrorist attack strategies and the capability of the terror organization. The dimension of terrorist attack strategies includes the minor dimensions that are the use of force multipliers and whether the terrorists use weapons of mass destruction (WMD) with or without a cocktail of weapons. The dimension of the terrorist group's organizational capability includes the minor dimensions of the group's operational capacity and the level of external support of the group given by outside sources.

The right half of Fig. 1 represents the terrorist impact from the attack. The impact is a function of the target vulnerability and the attack outcomes, which would be the extent of the terror resulting from the attack. The dimension of target vulnerability includes the minor dimensions of the target's systems preparedness response capacity and the target's organizational preparedness response capacity. The attack outcomes dimension in quadrant four contains the extent of terror produced on the target population and the actual casualties.

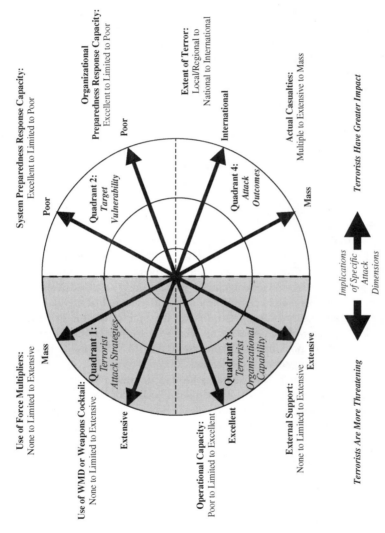

Fig. 1. Dimensions of Terrorist Attacks.

Terrorist Threat

Terrorist threat is the attack configuration and all its components used on the target nation. It is a function of the four minor dimensions of use of force multipliers, use of weapons of mass destruction, organizational capacity, and external support. Notice how the area of the circle, which represents the threat of an attack, is maximized if all of the areas are excellent or extensive.

Terrorist Strategic Intent

Strategy is defined as "an actor's plan for using armed forces to achieve military or political objectives" (Arrequin-Toft, 2001). Figure 1 indicates two dimensions for carrying a battle plan, the use of force multipliers, and the use of weapons of mass destruction. These two dimensions can be used in a variety of ways to produce a deadly cocktail of attacks on the target population.

Use of Force Multipliers

Use of force multipliers enhances a terrorist's attack by adding extra intricacies to the end product. There are wide varieties of force multipliers available for terrorist use. The use of deception serves the purpose of confusing the target by not allowing real information to reach intelligence officers, which causes the target population in some cases to make military mistakes. The use of media adds to the strategic effect of the terrorist by amplifying the fear in the target population. It also poses a problem to reporters who wish to disseminate information to the target population, by not allowing them the ability to differentiate from actual information from disinformation (Beelman, 2002).

The terrorist wants to provoke social control errors made by the enemy. These include massive quarantines and loss of civil liberties in the target population. Another way to enhance an attack is to use the targets own infrastructure against it. An example of this occurs with the 2001 anthrax attack in which a terrorist used the U.S. postal system to disseminate anthrax. Force multiplication of an attack can also occur through the terrorist striking certain locations or using replication within an attack.

Use of Weapons of Mass Destruction

The use of weapons of mass destruction can occur in various situations where the attackers are not concerned about public support for their cause or they may be used in situations where the terror group will deny responsibility for the attack. These organizations can use a single weapon or they can use a cocktail arrangement of multiple types of weapons. Terrorists receive weapons from either within the terrorist group or by rogue nations who support the political aims of a terror group. Some WMD, such as smallpox, allow a weaker country to attack the United States without the fear of retaliation. Many weapons of mass destruction are currently being disseminated to rogue nations by former nations that were once part of the U.S.S.R. (Lugar, 1999). These rogue nations include North Korea, Iraq, and Iran.

The difficultly for the United States occurs in determining what group or country made the WMD attack, in the same way it was difficult to determine who made the October, 2001 anthrax attacks. In any case, the WMD attack has a high likelihood of being very effective and thereby making the cost of continued battle for the dominant party costly. For example, a dirty bomb used by terrorist is easy to assemble, with the right materials, and the effects to the target population can be severely damaging (CDI, 2003).

Terrorist Organizational Capability
A terrorist group's organizational capability consists of its internal operational capability and its external support from other terror groups, rogue nations, and non-government organizations. Terror groups form informational and resource networks with other terror groups and rogue nations in order to enhance capabilities. Organizational capability is the level of sophistication that the terror group may employ in an attack.

Operational Capacity

Operational capacity is the level of sophistication of members in a terror group, excluding any support from outside organizations or countries. The level of sophistication determines the type of attack delivered by a terrorist group. The skills of the individual members of a terror group are one component of the operational capability of a terrorist organization.

An example of these types of capabilities present themselves in the terrorist attack on the World Trade Center. The terrorists needed to be highly sophisticated in flight mechanics so that they could fly the jets. This level of sophistication reflects extensive planning on the terrorist plot, and shows how no prevention occurred to stop the terrorists from acquiring this knowledge within the United States even after the FAA was alerted of the suspicious nature of one terrorist (ABC News, 2002).

External Support

External support comes from other terror groups and nations that sponsor terrorism. Terror groups network to increase their organizational capability just as executives network in order to improve their own and their firm's capabilities. "Firms are embedded in networks of cooperative relationships that influence the flow of information among them" (Gnyawali & Madhaven, 2001, p. 431).

External support comes in many different forms, which depends on the extent that a country or entity wishes to make their allegiance visible. Terrorists rely on external support in the forms of training, finance, and political support. Often, money is the most powerful tool that terrorist wish to receive. Monies can aid terrorists in buying weapons, bribing officials, paying operatives, or providing a social network that builds a popular base (Byman et al., 2001). The greater the extent of external support the stronger a terrorist organization or attack becomes.

Terrorist Impact

Terrorist impact is the effect on the target nation received from a terrorist attack. It is a function of the four minor dimensions of system preparedness and response capacity, organization preparedness and response capacity, the extent of terror, and the actual level of casualties occurred by the target population. Notice how the area of the circle, which represents the amount of terror caused by an attack, is minimized in area if the system response capability and target protection are excellent.

Target Vulnerability

Target vulnerability makes up of the minor components of system and organizational preparedness and response capacity. Note that target vulnerability is the only major dimension that the target government can control. If the target preparedness and response capacities are excellent, then the amount of terror inflicted is much smaller. Israel is a good example of how excellent system and organizational preparedness response capacity minimizes the impact of a terrorist attack. Israel has in place a good system to respond to transit bombings that occur within their population (Boyd & Sullivan, 1997).

System Preparedness and Response Capacity

System preparedness and response capacity is the personnel and resources necessary to protect the civilian population exposed to a terrorist attack. The system response capacity includes the number and level of training of first responder police and medical personnel, fire-fighting personnel and equipment and personnel needed to restore water and electrical services. Hospital services are a part of the system preparedness and response capacity.

Health care providers must set active communication lines with first responders in order to coordinate plans of action. Terrorist events have a dramatic effect on the system response capacity of the target population. It is generally agreed that response capacities must be in place in order for the health care system to navigate safely through a terrorist attack (Mahmoud & Lemon, 2002).

Organizational Preparedness and Response Capacity

Organizational preparedness response capacity is the resource that an organization has to alter the outcomes of a terrorist attack. Organization preparedness can range from poor to excellent. Depending on plans set up by the organization, response can adequately meet the challenges of terrorist attack. This is an essential area, which could mean the difference between life and death of a patient.

If a system has an overall adequate response capacity but the organization within the system does not, then the organization will be the weakest link in the chain. Organizations must take active measurements to combat any terrorist situation that could arise. This means preparing for either a physical attack or a bioterrorist attack. The organization must stockpile vaccines and make preparation plans for a surge of injured or worried well in order to effectively survive a bioterrorist attack (Institute of Medicine National Research Council, 2002).

Attack Outcomes

The outcomes of the terror attack are the actual level of casualties, which compare with the level of intended casualties to determine if the terrorist goals are realized or not. The attack outcomes also look toward the extent to which terror permeates throughout the target population. Health care providers should be cautioned against looking solely at actual casualties because they do not portray the entire picture of the terrorist attack, and what the overreaching outcomes will be on the organization.

Actual Casualties

The actual casualties are the number of killed or wounded in the terrorist attack. The number of actual casualties is always known, unlike the intended level of casualties. If there is a mass number of casualties, then the level of terror increases. Casualties occur from a variety of terrorist attacks ranging from a conventional attack to an unconventional one. The casualties are a definitive measure of the success of a terrorist attack in which inflicts the most damage to health care providers. The onslaught of casualties causes organizations to surge beyond capable serving capacities that can lead to the death of many patients.

Extent of Terror

The extent of terror may affect a local area, cover an entire nation, or be international in scope. The extent of terror in the model bases upon the geographical areas in which terror occurs. It does not refer to the extent of terror within a person. Terror has many psychological contributing factors that affect the level of terror within a geographical area, such as the number of people killed. These psychological factors thus dictate where the terror areas occur. The extent of terror caused by a terrorist attack can have drastic implication for any health care organization (Blair & Vlosich, 2003). It should also be noted that the extent of terror changes perceptions and policies in relation to terrorism as the more widely terrorist attacks occur within U.S. borders. These new perceptions and policies change the way in which terrorism is viewed by the public, and can have serious implications for health care policy.

TYPES OF TERRORIST
ATTACK CONFIGURATION

Overview

Figure 2 represents four possible types of terrorist attack configurations. These types describe the actual results of a terrorist attack in terms of casualties compared to the intended results. The goals of the terror attack may be realized or unrealized. If the goals of the terror attack realize, then the level of actual casualties and the extent of terror are at the same level as the intended casualties. The impact of the terror attack may be disproportionate or limited. When the impact of the attack is disproportionate, the effects far exceed the amount of effort expended in the attack. If the impact of the terror attack is limited, then the amount of terror inflicted is equal to or less than the amount of effort expended in the terror attack.

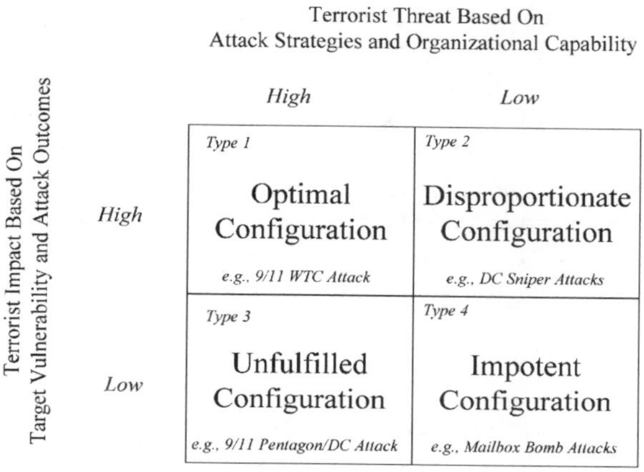

Terrorist Threat Based On
Attack Strategies and Organizational Capability

	High	*Low*
High	*Type 1* Optimal Configuration *e.g., 9/11 WTC Attack*	*Type 2* Disproportionate Configuration *e.g., DC Sniper Attacks*
Low	*Type 3* Unfulfilled Configuration *e.g., 9/11 Pentagon/DC Attack*	*Type 4* Impotent Configuration *e.g., Mailbox Bomb Attacks*

(Vertical axis: Terrorist Impact Based On Target Vulnerability and Attack Outcomes)

Fig. 2. Types of Terrorist Attack Configurations.

Description

We propose a model with a four-quadrant matrix with Terrorist Threat Based on Terrorist Attack Strategies and Terrorist Organizational Capability on the horizontal axis, and Terrorist Impact Based on Target Vulnerability and Attack Outcomes on the vertical axis (see Fig. 2). The type one configuration in the upper-left hand corner is the Realized Threat Configuration, which has a high terrorist threat level and a high terrorist impact level. The terrorists achieve their goal of inflicting their planned level of terror. The 9/11 World Trade Center attack is an example of this type configuration. A great deal of terror occurred by the attack. The actual casualties were as high as the intended casualties were, and the extent of terror extended to an international level.

The type two configuration is the Disproportionate Impact Configuration, which includes a high terrorist impact level and a low terrorist threat level. The outcome of the terror attack is above the level of terror that the terrorists expected to inflict. The Washington, DC sniper attacks are example of this configuration. The two attackers caused less than twenty deaths, yet they inflicted a large amount of terror at the regional level. They also inflicted some level of terror throughout the United States, because of the fear that possible accomplices of the snipers could strike anywhere.

The type three configuration is the Unrealized Threat Configuration, located in the lower left hand quadrant. It consists of a low terrorist impact level and high terrorist threat level. The realized level of terror is below the planned level of terror.

The casualties and the extent of terror are lower than the terrorists have intended. The 9/11 attack on the Pentagon represents this configuration. The terrorists planned to destroy the Pentagon and kill thousands of high-ranking military personnel. The actual casualties were below the level of the intended casualties.

The lower right hand quadrant, type four configuration, contains the Limited Impact Configuration. This configuration contains a low terrorist threat level and a low terrorist impact level. The Mailbox Bomb Attacks of late 2001 are examples of this configuration. The terrorist impact level and the terrorist threat level are both low. With this attack, the terrorists did not inflict a high level of terror in these attacks.

The Optimum Terrorist Configuration

The shaded area in Fig. 3 represents the effectiveness of the optimal terrorist attack. By placing, a point on each dimension at the outmost dimensional line desired by the terrorist creates the shaded area, and then shading the interior area. The terrorist organization's ability to threaten by force multipliers, weapons of mass destruction, operational capacity, and external support are high. The actual impact is also high because of the target has low system response capacity, organizational preparedness, along with the actual casualties and the extent of terror being high.

The configuration matrix allows one to study the threat from the terrorist's point of view and the threat from the target's view. The terrorists hope to maximize the "area" of the circle in the matrix, which indicates the amount of terror generated by the attack, by attacking unprepared targets and inflicting large numbers of casualties. Excellent terrorist operational capacity and external support increases in the level of terror. By increasing their organizational preparedness and system response capacity, the target government can decrease the amount of terror spread (or the size of the darkened area). These actions by the target government decrease the actual number of casualties and the amount of terror generated by the terrorist attack.

Terrorists intended casualties to high, since the terrorists in this optimum situation have the resources necessary to inflict large numbers of casualties. These resources include Weapons of Mass Destruction (WMD), such as nuclear, biological, chemical, and radiological weapons. Rogue nations that sponsor terrorism donate these weapons. These donations are made so that the rogue nation may strike at its enemies while at the same time allowing the rogue state to deny any involvement in the attack and avoid retaliation. There is also a possibility that the terrorist organizations develop WMD with their own resources.

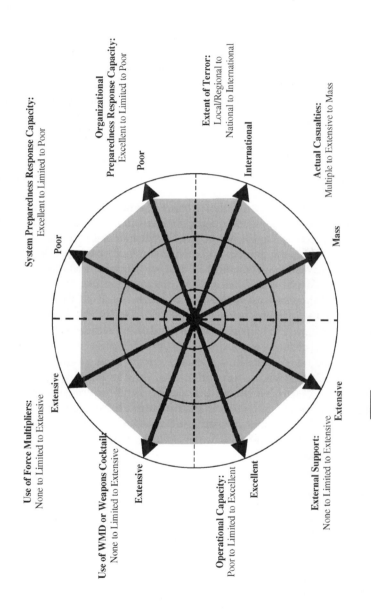

Pure "Optimal" Attack Configuration

Fig. 3. Optimal Terrorist Attack Configuration.

In the past, the intended casualties were often low to avoid worldwide outrage while still drawing attention to the terrorists' cause. As the goals of terror organizations change, the number of intended casualties changes as well. The terror group's goals may change from recognition of a cause to waging a holy war for example. In this situation, the terror group lacks concern for public opinion, and genuinely desires to inflict large numbers of casualties on an enemy nation.

The terror group's operational capacity is excellent in an optimum situation. The terrorist organization seeks knowledgeable experts in fields such as physics, medicine, computer science, and conventional warfare. In addition, terror groups look for external support from rogue nations and sympathizers in targeted countries. This external support possibly occurs from bogus charitable organizations, oppressed minorities in the target country, or financial and military support from nations that are "plausibly denied."

In the optimum configuration for the terrorists, the target country's response capacity and organizational preparedness are poor. Terrorists impose a greater level of terror against unprepared targets. System and organizational preparedness and response capacity include the number and effectiveness of first responders, such as police officers and emergency medical technicians. Number and effectiveness of hospitals affect the organizational capacity, including decontamination and inoculation services. Preparedness of these public services is an effective deterrent to terrorist attack. Israel, being a prime example of a country that expends great effort in establishing excellent response capacity and organizational preparedness, through capacity and preparedness decreases the level of terror.

Terrorists want their terror attacks to influence governments all over the world (Davis & Jenkins, 2002). They wish for their actions to intimidate countries not directly attacked. Terrorists believe that these neutral governments will not actively oppose them because of the fear of retaliation. Terrorists now hope for massive levels of actual casualties in order to deter attack from nations that have stronger conventional military forces than the countries that sponsor terrorism.

The optimum situation for the terrorist causes the greatest level of terror. The shaded area in the matrix is at its largest, which indicates a greater level of terror and intimidation. The optimum configuration for the terrorists is the least optimum for the targeted government. In this situation, the government infrastructure is not anticipating and unable to handle a WMD attack from a technically sophisticated and well-supported terrorist group. Preparation and planning by the target government decreases the size of the highlighted portion of the matrix and the level of terror caused by a terrorist attack.

The optimum situation for the target government is to have an attack on a target where the system response capacity and organizational preparedness level is excellent. The terrorists do not use WMD in the attack, the terror group has poor operational capacity and there is no external support for the terror organization. In this best-case scenario for the government, the amount of terror caused by the attack is minimal. It is even possible in this configuration to thwart terror attacks before they occur.

The World Trade Center Attack Configuration

Figure 4 represents the comparison between the optimal attack and the actual attack on the World Trade Center. The darker shaded area once again represents the optimal attack scenario while the lighter shaded area represents the actual attack on the Trade Center. The terrorists intended to inflict an extensive number of casualties in the World Trade Center attack. They hoped to kill and injure thousands of people, which worked in or near the twin towers. Although the terrorist managed to create a very large bomb out of the huge airline, no use of Weapons of Mass Destruction, such as nuclear or biological weapons, occurred (Posner, 2003). The use of these weapons could have killed or injured hundreds of thousands of people. This use of WMD increases the level of casualties into the mass category.

If the terror group's operational capacity increased, then they might develop and employ weapons of mass destruction. Given the terror group's external support amplified, it could have acquired nuclear or biological weapons from a state that sponsors terrorism. However, the terror group's organizational capacity rates as excellent, including the financial resources to support and train the pilots as they prepared for the attacks. The external support of the terror group is rated as extensive, including safe haven and training camps in Afghanistan. It also becomes clear that the terrorists received briefings from foreign intelligence officials (Jenkins, 2002).

The system response capacity of the United States was limited in many areas. Airline pilots were trained to accommodate hijackers in an attempt to minimize the loss of life. The terrorists took advantage of this policy in their planning of the attack. However, the emergency response capability of the United States is high in general terms with its large numbers of trained emergency response technicians and hospitals. The United States had adequate facilities to handle the injured, partly because of the large numbers of dead. In the terrorists' ideal scenario, terrorists hope to have large numbers of injured overwhelm the health care system. This is to show the average citizen that the government cannot protect him or her. The terrorists wish that the citizens would then demand that the government make concessions to the terrorists in order to stop the killing.

The organizational preparedness of the United States was limited. Air defenses were not as elaborate as they are today. None of the attacking aircrafts were

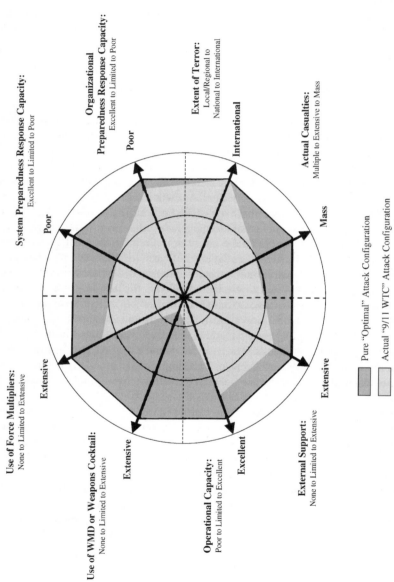

Fig. 4. Optimal Terrorist Attack Configuration: Contrasted With 9/11 WTC Attack.

brought down before hitting their targets, except for Flight 93 in Pennsylvania, which was crashed because of a struggle with the passengers on the flight with terrorists. This low level of preparedness is close to the level of preparedness desired by the terrorists in their optimum scenario.

The extent of the terror was international. This is the same as in the ideal terrorist scenario. It is unlikely that the extent of the terror might increase higher, unless the use of nuclear and biological weapons occurred. The actual casualties were extensive, but not at the mass level of the optimum terrorist attack. The actual casualties closely matched the intended casualties. In our terms, the terrorists realized their goals in terms of the number of casualties and the amount of terror caused.

The Pentagon Attack Configuration

Figure 5 represents the comparison between the World Trade Center attack and the attack on the Pentagon. The darker shaded area once again represents the World Trade Center attack, while the lighter shaded area represents the attack on the Pentagon. The intended casualties in the Pentagon attack amounted to thousands. The use of force multipliers is at the extensive level. The terrorists did not use WMD in this attack, although they may have desired to use them. The terrorists had excellent operational capacity, with trained pilots and excellent coordination between the separate groups conducting all of the 9/11 attacks. The terrorists also had sufficient funds to obtain entry visas, living expenses, and pilot training for the hijackers. The terrorists had extensive external support, with training camps in Afghanistan. The external support could have been greater if the hijackers obtained nuclear or biological weapons from some government that sponsors terrorism.

The target government's response capacity was poor in some areas. Airline pilots had not been trained for this scenario. The terrorists noticed the limitations in the airline policies for handling hijackers, and exploited these weaknesses. The air defense system around Washington did not provide an adequate defense in this attack, allowing the aircraft to hit the Pentagon (Jenkins, 2002). Structurally, however, the Pentagon was well prepared for the attack. The fire and medical response system performed fairly well, given that the number of injured was low.

The terrorists' goals were only partially realized. The extent of the terror was not as great as the World Trade Center attack, possibly because the numbers of casualties were low, and the Pentagon was not destroyed like the twin towers. In addition, the Defense Department operations were not significantly disrupted. The

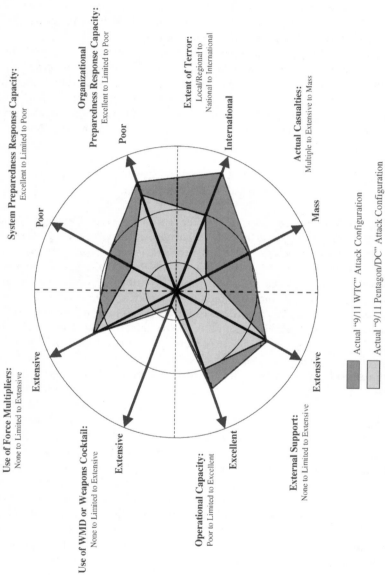

System Preparedness Response Capacity:
Excellent to Limited to Poor

Organizational Preparedness Response Capacity:
Excellent to Limited to Poor

Extent of Terror:
Local/Regional to National to International

Actual Casualties:
Multiple to Extensive to Mass

Poor

International

Mass

Poor

Use of Force Multipliers:
None to Limited to Extensive

Extensive

Extensive

Use of WMD or Weapons Cocktail:
None to Limited to Extensive

Extensive

Excellent

Extensive

Operational Capacity:
Poor to Limited to Excellent

External Support:
None to Limited to Extensive

Actual "9/11 WTC" Attack Configuration

Actual "9/11 Pentagon/DC" Attack Configuration

Fig. 5. 9/11 Pentagon Attack Configuration: Contrasted With 9/11 WTC Attack.

actual casualties in the Pentagon attack were less than hoped for by the terrorists. In their ideal scenario, casualties would have been in the thousands, and military operations would have been disrupted for months or years. The Pentagon would have been destroyed to add to the extent of terror.

The DC Sniper Configuration

The following, Fig. 6, represents the DC Sniper attack. The intended casualties in the DC sniper attacks were multiple. There was no intended use of WMD. The operational capacity of the two individuals was limited. The older sniper had been given basic marksmanship training in the military, but received no specialized sniper training. The younger sniper had not received no formal marksmanship training. There was no external support for the two. The weapon they possessed was legal to own and not difficult to obtain, even though it may have been stolen.

The outcome of these terror attacks was beyond the intended level. The level of terror was above any level that the pair could have expected. The level of terror was national even though the attacks occurred in the region of Washington, DC. In addition, the level of casualties was only at the multiple level of casualties. This high level of terror was achieved because of the fear that similar attacks could occur in other locations within the United States. This fear may have been worsened by the recent 9/11 attacks. The system response capacity in the Washington area was excellent in most ways, with professional law enforcement and medical services. However, the law enforcement officers were more accustomed to criminal activity rather than acts of terrorism. The suspects remained unapprehended for an extended amount of time.

The DC Anthrax Attacks

The darker area on Fig. 7 represents the actual attack configuration of the DC Anthrax attack. Of note in this configuration is the lack of Organizational Capability. However, by using a simple Weapon of Mass Destruction, namely the Anthrax Spore, the terrorist was able to generate a great deal of terror, especially internationally. This high degree of terror was in a great deal a reflection of the poor System Response Capacity that exists internationally. People all over the world recognized that little defense existed against this type of weapon and little could be done to respond to it.

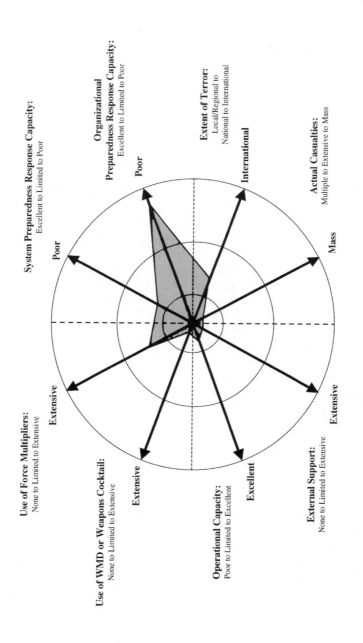

Fig. 6. Disproportionate Attack Configurations: DC Sniper Attack.

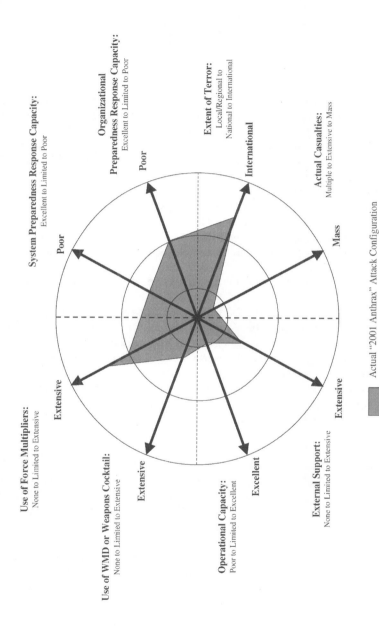

Fig. 7. Disproportionate Attack Configurations: Anthrax Attacks.

MODELING THE BIOTERRORIST
AND CYBER ATTACK SCENARIOS

Overview

Our model thus far demonstrates how the different dimensions can be employed in various ways in order to manipulate the magnitude of a terror attack. The magnitude of the terror is a function of the terrorists' and the target's capabilities and preparation. The ideal terrorist scenario covers the largest area of the model with large numbers of casualties, extensive use of Weapons of Mass destruction, and poor target response capability.

However, terrorist choices are not limited to the simple scenarios described above. Any terrorists who want to expand the shaded area of the model and thus increase the effectiveness of an attack has the option of simultaneously employing multiple types of targets, methods, and sites.

The different types of multiple attacks employed are described below.

- Multi-Method attacks are simultaneous attacks on the same target by different methods.
- Cocktail attacks are attacks on the same type of target with different types of weapons. An example may be a typical suicide bombing with conventional explosives, but in which biological agents are also released.
- Multi-Site attacks are simultaneous terror attacks at different locations such as were carried out in the September, 2001 attacks.
- Multi-Modal attacks simultaneously employ two or more of the attacks listed above. For example, multi-weapon and multi-site attacks such as Anthrax attack on humans while simultaneously attacking the Internet with a computer virus.

The following fictitious scenarios demonstrate how these multiple attacks can increase the effectiveness on all dimensions of the model. As shown in Fig. 8, multiple types of attacks require greater organizational capability and strategy. However, as they are introduced, the ability of the target to deal with them is decreased, and the terrorist's ability to inflict damage and casualties is increased.

The scenarios described above show how going from a single method on a single target to multiple methods on a single target, and then multiple methods on multiple targets can greatly increase the terror induced by a terrorist organization.

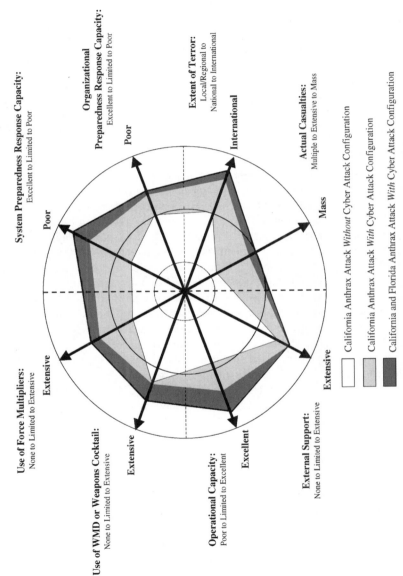

Fig. 8. Bioterrorist Attack Configuration Reflecting Hypothetical Scenario.

California Anthrax Attack

This scenario is an anthrax attack on California. The intended casualties are extensive. It assumes that the majority of the victims will be from the western United States. However, many victims will be from Asian countries. Because of the twelve-day inoculation period without symptoms, the victims will be dispersed all over the western United States and Asia. This will make diagnosis and treatment difficult. Also, people can become infected by contact with victims who have just returned from Disneyland. Note that one elderly woman in New York died because her letter went through a sorting machine that had already been through the decontamination process.

The use of WMD is extensive with four hundred kilograms of Anthrax agent used. Compare this to approximately one kilogram used in the 2001 anthrax attacks. The terrorist's operational capacity is excellent with its highly trained cells of dedicated experts including Ph.D.'s in computer science and the medical field. The terror organization has large amounts of cash, donated by or extorted from rich Saudi businessmen. The terrorist's external support is high with the "leased" expertise of bio-warfare scientists from the former Soviet Union.

The system response capacity is limited because the victims are scattered throughout the western United States and Asia. We assume the governments of these countries would demand assistance from the United States because their citizens were exposed on U.S. territory. This would stretch the limited American resources even further. The victim government's organizational preparedness level is limited because victims are scattered all over the western United States. In addition, the number of worried well that would appear throughout the western United States could overwhelm the medical system just by them selves. The actual casualties are extensive. However as illustrated in the third part of Fig. 9, this single type of attack on a single target is limited compared to introducing and additional method of attack on the same target.

California Anthrax and Cyber Attack

This scenario is an Anthrax attack on California including a nation-wide cyber attack. The intended number of casualties is massive. The use of WMD is extensive. The terrorists have excellent operational capacity. The terrorist organization has a high amount of external support. The victim government has a poor system response capacity. The government's organizational preparedness is poor. The extent of the terror reaches international proportions. The number of actual casualties is extensive.

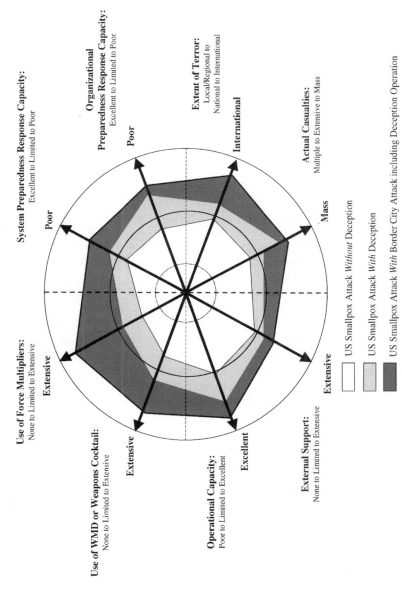

Fig. 9. Bioterrorist Attack Configuration for North Korea Scenario.

The Anthrax attack on California combined with a cyber attack on the government response system would delay any proper response. The cyber attack could prevent any coordinated response by local, state, and federal officials. A glut of vaccine and medical care in some areas would be matched by a shortage of vaccine and medical personnel in other areas. The ignorance among officials caused by the cyber attack would be obvious in the news media, and civilian panic would increase.

As Fig. 9 illustrates, the additional attack mode requires more organizational capacity on the part of the terrorists. It also means that the target is going to be less prepared, and more resources will be needed to deal with this additional attack.

California and Florida, Anthrax and Cyber Attack

This scenario includes Anthrax attacks in both California and Florida, as well as a cyber attack. In this attack the intended casualties are massive. There is extensive use of weapons of mass destruction. The terrorist group's operational capacity is excellent. Their external support is extensive. On the victim's side, vulnerability is high. The system response capacity is poor: organizational preparedness is poor. The extent of the terror reaches the international level. The number of actual casualties reaches the mass level.

We assume the Anthrax attack on California and Florida combined with a cyber attack would be perceived as a global attack. All United States citizens could feel that a close relative or neighbor has been exposed to Anthrax, even if the Florida attack was not on a massive scale. European tourists in the United States would feel exposed to the disease and cause panic in their home country. Along with Asian governments, European and African governments could demand immediate support from the United States and hold the U.S. responsible for the catastrophe. The panic and uncertainty that was felt in the United States on September 11 could be felt throughout the world.

As Fig. 9 again illustrates, the additional attack target requires more organizational capacity on the part of the terrorists. It also means that the country as a whole is going to be less prepared, and more resources will be needed to deal with this additional attack. The interaction of the dimensions added on to the interactions of the multiple attacks has the potential to drastically increase the terror inflicted on the target.

North Korean Smallpox Scenario

The scenario as described in the previous chapter, *Cocktails, Deceptions and Force Multipliers in Bioterrorism*, illustrates a situation in which North Korea

Table 1. Table for Attacks Presented.

	9/11 WTC Attack	9/11 Pentagon Attack	D.C. Sniper Attack	Anthrax Attack	Hypothetical Anthrax Attack	Hypothetical Smallpox Attack
Use of force multipliers None to limited to extensive	Limited	Limited	Limited	Extensive	Extensive	Extensive
Use of WMD or weapons cocktail None to limited to extensive	None	None	None	Limited	Extensive	Extensive
Operational capacity Poor to limited to excellent	Excellent	Excellent	Poor	Poor	Excellent	Excellent
External support None to limited to extensive	Extensive	Extensive	None	Limited	Extensive	Extensive
Actual casualties Multiple to extensive to mass	Extensive	Multiple	Multiple	Multiple	Multiple	Mass
Extent of terror Local/regional to national to international	International	International	Regional	International	International	International
Organizational preparedness response capacity Excellent to limited to poor	Poor	Poor	Poor	Limited	Limited	Limited
System preparedness response capacity Excellent to limited to poor	Limited	Limited	Excellent	Limited	Poor	Poor

plots with Al Qaeda to strike the United States. Figure 9 represents the diagram of the effect of the terrorist attack upon the United States spring break students. The figure represents the terrorist attack with smallpox, the terrorist attack with deception, and the terrorist attack with deception and replication.

The use of force multiplier was expensive, including use of deception, use of media, use of location, use of replication, and use of innovative tactics. Smallpox was the effective weapon of mass destruction sprayed upon the minor populations who have never been vaccinated for the disease. The operational capacity was limited, but the external support was extensive. The organizational and system preparedness were limited due to the fact that the bioagent would remained undetected while the target population disseminated the disease further.

The actual casualties were mass and the extent of terror reached international levels. The scenario shows that the United States is not prepared for such an attack. The U.S. must all prepare systems and organizations for such a mass epidemic, if it were to occur. It is the responsibility of all health care providers to stay current with these terrorists' agendas in order to ride the wave into the future.

The following table will provide a summary of all the attacks presented within this article. It shows in the table format each of the dimensions used to show the various attacks (Table 1).

Discussion

There is much ill-informed bland acceptance of the argument that the attacker needs three times the strength of the defender. That is rubbish. All he needs, given the advantage of deciding where and when, is three times at the point of attack. The defensive positions, even of a defender superior in numbers, are then turned and the defense begins to crumble, as at Shingmum (Thompson, 1989).

The model of terrorist attacks developed in this paper demonstrates that the terrorist real advantage is the flexibility to attack unprotected targets with multiple types of weapons without any time pressure. It demonstrates how the number of actual casualties need not be high in order to create a great deal of terror, if the public feels that they cannot be protected. In fact, extreme numbers of casualties may increase the resolve of the target to fight back.

The targets of terrorist attacks must battle the publics' perception of terror as well as the terrorist ability to inflict damage. Targets must not only try to fortify defenses, they must have response preparedness in order retain legitimacy within their own governments. Each dimension must be reduced so that the overall effectiveness of a terrorist organization can be minimized.

IMPLICATIONS FROM THE MODEL

The September 11 attacks clearly showed the susceptibility of organizations in the United States and across the globe to terrorist attacks. The U.S. economy lost billions of dollars in value in the months following the attack, and is still struggling to recover. Executives in every organization must minimize the threat of terrorism in order to preserve their organization. By understanding what makes an attack effective, and then limiting those factors can help an organizations survive during times of crises.

Healthcare executives carry all the responsibility of any other executive, but they also bear the additional burden of knowing that their response is one that may have dire consequences for the whole community. Healthcare organizations may become the primary targets of terrorist attacks if the terrorists believe that a stricken healthcare system would act as a force multiplier in creating panic.

The dimensions of terrorists attacks developed in this paper have implications for healthcare, both academically and commercially. Hopefully, these dimensions will allow us to be more efficient and effective as we develop our responses to the terrorists threat. Organizations in general must do all they can to limit the ability of terrorists to successfully implement an attack, and then they must have the ability to respond to the attack.

The dimensions displayed in Fig. 1 can be used as starting points to guide academics and practitioners. The left side of the model deals with characteristics controlled by the terrorists. However, both academics and practitioners should not ignore this side of the model.

Implications for Terrorism Dimensions

The ability for targets to have any control or impact on terrorist controlled dimensions is somewhat limited. Countries may be able to directly influence the terrorists' abilities but organizations are very limited to their own preparedness or the preparedness of the larger industry system. However, some focus should be developed on this side of the model.

Researchers should look at the terrorist controlled dimensions in order to determine what makes a target desirable (i.e. to what extend can casualties be expected) with resources controlled by the terrorist. The ability to predict a target must be developed before measures can be taken to protect. Researchers should develop models of the susceptibility of targets based on the desirability and hardness of the target. If targets can be predicted and hardened, then the terrorists may move on to other targets without ever inflicting damage.

Implications for Organization Dimensions

The right side of the model in Fig. 1, may prove to be of more immediate interest to both researchers and practitioners. The leaders of healthcare organizations may directly manipulate these dimensions.

Executives must try to harden their organization against attacks, thus making themselves less vulnerable to an attack. They must also consider the implications of an attack on the system that they operate within. This would include the whole business process of supplier through customer and might be thought of as an industry problem that must be dealt with on an industry basis. Researcher in this area should concentrate on how to employ human resources as well as technical resources to defend against terrorism.

Executives may want to consider building strategic alliances across their industry, including developing a plan with their competitors. We have seen that consumer confidence can be shaken over an attack in an area far from them. Thus, it is not one company that suffers, but all companies within an organization. Research in healthcare strategy should include studies on how security planning can become part of organizational competitiveness. Strategic alliances formed for economic competition must be modified to include alliances against terrorism.

Research needs to clarify how the effects of a terrorist attack differ from other types of disasters. The fear and panic created by a terrorist attack goes far beyond that created by a natural disaster if people believe that nothing can be done to stop the attack. Thus, the damage created goes beyond the actual damage as the victim's perceptions of events begin to form. Executives must find ways to minimize the terror created by these incidents. Control of perceptions must begin before any crises develops and must continue if a crises does occur. New dimensions must be found by researchers that will help practitioners control the aftermath of a terrorist attack.

In addition to developing preparations for the healthcare organizations, researchers must also find ways to link the private and public systems together. These systems often operate autonomously. As community preparedness plans are developed, it becomes painfully obvious that these systems must work in tandem. Communication and tactical plans must be developed by linking academic research, economic strategies, and community planning.

REFERENCES

ABC News (2002). FAA received alert about 9/11 hijacker. The Associated Press (May 13). Available at: http://abcnews.go.com/sections/us/DailyNews/homefront020510.html.

Arrequin-Toft, I. (2001). How the weak win wars: A theory of asymmetric conflict. *International Security, 26*(1), 93–128.

Beelman, M. (2002). The dangers of disinformation in the war on terrorism. *International Consortium of Investigative Journalists* (February 2).

Blair, J., Fottler, M., & Zapanta, A. (2003). The bioterrorism formula: Facing the certainty of the uncertain future. In: J. Blair, M. Fottler & A. Zapanta (Eds), *Bioterrorism Preparedness, Attack and Response, Volume 4, Advances in Health Care Management*. London: JAI Press/Elsevier.

Blair, J., & Vlosich, W. (2003). Cocktails, deceptions and force multipliers in bioterrorism. In: J. Blair, M. Fottler & A. Zapanta (Eds), *Bioterrorism Preparedness, Attack and Response, Volume 4, Advances in Health Care Management*. London: JAI Press/Elsevier.

Bolkcom, C., & Elias, B. (2003). Homeland security: Protecting airliners from terrorist missiles. *Report for Congress* (February 12).

Boyd, A., & Sullivan, M. (1997). *Emergency preparedness for transit terrorism*. Washington, DC: National Academy Press.

Byman, D. et al. (2001). *Trends in outside support for insurgent movements*. New York: RAND.

CDI (2003). Pascal's new wager: The dirty bomb threat heightens (February 4).

Chen, M. J. (1996). Competitor analysis and interfirm rivalry: Toward a theoretical integration. *Academy of Management Review, 21*(1), 100–134.

Cohen, W. (1993). *America in the age of Soviet power, 1945–1991*. New York: Cambridge University Press.

Cosgrove, M. (2002). Terrorist strategy and global economic implications. *International Business & Economics Research Conference*. Las Vegas, NV.

Davis, P., & Jenkins, B. (2002). *Deterrence and influence in counterterrorism: A component in the war on Al Qaeda*. Santa Monica, CA: RAND.

Dolnik, A. (2003). Die and let die: Exploring links between suicide terrorism and terrorist use of chemical, biological, radiological, and nuclear weapons. *Studies in Conflict & Terrorism, 26*, 17–35.

Gnyawali, D. R., & Madhaven, R. (2001). Cooperative networks and competitive dynamics: A structural embeddedness perspective. *Academy of Management Review, 26*(3), 431–445.

Gray, C. (2002). Thinking asymmetrically in times of terror. *Parameters* (Spring), 5–14.

Greenfield, V. (2002). *The role of the office of homeland security in the federal budget process: Recommendations for effective long-term engagement*. Santa Monica, CA: RAND.

Hoffman, B. (1998). *Inside terrorism*. New York: Columbia University Press.

Institute of Medicine National Research Council (2002). Countering bioterrorism. Washington, DC: National Academic Press.

Jenkins, B. (2002). *Countering Al Qaeda*. Santa Monica, CA: RAND.

Kreighbaum, J. M. (1997). An indirect approach to warfare: Attacking an enemy's moral forces. In: *Office of History*. Maxwell AFB, AL: Headquarters Air University.

Levine, D., & Hoffer, W. (1991). *Inside out: An insider's account of wall street*. New York: Putnam Publishing Group.

Liddell-Hart, B. H. (1967). *Strategy*. New York: Praeger.

Lovelace, D., Jr. (1997). The evolution in military affairs: Shaping the future of U.S. armed forces. *Strategic Studies Institute* (June 16).

Lugar, R. (1999). The threat of weapons of mass destruction: A U.S. response. *Nonproliferation Review* (Spring-Summer).

Mahmoud, A., & Lemon, S. (2002). Summary and assessment. In: S. Knobler, A. Mahoud & L. Pray (Eds), *Biological Threats and Terrorism* (pp. 1–19). Washington, DC: National Academic Press.

McKenzie, K. F. (2000). *The revenge of the Melians: Asymmetric threats and the next QDR* (pp. 1–64). Washington, DC: Institute for National Strategic Studies, National Defense University, 2000.

Metz, S., & Johnson, D. V., II (2001). *Asymmetry and U.S. military strategy: Definition, background and strategic concepts*. Carlisle, PA: Strategic Studies Institute, U.S. Army War College.

Myers, R. (2002). The U.S. military: A global view of peace and security in the 21st century. *Electronic Journal of the Department of the State, 7*(4).

The Office of the Department of Defense (2002). Review of reserve component contributions to national defense (December 20).

Perl, R. (1997). Terrorism, the media, and the government: Perspectives, trends, and options for policymakers. *CRS Issue Brief* (October 22).

Posner, G. (2003). *Why America slept: The failure to prevent 9/11*. New York: Random House.

Prins, G. (2002). *The heart of war: On power, conflict and obligation in the twenty-first century*. London and New York: Routledge.

Segarra, A. (2002). Agroterrorism: Options in congress. *Report for Congress* (July 17).

Smith, J., & Thomas, W. (2001). The terrorism threat and U.S. government response: Operational and organizational factors. U.S. Air Force Academy, Colorado: USAF Institute for National Security Studies.

Thompson, R. (1989). *Make for the hills: Memoirs of far Eastern wars*. London: Leo Cooper.

Vandenbroucke, L. (1993). *Perilous options: Special operations as an instrument of U.S. foreign policy*. New York: Oxford University Press.

PART II:
CHAOS, COMPLEXITY
AND CHANGE

CHAOS AND COMPLEXITY IN A BIOTERRORISM FUTURE

Reuben R. McDaniel Jr.

ABSTRACT

Preparing for a potential bioterroism is a difficult task for health care leaders because of the fundamental unpredictability of bioterroist acts. Complexity science thinking is presented as an approach that can help in this task. Basic concepts from complexity science, especially the role of relationships, are presented. Specific recommendations for action including sensemaking, learning, and improvisation are made. A case study is used to illustrate the power of complexity science thinking in assisting health care leaders addressing potential bioterroism. Questions for further research are presented.

Bioterrorism is a real and present danger in the modern world. Understanding how to respond to this danger is difficult at best and impossible at worst. Perhaps the most we can do is to think seriously and deeply about plausible situations and circumstances in which bioterrorism might occur, think about how we might avoid this occurrence, and consider possible responses if bioterrorist acts do occur. Clearly, we will need all of our intellectual resources and wisdom if we are to do this in a reasonable way. While there are many possible paradigms and frameworks that will be useful in this endeavor, the complexity sciences offer some helpful perspectives, as well as some ways to possibly see things that might be missed using other perspectives. The purpose of this chapter is to bring complexity science thinking to the discussion of the role of health care organizations

Bioterrorism, Preparedness, Attack and Response
Advances in Health Care Management, Volume 4, 119–139
ISSN: 1474-8231/doi:10.1016/S1474-8231(04)04005-4

and health care leadership, in a world confronted with potential bioterroism. In particular, the chapter will consider how health care leaders might use some ideas from considerations of chaos and complexity to think more creatively about the meaning of bioterrorism for their endeavors. This chapter is written as an essay because it is fundamentally different from a traditional report of empirical research or theory development. Instead, it is an attempt to share thinking, informed by complexity science, about health care leadership in an age of bioterrorism.

Considerations of institutional responses to the threats of bioterrorism almost invariably admonish institutions and their leadership to be prepared. There is agreement that health care organizations have an obligation to be prepared, or to get prepared, for bioterrorist attacks. The difficulty comes in trying to answer the question, "be prepared for what?" Should one be prepared for an attack of anthrax through the mail? Or perhaps one should be prepared for a smallpox epidemic. Should one be prepared for a local attack to panic, or a mass attack to destroy? Is preparation for bioterrorism attacks on the food and/or water supply a reasonable approach, or will the attacks likely be more direct? What is the comparative likelihood of a biological attack vs. a chemical attack, and how does one prepare differently for each? These questions, and many others like them, confront health care leaders as they think about questions surrounding bioterrorism issues. Complexity science suggests some strategies for addressing the question; "be prepared for what?" and this chapter will attempt to use complexity science to shed light on this question.

The essay is organized as follows. It begins with a brief discussion of some basic concepts from complexity science, and this forms the basic framework that guides the heart of the discussion of the question, "be prepared for what?" The conventional wisdom that one might follow in thinking about preparing for a bioterrorist attack is identified. This is followed with ideas from complexity science that suggest alternative approaches to the conventional wisdom as well as an attempt to specify these approaches in a clear and comprehensive manner, paying particular attention to the role of relationships. This section of the essay is an attempt to be specific in understanding the recommendations that flow from complexity science when thinking about how health care leaders and their organizations can be effective in an era of bioterrorist danger. It briefly suggests ways people might engage in effective action to be more successful in responding to bioterrorism, and closes with some general research questions suggested by the analysis.

Be assured, there is no universal approach or universal wisdom regarding bioterrorism. This chapter presents a complexity science approach that opens the door for more creative and innovative methods for leading health care organizations in the face of this danger. Health care organizations will clearly be fundamental in any societal response to bioterrorism and therefore, any assistance that can be provided to health care leadership as they face this reality is extremely important.

COMPLEXITY SCIENCE

Complexity science (Anderson, 1994; McDaniel & Driebe, 2001) is a loosely connected set of inquiries targeted at understanding two interesting phenomenon. First, how can fairly simple systems generate very complex behaviors? This is normally seen as the domain of chaos theory (Gleick, 1987). Secondly, how can fairly complex systems generate simple patterns of order? This is the domain of what is normally referred to as complexity theory (Waldrop, 1992). These two questions, how do we generate complexity from simplicity and how do we generate simplicity from complexity, have become very important to sciences at all levels (Capra, 1996). Science traditionally has assumed that complex things were generated by complex patterns of interaction and that simple patterns of interaction generated simple things. But complexity science suggests that the world may operate in more fascinating ways. Complexity science differs from "normal" science in its attention to these intriguing relationships between the simple and the complex (Cowan, Pines & Meltzer, 1994; Driebe, 2000; Ford, 1989).

There are many domains of complexity science. To name just a few, there is the study of dissipative systems (Prigogine & Stengers, 1984), the study of self-organized criticality (Bak, 1996; Jensen, 1998) and the study of cellular automata (Wolfram, 2002). Of particular interest to the present discussion is the study of complex adaptive systems (Cilliers, 1998; Gell-Mann, 1994; Holland, 1995; McDaniel & Driebe, 2001; Plesk, 2001). Complex adaptive systems were, perhaps, most notably studied by the group of scientists from the Santa Fe Institute in their early explorations of complexity (Waldrop, 1992) but these systems are now the subject of considerable study in areas including biology (Camazine et al., 2001) and economics (Arthur, Durlauf & Lane, 1997). There is an emerging consideration of these systems with respect to organizations (McDaniel & Driebe, 2001; Stacey, 1992) and this consideration forms the basis of this discussion. Complexity science had caught the attention of the health care community. The Institute of Medicine in its recent report on quality issues in health care suggests that the application of complexity science to health care offers one of the best paths to a better health care future (IOM, 2001).

COMPLEX ADAPTIVE SYSTEMS (CAS)

Health care organizations are complex adaptive systems (CAS) and as such, share the common properties of these systems (McDaniel & Driebe, 2001; Plesk, 2001; Zimmerman, Lindberg & Plesk, 1998). A readily recognized property of CAS is that they are composed of separate and diverse elements often know as agents (Cilliers, 1998). When we think of organizations, we often think of these agents as people; certainly people are often agents in health care organizations conceptualized as CAS. However, agents may be subsystems within the larger organization, such as the accounting department or the surgery department. Certain processes such as inventory control, or infection control may also be agents in a CAS. These diverse agents have the capacity to exchange information and energy with other agents, and to change their behavior based on these exchanges. They have the capacity to learn; to gather information and use this information to guide future behavior (Capra, 2002; Kaufman, 1995).

The most critical characteristic of CAS is that the relationships among agents are nonlinear and dynamic (Axelrod & Cohen, 1999; Capra, 1996, 2002; Kaufman, 1995). This suggests that outputs of system behavior may well not be proportional to inputs. A small effort at correcting a nurse's behavior may result in a large disruption in a department. Likewise, a massive effort to establish a quality improvement system may result in almost no change in the day-to-day behavior of people. It also suggests that the interactions among agents exist over time and may change over time, independent of intentions. The positive effects of a reward for good performance may decay over time as the recipient comes to see that reward as a right rather than a motivator. It is the dynamic, nonlinear interactions among diverse agents with the capacity to change because of these interactions, which leads to complexity (Cilliers, 1998; McDaniel & Driebe, 2001; Plesk, 2001).

CAS self-organize, using their web of connections to configure and reconfigure themselves, independent of any hierarchical control. Many companies are utilizing this property to develop solutions to complex problems (Bonabeau & Meyer, 2001). Physicians form informal groups that have a "mind of their own" as they discuss hospital policies for cost control. Patterns of relationships within a surgical team create a capacity to learn (or not learn) as they attempt to incorporate a new procedure into their practice (Edmonson, Bohmer & Pisano, 2001). Health care organizations must maintain respect for self-organized sets of relationships and not simply discount them as informal and, therefore, unimportant. It is the set of self-organized relationships that is likely to be most able to respond to new occurrences and to help leaders cope with bioterrorist attacks. Organizations in which there are a large number of high quality interactions among agents are more capable of effective self-organization that is supportive of overall systems behavior (Camazine et al., 2001; Capra, 2002; Waldrop, 1992).

CAS have emergent properties, and therefore one cannot understand a CAS by simply understanding the agents (Holland, 1998; Johnson, 2001). These emergent properties of CAS are what really give each health care organization its distinctive characteristics. Capra (2002) puts it very directly:

When carbon, oxygen and hydrogen atoms bond in a certain way to form sugar, the resulting compound has a sweet taste. The sweetness resides neither in the C, nor in the O, nor in the H; it resides in the pattern that emerges from their interaction. It is an emergent property. Moreover, strictly speaking, the sweetness is not a property of the chemical bonds. It is a sensory experience that arises when the sugar molecules interact with the chemistry of our taste buds, which in turn causes as set of neurons to fire in a certain way. The experience of sweetness emerges from that neural activity (pp. 41–42).

The quality of health care for a given hospital is an emergent property, not simply because of the doctors or the nurses or the food service, but because of the interaction of these agents with each other and with the patient that creates quality health care. Health care quality is an emergent property of the hospital. Likewise, when a hospital board develops a lack of responsible oversight the irresponsibility is usually an emergent property of the board rather than the result of a single member's attitude or behavior. Although a single member, such as the chairman of the board or the head of the audit committee might be held accountable, the characteristic of irresponsibility is one that emerges from the patterns of interactions within the board and between the board and its environment.

The quality of these connections is a key determinate of the quality of the emergent properties of the system. Emergent properties are not static in nature but are dynamic (Goldstein, 1999). This means that the emergent properties of a system require continuous attention. When connections are good and relationships are well developed emergent properties are congruent with each other and consistent with each other over time (Johnson, 2001). One can ask "what are the emergent properties of a health care organization that are likely to be most helpful in a bioterrorist attack?" Among such emergent properties is certainly a willingness to do whatever is necessary to make things work, calmness when unexpected things happen, and a strong understanding of the intersection of psychological and physiological needs of people in crisis. The development of these kinds of emergent properties requires attention from organizational leadership, and especially attention to relationships.

We often think of the need for systems to adapt to their environment. This is seen as a prerequisite for survival. The study of complex adaptive systems shows us that organizations truly co-evolve with their environment rather than simply adapt to their environment (Boisot & Child, 1999; Stacey, 1995; Wheatley, 1992). Each move by an organization elicits a response from the environment. Therefore, each action taken to "adapt" to a particular environmental circumstance will change that circumstance and, thereby, make the adaptive action inappropriate. An insurance company establishes a procedure-by-procedure payment system and, lo and behold, the hospital does more procedures. So you try again, only to find that you face new problems as co-evolution occurs.

The notion of fitness landscapes (Kaufman, 1995; Levithal & Warglien, 1999) is helpful here, "Organisms, artifacts and organizations all evolve and co-evolve on rugged, deforming fitness landscapes.... Tracking peaks on deforming landscapes is central to survival" (Kaufman, 1995). The CAS not only responds to a fitness landscape, but it shapes that landscape and that shaping is a function of the interrelationships and connections in the system and between the system and its environment. The notion of co-evolving fitness landscapes suggests that leaders must be willing to risk doing worse in order to do better; you may have to go down one peak in order to position yourself to go up another, higher peak. Hospitals buy primary care practices and the pattern of primary care in the community changes. Cardiologists set up a heart hospital and new cardiologists are brought into town by the old hospital so that the old hospital can maintain its line of business.

The effort to adapt to the environment is, in a real sense, endless and full of surprises. Co-evolution continues to occur as CAS continue to seek to improve their position on an ever-changing landscape (Johnson, 2001; Kaufman, 1995; Levithal & Warglien, 1999). When health care organizations respond to a bioter-rorist attack, they will, in fact, shape the nature of events as well as be shaped by the nature of events. Leadership must be aware of this interaction and develop strategies to monitor the unfolding situation. It is not adequate to do something in response to a situation and then expect that response to have no effect on the situation itself.

Health care organizations are CAS and the properties of CAS, taken together, lead to fundamental uncertainty in the dynamic unfolding of health care organizations and of the environment in which they operate (McDaniel & Driebe, 2001; Plesk, 2001). Plesk puts the state of affairs this way,

> CAS science suggests that we cannot hope to understand a priori what a CAS will do or how to optimize it. A design cannot be completed on paper. Past attempts to do this in health care have not succeeded in part because they may not have been satisfactory designs, but mainly because a new understanding of "design" is needed (pp. 327–328).

The learning capacity of agents and the nonlinear interactions among agents results in self-organization, emergence and co-evolution, all unfolding in an unpredictable fashion. While natural aversion to uncertainty, coupled with the historical reliance on the certainties implicit and explicit in Newtonian thinking (Capra, 1983; Wheatley, 1992), leads us to want to avoid surprise and fundamental uncertainty, we cannot. "Perhaps the most important thing we are learning today about organizations is that if they are going to succeed, they must give up their obsession with control, knowing what is going on, and seeking stability" (Bergquist, 1993; Stacey, 1992; Wheatley, 1992) (McDaniel & Walls, 1997,

p. 374). This awareness can inform our thinking as we prepare for bioterrorism. But first let us touch on the conventional wisdom about preparedness.

CONVENTIONAL WISDOM

The conventional wisdom about how to prepare to deal with bioterrorism derives from the conception of the world as a Newtonian place, and this historical view dominated scientific thought well into the 20th century (Capra, 1983; Driebe, 2000; Wheatley, 1992). This view suggested that the ideal of science should be an objective description of the world that would lead to a command and control approach to managing situations (Weber, 1971). Out of this approach, processes for rational decision-making have been developed (Klein, 2001) and these form the conventional wisdom about developing an appropriate capacity for responding to bioterrorism.

Start by figuring out the alternative possible scenarios and then figure out the probability of each. Compute the cost of each. Probability times cost equals the seriousness of the threat. The values of variables might be determined in a variety of ways, but one often suggested is to appeal to experts for their vision of possible scenarios and for estimates of cost and probabilities. Another is to allow some kind of political process to set the parameters, as is usually the case when legislative decisions are made about the allocation of resources to combat terrorist acts. In any case, according to conventional wisdom, the projected seriousness of alternative threats must be determined.

Then, because resources are never limitless, one must make plans for dealing with the most serious threats. This means calculating the resources needed and determining how these are to be deployed. There must be an organization for executing the plan. This includes defining roles and responsibilities of various units. The right people must be hired and trained. Careful assessments of readiness must be conducted. Audits of preparedness for each serious threat must be done and appropriate adjustments must be made. A variety of simulations are often conducted for purposes of assessment, training and development of a knowledge base for action. The result of this activity, within this viewpoint, is preparedness.

Conventional wisdom regarding approaches to bioterroism preparedness depends on organizational capacity for forecasting and calculation, and is a reductionist view of the world. The proper way to solve (or resolve) a problem is to break it down into manageable segments and deal with each segment. Then reassemble these segments into a response. This is considered to be logical thinking for a Newtonian world (Capra, 1996; Wheatley, 1992).

A COMPLEXITY SCIENCE APPROACH

Modern science, and particularly complexity, science suggests that reductionist processes, despite their apparent logical framework, may not be the only or even the best way to cope with environments, such as the threat of bioterrorism, that are far from equilibrium (Anderson, 1999). Reasons for this view of the inadequacy of traditional reductionist thinking lie in the very characteristics of complex adaptive systems; learning agents, nonlinear interdependencies, self-organization, emergence and co-evolution. These characteristics, which lead to fundamental unknowability, cannot be ignored. They suggest that our capacity to forecast and plan, at least in the conventional view of planning, is minimal at best.

A complexity science approach to the question, "be prepared for what" is to be prepared to respond to the unanticipated using systems that are themselves complex adaptive systems and, therefore, unpredictable in their unfolding over time. The problem becomes to get ready to execute reasonable behavior in the face of events when you have no way to prepare for a specific, predictable event. Do not try to outguess the dynamic unfolding of events because the ongoing dynamic is unpredictable. Pietro D. Marghella (2002), Medical Plans Operations Officer for the Joint Chiefs of Staff put the issue this way:

> No plan withstands contact with the "enemy." Flexibility will always be the key to success. The events of 9/11 altered *all* existing plans: the lesson is, be prepared. Each event will be unique and unpredictable. The ability to rapidly assess and adapt to very fluid situations will be critical. Therefore, assess, adapt, tailor available resources and respond.... Interagency trust and coordination is therefore of paramount importance, as you will invariably rely on others to complement or complete your resources. In the end, perhaps the best weapon the professional planner can field is his or her Rolodex, as effective communications with appropriate respondents is critical to the successful consequence management mission (p. 22).

As one can ascertain from the above quote, and consistent with our understanding of complex adaptive systems, response to the unanticipated requires that we pay attention first of all to issues of relating. This is so that self-organization, emergence and co-evolution can take place in a manner that leads to appropriate sensemaking, learning and improvisation in the face of unfolding events of bioterrorism. Complexity science suggests that in addition to attempting to forecast what might happen in a bioterrorist attack and planning to meet the expected conditions, we might well focus on the development of health care organizations with systems of relationships that will enable the organizations to be better prepared to act effectively in an uncertain world.

RELATIONSHIPS

As one considers the development of relationships in organizations or in systems of organizations one must carefully consider the role of leaders. (For economy of expression the term "leader" will refer to all of those who play a role in the shaping of CAS, including those who might call themselves managers or administrators and also including political actors such as members of legislative bodies or executive branches of government.) The leader is not removed from the system, overseeing its development but is an integral part of the system, both effecting and being effected by the patterns of relationships that develop over time (Stacey, Griffin & Shaw, 2000). Therefore, the development of positive systems of relationships is not a process of a leader telling others how they should behave with respect to each other, but it is the result of an emerging property of the CAS. However, leaders do have the capacity to influence the way patterns of relationships develop because of the legitimate or position power normally invested in them. Relationships in CAS are nonlinear (Cilliers, 1998) and therefore one cannot establish with precision what a relationship will be over time. However, one can influence relationships so that they facilitate, rather than hinder, positive system development.

Relationships can be characterized by either strong or weak ties (Granovetter, 1973). Because we usually enjoy the good feelings that derive from strong ties, there is often an emphasis in relationship systems on cohesiveness and closeness. However there are significant benefits to be derived from weak ties. Weak ties can lead to more effective bridges between subgroups, richer sets of interactions within and outside of the CAS, better diffusion of information and a higher probability of the discovery of critical information (Putnam, Phillips & Chapman, 1996). It is also possible that people with weak ties may be more willing to share ideas, to be early adopters of new technologies and be less prone than those with strong ties to be subject to social pressures and group think (Papa, 1990). This suggests that preparations for bioterrorism should not attempt to simply establish strong relationships among all of the actors. Response is likely to be more effective when there is a balance of weak and strong ties in the system.

In a similar vein, communications within systems can be either rich or lean (Daft, 1989). Rich communication relationships can carry high levels of information and are most useful when problems are ambiguous. Lean communication relationships carry lower levels of information and are most useful when problems are highly structured with clear lines to problem solutions. While organizations certainly deal with both kinds of information, complexity science suggests that ambiguity is more likely to be a characteristic of dynamic situations than might be supposed. Certainly, situations surrounding bioterrorism will be unclear and present complex problem solving challenges. This suggests the need to develop rich linkages. When information transmittal is face-to-face, the information transfer is much richer than when information transmittal is through formal written and numeric documents (Daft & Lengel, 1984). But many health care organizations have developed the habits of "everything in writing" and treating problems as though they were well structured. What is needed is further development of rich information mediums so that they will be ready and available. Systems should practice face-to-face discussion and telephone conversations so that they will be prepared to use these relatively rich processes when they are needed. Simply getting everything in writing and everything in the book of standard operating procedures won't do when faced with the uncertain unfolding of events in CAS. It is unlikely that practice guidelines will be much use when faced with a bioterrorist attack. Health care organizations need to pay careful attention to the development of a capacity for rich relationships rather than to focus all of their attention on highly structured and formalized systems of communications.

Interdependencies in systems may be characterized as pooled, sequential or reciprocal (Daft, 1989; Thompson, 1967). In pooled interdependence each element of a system makes a discrete and independent contribution and rules and procedures should be used to standardize activities across elements. In sequential interdependence, contributions take a serial form and the order of contribution can be specified and coordination is required. Sequential interdependence places a heavier burden on leadership than does pooled. The heaviest leadership burdens are created when a system is characterized by reciprocal interdependence. In reciprocal interdependence, the outputs of each element become inputs for others and mutual adjustment is required for proper functioning. The organizational structure must allow for frequent communication among elements. The nonlinear relationships among agents often create a situation of reciprocal interdependence and certainly this is true for unique and unpredictable events such as those present in a bioterrorist situation. In CAS, special attention must be paid to tactics and strategies for managing reciprocal interdependencies.

Health care organizations preparing for a bioterrorist attack must strive to develop high-level capacities for maintaining and managing loose ties, for engaging in rich information exchanges and for enabling reciprocal interrelationships. Complexity science suggests that the rational for this is the need to facilitate positive self-organization, constructive emergence and effective co-evolution.

What is true when systems of highly developed relationships have facilitated positive systems characteristics? The organization will be characterized by respectful interactions, mindfulness and the development of a collective mind (Weick, 1993; Weick & Sutcliffe, 2001; Weick, Sutcliffe & Oberstfeld, 1999; Weick & Roberts, 1993). Weick and his colleagues have developed these ideas quite fully and, therefore, no attempt is made to develop them here. Rather, a short definition will be provided for each concept from which it will be clear that these characteristics are critical to health care organizations in a bioterrorist future. You are engaged in respectful interaction when you,

(1) Respect the reports of others and (are) willing to base beliefs and actions on them (trust); (2) Report honestly so that others may use your observations in coming to valid beliefs (honesty); and (3) Respect your own perceptions and beliefs and seek to integrate them with respect of others without depreciating them of yourselves (self-respect) (Weick, 1993, pp. 642–643).

Mindfulness is the "joint capacity to induce a rich awareness of discriminatory detail and a capacity for action" (Weick, Sutcliffe & Oberstfeld, 1999, p. 88). CAS that manage mindfully, "organize themselves in such a way that they are better able to notice the unexpected in the making" (Weick & Sutcliffe, 2001, p. 3). Collective mind is intimately related to the achievement of reliable performance, "Reliable performance may require a well-developed collective mind in the form of a complex attentive system tied together by trust" (Weick & Roberts, 1993, p. 378). "Contributing, representing, and subordinating, actions that form a distinct pattern external to any individual, become the medium through which collective mind is manifest" (Weick & Roberts, 1993, p. 364). These attitudes each require paying attention in a world where attention is likely to be in short supply.

The contribution of complexity science to this line of thinking is to tie these ideas together in terms of the basic characteristics of complex adaptive systems. Respectful interaction, mindfulness and collective mind are all social acts, grounded in the nonlinear interactions among agents and leading to positive self-organization, emergence and co-evolution. These ideas, therefore, are basic in a world of fundamental uncertainty.

Two other ideas must be considered when examining issues of relating in CAS. The first to these is the idea of participation. When health care organizations are considered as complex adaptive systems, a strong case can be made for making organizations more complex internally through a fairly simple managerial rule – using participative decision making (Ashmos, Duchon, Huonker & McDaniel, 2002). Participation is a strategy for enhancing connectivity in organizations. This enables more effective self-organizing, emergence and co-evolution than when there is an autocratic management system that discourages connectivity. Recent research has shown that organizations with greater participation, well managed, out-perform those with less participation (Ashmos, Duchon, Huonker & McDaniel, 2002). Because respectful interaction, mindfulness and collective mind are all social acts, they can be considerably enhanced by a participative style of leadership. Participation requires practice. If leaders want people to be able to contribute in the time of crises then people must be given practice at participative behavior before crisis occurs. When they have such practice they are much more likely to respond positively when faced with unexpected events.

The second key idea is trust. Effective CAS are not built on bureaucratic structures that create expectations and obligations. They are built of trust and on systems of trust. Trust is clearly a relationship-centered phenomenon and the social aspect of trust is critical (Rousseau, Sitkin, Burt & Camerer, 1998; Tyler & Kramer, 1996). McGuire and Anderson (1999), in discussing the nation's health care dilemma speak clearly about the role of trust in the establishment of a functioning health care system. "A provider may occasionally scare a patient into a new lifestyle, yet most effective and lasting behavior changes occur when provider and patient trust each other and both understand that the patient's survival is the basis for their interaction" (p. 114). Trust is seen as fundamental to the effective functioning of health care systems and yet, trust is in jeopardy in the modern health care system (Annison & Wilford, 1998; McGuire & Anderson, 1999). Recent research on the utilization of telemedicine systems shows that a primary reason that these systems are not used is the lack of trust between the clinicians on the two ends of the virtual connection (Paul & McDaniel, 2002). Trust is a key to the development of respectful interaction, mindfulness and collective mind. Thus, it can be seen that for a health care system to be effective in developing the relationship systems required for bioterrorism readiness, that system must be effective in developing trust within the system and between the system and environmental elements.

Participation in decision making and development of interpersonal trust are key elements in developing meaningful relationships. Without these, the system of relationships in the organization is likely to be sterile and unproductive.

EFFECTIVE ACTION

Given the scheme of characteristics and properties that describe effective CAS, what are some of the specific actions that might be required? The first of these is sensemaking. Sensemaking is often seen as the human response to bounded rationality. However, in an uncertain world, sensemaking is a strong positive response that enables action (Thomas, Clark & Gioia, 1993; Weick, 2001). Sensemaking requires intensive interaction as people try to discover who they are, why they are there, and what is happening to them. Leaders must recognize that sensemaking is a social act, and must create opportunities for the necessary interactions to occur. Sensemaking is enhanced through diversity among agents (McDaniel & Walls, 1997). Therefore, there is a need to have wide representation at the decision-making tables (Ashmos et al., 2002). Preparation for bioterrorism is too important to be left to the experts because making sense of the situation is necessary for action at all levels and in all domains.

A second focus of attention for action is learning. However, the most appropriate learning may not be learning in preparation for action, but learning while action is going on. As noted by Stacey (1995), "The most important learning we do flows from the trial-and-error action we take in real time and especially from the way we reflect on these actions as we take them" (p. 17). Healthcare organizations, as they attempt to prepare for a bioterrorism future, must become skilled at learning from samples of one (March, Sproull & Tamuz, 1991). This suggests that they attend to multiple interpretations of unfolding events and pay careful attention to various ways that stakeholders might perceive the world. What looks to one like a satisfactory adjustment to the need for more space in an emergency room, may look to another as the need for more aggressive, and accurate triage systems. This may lead to conflict over resource utilization. If attention is paid to both views, it is more likely that learning about the situation will be achieved and a satisfactory resolution attained. In traditional views of organizational life, knowledge is the key, but in a complexity view the capacity to learn is the key. Certainly this capacity to learn is built on existing knowledge, but the focus shifts. In a bioterrorist attack we want systems that can learn what to do rather than systems that know what we did right (or wrong) in the last terrorist attack.

A third focus of attention for action is improvisation. When the unfolding of events is uncertain at best, improvisation becomes key to success. Crossan and Sorrenti (1997) define improvisation as "intuition guiding action in a spontaneous way" (p. 156). The capacity of a system to improvise is dependent on the degree to which the system understands its resources and the degree to which people can work together to exploit these resources (Crossan, White, Lane & Klus, 1996). Improvisation is not a causal activity that "kind of happens." It´is a disciplined craft that can be learned and applied to situations as they arise (Weick, 1998). When CAS are characterized by healthy patterns of relationships, their capacity for improvisation goes up. When physicians and nurses communicate freely about the clinical situations they face, their potential for improvisation in the face of the unexpected expands. They are better able to invent, on the spot, the behaviors that are likely to lead to improvements in the situation. This aptitude is critically important when events are unfolding rapidly in ways that cannot be predicted.

Effective action is characterized by sensemaking, learning and improvisation. These are difficult to achieve. Each, however, is critical to health care leadership in an era of bioterrorism.

A CASE IN POINT

In order to gain a better understanding of the concepts discussed above, it would be instructive to consider a specific case in point. The Fall 2002 issue of *Frontiers* was devoted to a discussion of "When Disaster Strikes: Healthcare's Response." One of the lead articles in that discussion was by David J. Campbell, FACHE, "9/11: A Healthcare Provider's Response" (pp. 3–13) and this article outlines the response of Saint Vincent Catholic Medical Centers' (SVCMC) eight hospitals to the attack on the Word Trade Center. St. Vincent's Manhattan was the nearest level 1 trauma center to the disaster and, as such, was centrally involved in the immediate aftermath of the bombings. There is no evidence in the article that Mr. Campbell was using any concepts from complexity science to describe the events or to analyze the unfolding state of affairs. However, when one reads his account through the lens of complexity science, the situation at St. Vincent's certainly illustrated the applicability of the concepts discussed here (Campbell, 2002).

St. Vincent's had a diverse set of agents that interacted to respond to various events. The team of clinicians that was called upon to speak to the press represented a wide variety of specialties, and was able to create a comprehensive and trusting relationship with the media (p. 11). Volunteers flooded the hospital to offer their services (p. 4). The situation demanded not only acute care physicians, but also behavioral specialists to deal with the multiple effects of the tragedy (p. 5). Nonlinear interactions among the hospital staff, volunteers, city government and other hospitals had to be managed (p. 11).

Self-organization within the system was demonstrated in several ways, such as when the role of the family center shifted from a source of information to providing counseling to families and friends of victims (p. 5). The behavioral staff was self-organizing when it found itself serving, not only the direct victims of the tragedy but hospital employees as well (p. 5). People from a wide variety of hospital sections were self-organized to meet the emerging communication needs on a 24 hour-a-day basis (p. 11). Systems properties emerged as the overwhelming need of people for information and relief from horrors that had been witnessed became a priority (p. 5). SVCMC's computers emerged as the consistent communication link when telephone service was disrupted (p. 4). Co-evolution took place as the city provided water to compensate for broken water mains (p. 4) and as the New School University provided space to relieve an impossibly crowded family center (p. 5).

Relationships were the key for much of St. Vincent's success in responding to the disaster. The first "lesson learned" reported in the article had to do with the need for regional planning for regional response to disasters, and the critical need to have appropriate liaison strategies developed before they are needed (p. 6). The Greater

New York Hospital Association is in the process of developing a comprehensive emergency contact directory so that organizations that do not have tight ties with each other can be of assistance to each other (p. 8). SVCMC has built on "its history of community outreach, its preexisting relationships, and the skills, ability and dedication of its staff to address [the issues of behavioral trauma]" (p. 10). Characteristics of respectful interaction, mindfulness and collective mind can be clearly seen in SVCMC's response to the psychological needs of those directly affected by the tragedy (p. 5). They can also be seen in the response to the need to provide accurate and timely information to the media (p. 11). Improvisation was the order of the day as surprises; some positive such as the influx of volunteers and donations that flooded the system (p. 4), and others more problematic, such as the torrent of families and friends searching for loved ones that threatened to overwhelm the hospital (p. 5), had to be managed. In both cases, the capacity of SVCMC to improvise and develop solutions on the spot was key in their ability to cope with the situation.

Complexity science calls our attention to aspects of the situation at St. Vincent's after the 9/11 disaster that might otherwise be missed. These aspects provide significant opportunities for other health care organizations to learn from the St. Vincent's experience.

In particular, complexity science calls our attention to the relationship system that had been built over time and that enabled the hospital to respond in unplanned ways can easily be overlooked if one is limited to traditional perspectives of organizational life. Also the capacity of the staff to self-organize to solve new problems can be missed from traditional perspectives. Complexity science calls our attention to the fact that things did not necessarily work according to plan but that the existence of a system characterized by respectful interaction, mindfulness and collective mind has tremendous capacity for sensemaking, learning and improvisation. This suggests that the nature of CAS should be fully recognized when planning for bioterrorism attacks, qualities that lead to more effective behavior should be nourished, and attention should be paid to the level of preparedness that is present to deal with the unexpected. SVCMC was prepared to respond to the unanticipated.

Certainly, one can easily envision things that might have been better and certainly SVCMC should change to improve its response to terrorism. But the 9/11 disaster cannot be seen as a signal to prepare for planes crashing into the World Trade Center. That is unlikely to happen again. The problem is not to learn from 9/11 how to respond to 9/11, but to learn how to respond to some different terrorist activity. Complexity science suggests that the events of 9/11 must be seen as a signal to strengthen the capacity of the system to self-organize, utilize emergent properties and to co-evolve with the environment.

RESEARCH QUESTIONS

When we consider the arguments developed in this essay, several research questions come to mind. These are questions that face leaders as they contemplate the value of a complexity science approach to leadership in a health care organization.

(1) How rapidly can self-organization, emergence, and co-evolution take place? Surge effects are likely to overwhelm a system in a terrorist attack. What does complexity science have to say about surge effects and how can self-organization, emergence and co-evolution take place in the presence of surge effects or environmental shocks?

(2) How can technology facilitate self-organization, emergence and co-evolution? Much of technology development in the past has focused on the value of information in a world where information was the source of predictability. Complexity science suggests that much unpredictability is fundamental, not a result of inadequate information. Can technology make a positive contribution to the development of positive relationships in CAS? What is the role of technology in helping systems cope in an unpredictable world?

(3) What are the respective roles of clinicians and administrators in developing a sense of respectful interaction, mindfulness and collective mind throughout the health care organization? Complexity science suggests that diversity in agents is a very positive factor when the relationship system among the agents is well managed. In health care organizations the disparate roles of clinicians and administrators sometimes makes relationship management difficult. How can the groundwork be laid in a heath care organization so that the necessary positive attitudes are developed and maintained?

(4) How do you train people to be effective in complex adaptive systems; how do you train people to be ready for the unexpected? What is the meaning of the word "practice" in a world where we must learn from samples of one? What is the set of experiences that will prepare those who must be first responders to grasp patterns and intuit appropriate responses? How can simulation exercises and computer simulations be used most effectively when neither can simulate the actual course of events?

(5) There needs to be considerable research on the emergent properties of health care organizations that are likely to be useful in crisis situation. In addition, little is known about how to guide the development of emergent properties. What is the set of member and leadership behaviors that will elicit emergent properties of health care organizations that are necessary for satisfactory responses to bioterrorist attacks?

Complexity science demands a dynamic approach to research rather than the static approach that often characterizes organizational research. It is necessary to look at things "over time" in order to understand the phenomenon in question. We must identify and, where necessary, invent research strategies that produce a depth of understanding in circumstances where prediction, the normal indicator of understanding, is not possible. These research designs and methodologies may not be characteristic of traditional research in heath care organizations. More creative use of case study designs including the imaginative use of multiple case designs will likely be required. Ethnographic studies that enable very deep probes into organizational life will be required. Research into complexity science inspired research questions will require the use of the whole range of existing qualitative techniques and the invention of new techniques. New quantitative designs will also be required in order to capture the nonlinear dynamics of complex adaptive systems. Simulations including both computer simulations and "live action" simulations must be utilized in order to gain some insights into phenomenon, such as bioterrorism attacks, that cannot be studied in real life. The challenges for the research community are daunting when we recogniz the nature of health care systems as complex adaptive systems and the fundamental characteristics of these systems. These challenges, however, must be confronted if the research community is to effectively assist health care leadership in facing the possibility of bioterrorism while

CONCLUSIONS

The role of complexity science in thinking about leadership issues facing health care organizations in a world confronted with the threat of bioterrorist attack is to bring a different set of ideas to the table. Traditional thinking suggests that people should respond to rules; do what they are supposed to do perfectly. Complexity science suggests that people must be prepared to respond to unpredictable, dynamic situations. Clearly both things are true at the same time. But we are used to thinking about health problems as naturally occurring events not as planned bioterrorist attacks on our well-being, and complexity science may have some particularly important contributions to make as we try to think about this new situation.

Complexity science, and in particular theories of Complex Adaptive Systems (CAS), suggest perspectives that focus leadership attention on several items that might be overlooked if only a conventional approach to organizational analysis is used. From a CAS perspective, health care organizations self-organize, emerge and co-evolve. This suggests that in order to prepare organizations for the unexpected particular attention must be paid to the nonlinear interdependencies among organizational elements. Leaders must give consideration to the value of loose ties, rich communication linkages and reciprocal relationships because these can enable an organizational to respond in reasonable and helpful ways when the unexpected occurs. CAS are most effective when the members have attitudes that promote respectful interaction, mindfulness and the development of collective mind and when leadership is participative and built on a bed of trust. In a crisis situation, these attitudes are critical because when health care organizations nurture these factors, they will be able to take effective action that includes the capacity for sensemaking, learning and improvisation. In this fashion, health care organizations and their leaders can be prepared to cope with unfolding events that could not be planned for in the traditional sense.

The problem facing health care leadership in a world where bioterrorism is a real possibility is to get ready to execute reasonable behavior in the face of events when you have no way to prepare for a specific, predictable event. Complexity science, and in particular, the theory of complex adaptive systems, can help in this endeavor. It has been the purpose of this chapter to suggest how complexity science can help and to point to elements that must be considered when the view of complexity science is used to examine the problems presented by bioterrorism.

REFERENCES

Anderson, P. (1999). Complexity theory and organization science. *Organization Science, 10*(3), 216–232.

Anderson, P. W. (1994). The eightfold way to the theory of complexity: A prologue. In: G. A. Cowan, D. Pines & D. Meltzer (Eds), *Complexity: Metaphors, Models and Reality* (pp. 7–16). New York: Addison-Wesley.

Annison, M., & Wilford, D. (1998). *Trust matters*. San Francisco, CA: Jossey-Bass.

Arthur, W. B., Durlauf, S. N., & Lane, D. A. (1997). *The economy as an evolving complex system II.* New York: Addison-Wesley.

Ashmos, D., Duchon, D., Huonker, J., & McDaniel, R. (2002). What a mess! Participation as a simple managerial rule to "complexify" organizations. *Journal of Management Studies, 39*(2), 189–206.

Bak, P. (1996). *How nature works*. New York: Springer-Verlag.

Bergquist, W. (1993). *The postmodern organization: Mastering the art of irreversible change.* San Francisco: Jossey-Bass.

Boisot, M., & Child, J. (1999). Organizations as adaptive systems in complex environments: The case of China. *Organization Science, 20*(3), 237–252.

Bonabeau, E., & Meyer, C. D. (2001, May). Swarm intelligence: A whole new way to think about business. *Harvard Business Review*, 107–114.

Camazine, S., Deneuborg, J., Franks, N. R., Sneyd, J., Theraulaz, G., & Bonabeau, E. (2001). *Self-organization in biological systems*. Princeton, NJ: Princeton University Press.

Campbell, D. (2002). 9/11: A healthcare provider's response. *Frontiers of Health Services Management*, *19*(1), 3–13.

Capra, F. (1983). *The turning point: Science, society and the rising culture*. Hammersmith, London: Flamingo.

Capra, F. (1996). *The web of life*. New York: Anchor Books Doubleday.

Capra, F. (2002). *The hidden connections*. New York: Doubleday.

Cilliers, P. (1998). *Complexity and postmodernism: Understanding complex systems*. New York: Routledge.

Cowan, G. A., Pines, D., & Meltzer, D. (Eds) (1994). *Complexity: Metaphors, models and reality*. New York: Addison-Wesley.

Crossan, M. M., & Sorrenti, M. (1997). Making sense of improvisation. *Advances in Strategic Management*, *14*, 155–189.

Crossan, M. M., White, R. E., Lane, H. W., & Klus, L. (1996, Spring). The improvising organization: Where planning meets opportunity. *Organizational Dynamics*, 20–35.

Daft, R. L. (1989). *Organization theory and design* (3rd ed.). St. Paul, MN: West Publishing Company.

Daft, R. L., & Lengel, R. H. (1984). Information richness: A new approach to managerial behavior and organization design. In: B. Staw & L. L. Cummings (Eds), *Research in Organizational Behavior* (Vol. 6, pp. 191–233). Greenwich, CT: JAI Press.

Driebe, D. J. (2000). Complexity, chaos and our conception of nature. *Proceedings of the International Expo 2000*, 1–8.

Edmonson, A. C., Bohmer, R. M., & Pisano, G. P. (2001). Disrupted routines: Team learning and new technology implementation in hospitals. *Administrative Science Quarterly*, *46*, 685–716.

Ford, J. (1989). What is chaos that we should be mindful of it? In: P. Davies (Ed.), *The New Physics*. Cambridge: Cambridge University Press.

Gell-Mann, M. (1994). Complex adaptive systems. In: G. A. Cowan, D. Pines & D. Meltzer (Eds), *Complexity: Metaphors, Models and Reality* (pp. 17–45). New York: Addison-Wesley.

Gleick, J. (1987). *Chaos: Making a new science*. New York: Penguin Books.

Goldstein, J. (1999). Emergence as a construct: History and issues. *Emergence*, *1*(1), 49–72.

Granovetter, M. S. (1973). The strength of weak ties. *American Journal of Science*, *78*(6), 1360–1380.

Holland, J. H. (1995). *Hidden order: How adaptation builds complexity*. New York: Addison-Wesley.

Holland, J. H. (1998). *Emergence: From chaos to order*. New York: Addison-Wesley.

Institute of Medicine (2001). *Crossing the quality chasm: A new health system for the 21st century*. Washington, DC: National Academy Press.

Jensen, H. J. (1998). *Self-organized criticality: Emergent complex behavior in physical and biological systems*. New York: Cambridge University Press.

Johnson, S. (2001). *Emergence: The connected lives of ants, brains, cities, and software*. New York: Addison-Wesley.

Kaufman, S. A. (1995). *At home in the universe*. New York: Oxford University Press.

Klein, G. (2001). *Sources of power: How people make decisions*. Cambridge, MA: MIT Press.

Levithal, D. A., & Warglien, M. (1999). Landscape design: Designing for local action in complex worlds. *Organization Science*, *10*(3), 342–357.

March, J. G., Sproull, L. S., & Tamuz, M. (1991). Learning from samples of one or fewer. *Organization Science*, *2*(1), 1–13.

Marghella, P. (2002). Medical planning considerations in consequence management. *Frontiers of Health Services Management*, *19*(1), 15–23.

McDaniel, R. R., & Driebe, D. J. (2001). Complexity science and health care management. *Advances in Health Care Management, 2*, 11–36.

McDaniel, R. R., & Walls, M. (1997). Diversity as a management strategy for organizations: A view through the lenses of chaos and quantum theories. *Journal of Management Inquiry, 6*(4), 371–383.

McGuire, M. T., & Anderson, W. H. (1999). *The U.S. health care dilemma: Mirrors and chains.* Westport, CN: Auburn House.

Papa, M. J. (1990). Communication network patterns and employee performance with new technology. *Communication Research, 17*(3), 344–368.

Paul, D. L., & McDaniel, R. R. (2002). *A field study of the effect of interpersonal trust on virtual collaborative relationship performance.* Unpublished manuscript, University of Denver.

Prigogine, I., & Stengers, I. (1984). *Order out of chaos: Man's new dialogue with nature.* New York: Bantam Books.

Putnam, L. L., Phillips, N., & Chapman, P. (1996). Metaphors of communication and organization. In: S. R. Clegg, C. Hardy & W. R. Nord (Eds), *Handbook of Organizational Studies* (pp. 375–408). Thousand Oaks, CA: Sage.

Rousseau, D. M., Sitkin, S. B., Burt, R. S., & Camerer, C. (1998). Not so different after all: A cross-discipline view of trust. *Academy of Management Review, 3*, 393–404.

Stacey, R. D. (1992). *Managing the unknowable.* San Francisco, CA: Jossey-Bass.

Stacey, R. D. (1995). The science of complexity: an alternative perspective for strategic change processes. *Strategic Management Journal, 16*(6), 377–495.

Stacey, R. D., Griffin, D., & Shaw, P. (2000). *Complexity and management: Fad or radical challenge to systems thinking?* New York: Routledge.

Thomas, J. B., Clark, S. M., & Gioia, D. (1993). Strategic sense-making and organizational performance: Linkages among scanning, interpretation, action and outcomes. *Academy of Management Journal, 36*, 239–270.

Thompson, J. D. (1967). *Organizations in action.* New York: McGraw-Hill.

Tyler, T., & Kramer, R. M. (1996). Whither trust. In: R. Kramer & T. Tyler (Eds), *Trust in Organizations: Frontiers of Theory and Research* (pp. 1–15). Thousand Oaks, CA: Sage.

Waldrop, M. M. (1992). *Complexity: The emerging science at the edge of order and chaos.* New York: Touchstone.

Weber, M. (1971). The ideal bureaucracy. In: B. L. Hinton & H. J. Reitz (Eds), *Groups and Organizations: Integrated Readings in the Analysis of Social Behavior* (pp. 452–457). Belmont, CA: Wadsworth.

Weick, K. E. (1993). The collapse of sense making in organizations: The Mann Gulch disaster. *Administrative Science Quarterly, 38*, 628–652.

Weick, K. E. (1998). Improvisation as a mindset for organizational analysis. *Organization Science, 9*(5), 543–555.

Weick, K. E. (2001). Leadership as the legitimation of doubt. In: W. Bennis, G. M. Spreitzer & T. G. Cummings (Eds), *The Future of Leadership* (pp. 91–102). San Francisco, CA: Jossey-Bass.

Weick, K. E., & Roberts, K. H. (1993). Collective mind in organizations: Heedful interrelating on flight decks. *Administrative Science Quarterly, 38*, 357–381.

Weick, K. E., & Sutcliffe, K. M. (2001). *Managing the unexpected: Assuring high performance in an age of complexity.* San Francisco, CA: Jossey-Bass.

Weick, K. E., Sutcliffe, K. M., & Oberstfield, D. (1999). Organizing for high reliability: Processes of collective mindfulness. *Research in Organizational Behavior, 21*, 81–123.

Wheatley, M. (1992). *Leadership and the new science.* San Francisco: Berrett-Koehler.

Wolfram, S. (2002). *A new kind of science.* Champaign, IL: Wolfram Media.

Zimmerman, B., Lindberg, C., & Plesk, P. (1998). *Edgeware: Insights from complexity science for health care leaders.* Irving, TX: VHA Inc.

ENVIRONMENTAL JOLT OF BIOTERRORISM

Leonard Friedman and Peter Marghella

ABSTRACT

Health care organizations are accustomed to rapid and often discontinuous environmental change. Even when contemplating large scale change including the decisions to merge or integrate operations, health care managers can draw upon the expertise and advise of peers who have gone through similar experiences. However a bioterror event is a class of change that represents something totally unplanned and for which the industry has little or no experience in confronting. The objective for health care organizations is to mitigate the effects of this type of an event. Specific ideas for taking systems oriented, network-centric approach to disaster planning are provided.

To be prepared against surprise is to be *trained*. To be prepared for surprise is to be *educated*. Education discovers an increasing richness in the past because it sees what is unfinished there. Training regards the past as finished and the future as to be finished. Education leads toward a continuing self-discovery. Training leads toward a final destination. Training repeats a completed past in the future. Education continues an unfinished past into the future.

James P. Carse from *Finite and Infinite Games*

The objective of this chapter is to help current and prospective health care executives think the unthinkable as they plan for a post September 11 world. We have three goals for this section. The first is a brief review of some of the important theory and practice associated with managing change in health care organizations.

Bioterrorism, Preparedness, Attack and Response
Advances in Health Care Management, Volume 4, 141–162
Copyright © 2004 by Elsevier Ltd.
All rights of reproduction in any form reserved
ISSN: 1474-8231/doi:10.1016/S1474-8231(04)04006-6

The focus will be on the type of change that Meyer, Goes and Brooks (1990) describe as turbulence that may even shift into hyperturbulence. While trying to manage in a turbulent environment is challenging, the events faced are familiar and reasonably commonplace. There is certainly a high degree of difficulty associated with decision making in these environments and a series of poor decisions can lead to poor organizational performance. However, the environment while stressful is not dangerous per se. The second goal of our chapter is examine how health care organizations react and respond to specific environmental jolts, particularly when the jolts represent that which is completely new and outside of the frame of reference of the decision maker. These jolts serve as catalysts for action that prompts stakeholders to question the status quo and potentially bring down existing structures, or – at a minimum – serve to uncover alternative arrangements (Meyer, Brooks & Goes, 1990; Sine & David). In this case we believe that the threat of attack by one or more weapons of mass destruction (WMD) with a potential for mass casualties is the sort of environmental jolt that will cause health care executives to seriously consider alternative ways of planning and organizing. Finally, we hope to give the reader a series of suggestions and recommendations that they might use as they prepare to confront a frightening but real threat that none of us dare to ignore. In order to provide the context for our approach to these questions, we present the following fictitious but all too plausible scenario.

In a major metropolitan area, a large number of theater goers have gathered for the opening of an important new play. Every seat in the 2,000 person capacity theatre is filled when in walks the President of the United States, accompanied by her entourage and security detail to take their places. All is going well up until the start of the final act when a group of approximately 100 persons dressed entirely in black burst into the theatre and announce that in the name of all the oppressed peoples of the world, they are taking everyone in the theatre hostage. All the exits are blocked by the intruders who set about placing small metal canisters at various locations in the building. The leader of the group grabs a microphone and explains that the canisters contain a "cocktail" of Saxitoxin in addition to other, longer lasting nerve agents.

After reading a manifesto describing the group's political beliefs and the members willingness to die in order to advance their cause, the leader pulls out a small radio transmitter and presses a button. A valve on each of the metal canisters opens, releasing its contents into the room. The air conditioning system does an effective job spreading the gas to every corner of the theatre and the ventilation system engineered to draw in fresh outside air and purge inside air also operates as designed, beginning to diffuse the gas into the night air. Those not instantly overcome immediately try to flee the building, bursting through the doors and running into the street. Police and fire personnel soon arrive. There are five hospitals within a five-mile radius of the theater. Only one is a designated Level 1 trauma unit. Within 30 minutes of the detonation of the gas canisters, those who are still living begin arriving at the nearest emergency rooms.

Let us assume that you, the reader, are the president and CEO of the hospital nearest the theatre. This evening, you are out having dinner with your family when suddenly your cell phone rings. On the other end of the line is the ED manager who is screaming into the phone in order to make himself heard over what appears to be a scene of utter chaos. People are pouring into the ED by ambulance, personal vehicle, or taxi. People who are displaying a whole host of symptoms take up every available bed, gurney, chair, and sofa but in a number of cases, patients are going into respiratory arrest and can not be revived. Calls have gone into every available physician and nurse listed on the on-call list but their pages are not being returned. There seems to be no end of people streaming in either sick or wanting to know what is happening. The ED manager has managed to find a copy of the hospital's disaster plan stuck away on some back shelf but it is of little use unless the disaster included fire, flood, earthquake, or power outage. Not knowing the source of the problem, you tell the manager that you will be there as soon as you can but to implement the triage system that was part of the overall disaster plan.

You send your family home and hail a cab to take you to the hospital. You now hear the omnipresent wail of police and fire sirens. Traffic around your hospital has degraded into total gridlock with more rescue vehicles trying to get to the ED entrance and people becoming sick on the streets. You jump out of the cab and sprint the remaining six blocks to the hospital. A scene that reminds you of the time you served in a front line unit during wartime assaults your eyes. The dead and dying appear to be everywhere. CDC mandated universal precautions are nowhere to be found as every available person with even the slightest bit of clinical knowledge is working on what seems to be a flood of humanity. As you enter the doors to the hospital, the first reporters from the local television station arrive. The reporter asks you what your plans are to treat the victims of the explosion at the theatre and what of reports that some type of gas was venting from the air conditioning system.

As you brush by the reporter with a rapid, "No comment," your thoughts are on the people directly affected by this incident, whatever it was, yet you can not help but think about the condition of your hospital and its ability to handle the sick and injured. When you left your office this evening, after another overly long and stressful day, you were concerned that the hospital was running at just under 90% of its licensed capacity. You spent a good part of the day with your Vice President for Patient Services trying to develop strategies to cope with both the continuing high number of nursing vacancies coupled with the announcement that the current nurses had requested union representation. Physicians assigned to cover the Emergency Department were unhappy with both the hours required and the level of compensation. You were dealing with the news that the largest underwriter of medical liability insurance in the state had just decided to drop their malpractice book of business due to loss ratios that had turned upside down in recent years. As though this were not enough, the information system team was preparing to roll out a new software package designed to link together all clinical and administrative parts of the organization. In order to make the rollout happen, the current IS needed to be taken off-line for most of the night. Only essential functions would be left operating but your IS manager suggested paper backups just in case. In the morning, your first meeting was with your Board chair to discuss a potential merger with one of the other hospitals in town.

The level of change and uncertainty you were facing just in the day-to-day operations of your hospital seemed at times to be more than you could manage, yet somehow, you were able to break problems and activities into manageable pieces and more importantly, depend on a skilled and trusted staff. And now, looking at the whirl of activity surrounding you in the ED, these other concerns and uncertainties seem to be completely irrelevant. The environmental jolt that you and your hospital have sustained cause you to question whether you have anything near the capability to deal with the disaster at hand. How could you have prepared for this unthinkable event? Will the hospital be able to survive and provide care for the community that has come to depend on its constant and reassuring presence?

The events described in the previous scenario are fictional. However, world events over the past 18 months suggest that they may not be as farfetched as one might have imagined not so long ago. The sentinel event of the September 11, 2001, attack in the United States, followed in 2002 by the bombing of the night club in Bali and most recently, the attack on the Madrid rail lines suggests that we have not seen the last of these horrific acts. It is ironic that in all three events, weapons of mass destruction (WMD) including chemical, biological, radiological, nuclear, and high explosive (CBRNE) were not used by the perpetrators but that gas pumped into the theatre in Moscow by Russian security forces effectively killed both combatants and civilians. It can be argued that those who commandeered the aircraft on September 11 converted these planes loaded with highly flammable J-1 fuel into flying missiles thereby creating a WMD. Semantics aside, "the events of 9/11 have ushered in the era of the asymmetrical threat" (Marghella, 2002). These unconventional attacks suggest that in order to effectively deal with their disruptive nature, all stakeholders associated with health care organizations will need to devise new ways of thinking about, planning for, and managing change.

MANAGING NORMAL CHANGE

In a recent issue of *Advances in Health Care Management*, the contribution by Goes and his colleagues (Goes, Friedman, Seifert & Buffa, 2000) dealt with the objective of attempting to speak to and classify the wide variety of organizational change that takes place in health care organizations. According to that model, change is conceptualized as occurring along three dimensions including level, type, and mode. The level of change is seen as taking place at either the organizational or the industry level. There is a growing appreciation of change that takes place at the level of groups or teams of individuals particularly in the context of quality improvement efforts and team performance (Edmonson, Bohmer & Piscano, 2001). One criticism of the matrix model is that it fails to take into account this type of group based change, but for our purposes, we will blend change at the group level into organization level change. Despite the criticism, we use this model since it allows us to visualize change that occurs along three discrete axis, hence providing a richness and complexity that would be unavailable by only looking at one or two dimensions.

Type of change is conceptualized as taking place along a continuum beginning with incremental change progressing to change commonly described as radical or discontinuous. Incremental change is that which is most common to managers and administrators regardless of what industry they are connected to. At the other end of the continuum are those revolutionary (may we say paradigm shifting) changes that Kuhn (1996) so eloquently spoke of over 30 years ago. Finally, Goes and colleagues suggest that change needs to be categorized along a third dimension, termed mode of change. Along this axis, change is classified as either deterministic or volunteristic. Deterministic change is imposed on the organization or industry by outside entities and is such that there is little option other than to respond to the change in very much a reactionary mode. Examples of this change are the current HIPAA privacy regulations, Stark rules, JCAHO accreditation requirements, or even changes to billing forms when third party payers modify their procedures. There are certainly options on how to respond to these deterministic changes but for most health care organizations, there is no option other than to make the changes specified. Alternatively, volunteristic change is driven exclusively by the organization or industry. In this case, we see examples that include the decision to implement a new telecommunications system, begin a series of rapid cycle improvements, or expand into a new service area. In every case, volunteristic change is proactive and results from (hopefully) a significant amount of thought, deliberation, and intentional consultation with key stakeholders.

The resulting $2 \times 2 \times 2$ matrix yields an eight celled cube in which various change activities and processes are categorized (see Fig. 1). Each of these cells

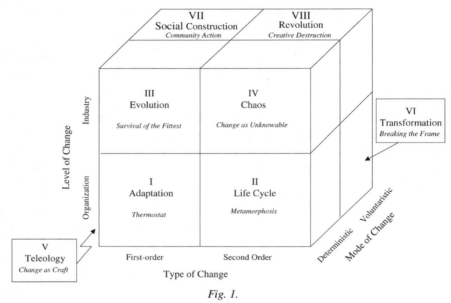

Fig. 1.

is identified with a dominant metaphor that serves to express the theory that underscores the type of change contained within. As might be expected, Cell I, Adaptation, dominates the change processes that are described in both the general business and health care literature. Incremental, deterministic, and organization based change seems to be the most frequent change process conducted and described. Life cycle is the metaphor in Cell II where more radical and deterministic change takes place at the organization level. A current example of this phenomenon is in the managed care industry where HMO products have been steadily declining in popularity over the past few years and are being replaced by PPO options (Ginter, Swayne & Duncan, 2002). Cell III speaks to an evolutionary process where at the industry level change occurs in a deterministic and (mainly) incremental level. If we use the natural selection metaphor, the environment exerts pressures such that organizations best suited to their environments have the capacity to survive. Mimetic isomorphism is a common response to these pressures as was seen in the rapid adoption of magnetic resonance imaging in the late 1980s and 1990s (Friedman & Goes, 2000). The final deterministic change Cell (IV) involves radical change at the industry level. These sorts of changes fall in the realm of chaos where change is often unknowable and uncontrollable. Chaos theory and complex adaptive systems are categorized in this part of the matrix. Recent work by Reuben McDaniel and his colleagues (2000) provides some of the best application of this theory to change in healthcare organizations.

The second half of the change model posits that change occurs voluntarily in either organizations or industries. Cell V is uses the metaphor of change as craft in order to describe first order, organization based change. Much as Mintzberg (1987) described strategy as a process of crafting, organizations can craft change in an intentional and purposeful manner. Quality improvement is a clear example of this form of change. Transformation (Cell VI) is the type of change that occurs when organizations undergo radical, second order change. As originally envisioned, reengineering is the clearest example of this form of change process. At the industry level, Cell VII is representative of social construction where change is organized and takes place through a community of organizations that share interests and an environment. It is ironic that there is a significant tension among health care delivery organizations today particularly when anti-trust laws interfere with the ability of these organizations to cooperate with one another to either generate revenue or decrease costs by sharing technology. Finally, Cell VIII refers to the phenomenon of an industry revolution where radical change is voluntarily spread across an entire industry. A revolution waiting to take place will center on the technology made available through the efforts of the human genome project. While the Internet has profoundly affected the ways in which all stakeholders in health care delivery interact with one another, the implications inherent in manipulating the human genome are truly profound. How will our industry at all levels choose to respond when it becomes possible to identify and modify the genes that trigger heart disease, cancer, or Alzheimer's disease? What sorts of services shall we provide when the genes that control aging are discovered and can be altered?

The point to discussing the Goes et al. model of organizational change is that we make an a priori assumption. That assumption is that health care organizations deal with change that in all but a few instances are both predictable and within the context of what has been seen before. These two notions are interrelated. In every cell of the Goes change matrix except for numbers II and IV, change is knowable and understandable. That is, even if we did not choose to create the HIPAA privacy regulations (a deterministic, first order, industry wide change), health care organizations know and understand what is required in order to operationalize the changes required. The Federal government has specified what is necessary for covered entities to comply with the new requirements and there is no shortage of consultants to help health care organizations meet the HIPAA standards. While protection of patient data and the imposition of stringent privacy standards may mean a significant cost to the organization and might even be considered something other than incremental, we know what needs to take place and by when. All of the cells in the volunteristic half of the matrix (Cells V–VIII) presume that change is initiated intentionally and that decision-makers know (or at least think they know) what sort of outcome resulting from the change process is desired.

Coupled with knowing what is expected is the even more important attribute of understanding that the context of change resides within the familiar. Whether change is something as simple as the imposition of a new process for answering to telephone to a complex and lengthy event such as the merger of several organizations to form an integrated delivery network, the events required to make the change happen include those things that are familiar and (relatively) common to the health care industry. Even the introduction of TQM (which was based on the manufacturing model developed by W. Edwards Deming in 1988) was made both palatable and acceptable for health care by Donald Berwick (1989). Second order change processes not routinely encountered by health care executives have to be considered on the basis of both context and prior experience with similar events. Gharajedaghi (1999) observes that without careful consideration of the context, solutions to problems or approaches to change are likely to be ineffective. In fact, "the ability to define context requires purposeful behavior, a holistic orientation to seeing the bigger picture and putting the issues in proper perspective." The problem and the solution are necessarily linked to one another. Without the context of what is commonly seen and recognized in health care settings, change processes become potentially mindless and worthless.

While the persons within health care organizations have the capability of conceiving of and dealing effectively with the vast majority of change processes, what happens when change is both unknowable and out of context? While chaos (the metaphor for Cell IV) might provide some clues, the work of McDaniel and others who study chaos and complexity theory recommend that we think about change as occurring within a larger system, made up of large number of linkages and loops. All of these loops and linkages occur as part of systems that while complex, are coherent. Again, Gharajedaghi (1999) provides a very useful perspective when he states, "the imperative in interdependency, the necessity of reducing endless complexities, and the need to produce manageable simplicities require us to proceed from a basic frame of reference and focus on the relevant issues." Our question is what happens when these conditions are violated? What happens to decision makers when events occur that are totally out of context and are not within the knowledge realm of the executive? These traumatic events in the immediate environment require a completely different way of sense making and reacting. How shall we anticipate and effectively deal with change that is both deterministic, radical and out of context? It is this fourth dimension of change that we address in the next part of the paper.

RESPONDING TO ENVIRONMENTAL JOLTS

The idea of environmental jolts is a familiar concept in the language of strategic management. An important paper by Meyer and colleagues (1990) carefully reviewed the literature of discontinuous change generally and used the hospital industry in the San Francisco Bay area as the model through which they tested their assumptions. As part of their theoretical model they needed to explain unexpected high intensity change that was observed during their field study. In the hospitals they studied, change did not occur in a continuous manner. Rather, there were periods of relative stability interspersed by bursts of high-velocity change. These periods of discontinuous change were posited to be a necessary condition in order to allow for innovation and opens up opportunities for entrepreneurs. The paleobiologist Steven J. Gould first coined the term "punctuated equilibrium" in 1977 as a way of explaining a similar observation in the fossil record. These dramatic changes in the normal equilibrium of either biological organisms or the constructions we term organizations, allow for significant and lasting change to occur.

It is at this point that another important assumption must be made as we continue forward. We assume that health care organizations are but a part of a much larger health care system. Do not confuse the use of the term system with what many think of when using this particular phrase. For many people, a health care system is one that has been rationally and intentionally constructed in such a way that the many parts relate to one another to express the objectives of those who designed the system. The key element here is intentionality. In the language of systems thinking, components of the system relate to one another in a multitude of both planned and emergent ways. Cause and effect relationships exist but are overwhelmed by multiple interrelationships between and among the various components of the system. Because of this interrelationship, attempting to "fix" a problem in one part of the system frequently results in certain unintended consequences in another part.

A metaphor for this is contained within a cellular telephone. Suppose for a moment that a visitor from another world were to come across this almost ubiquitous piece of technology. In order to make sense of this cell phone, the visitor pulls out a small case of very small tools and begins to disassemble the device, laying out all the parts on a table. After a short time, every wire, circuit board, and piece of metal and plastic lays spread out in front of the extraterrestrial. Unfortunately, our visitor is more confused now than before. Why? Because the cellular telephone is part of a much larger system – something that we call a communications system. That system is made up of multiple subsystems including (but not limited to) manufacturers, computer systems, senders and receivers, antennas, and national telephone companies. Simply by breaking down one component of the system (albeit an important part) to it's smallest units does not for a moment help us understand the larger system.

This begs the question of why we work so hard at optimizing the various parts of the system in the hope that by doing so, we will improve the larger system. In the words of Gharajedaghi (1999) in order to understand systems, we need to appreciate the structure, function, and process of the system. When we work at optimizing the parts, we do so without considering the underlying structure and processes. The discipline of organization design concentrates on structure without regard for function or process. When we attempt to map the particular processes (much like we do in an engineering flowchart) in an attempt to understand process, we lose sight of both structure and function. For Ghargjedaghi, structure defines components and their relationships. Function defines the outputs or results. Process describes the activities and expertise required producing the outputs. All three of these activities must be considered together in a holistic manner in order to fully understand a particular system.

Failure to carefully think about structure, function, and process can have fatal consequences. In his masterful book, Perrow (1984) carefully traces the origins of a number of "normal" accidents that occurred because those who designed a particular system failed to understand and appreciate the complexity of the system they built. In his examination of catastrophic failures of carefully designed and engineered complex systems, Perrow comes to two important conclusions. In complex systems, subsystems are highly interactive with one another. Additionally, the various subsystems are tightly coupled so that when one part of a system fails, it is likely to have a dramatic effect on a number of other parts of the system. In his examination of the failure of the Three Mile Island nuclear power plant and accidents involving aircraft, Perrow can point to the normal (and even expected) catastrophic accidents that occur due to a small failure somewhere in the system.

What is it about an environmental jolt that has the potential to cause even well designed systems to fail? In some instances, environmental jolts, while disruptive, provide new opportunities for organizations to take advantage of the uncertainty being experienced by other firms. One explanation may have something to do with the strategy selected by an organization. It might be assumed that an environmental jolt causes decision makers to choose a strategy different than the one currently being pursued. However, in their study of firms going through "boom and bust" cycles, Grønhaug and Falkenberg (1989) determined that the strategies do not change as a result of the jolt. Their explanation is that the core competencies of the organization dictate the reaction to even the most difficult and challenging environment.

Let us now consider what happens to systems when confronted with environmental jolts so large as to result in a breakdown of large parts of the system. What sorts of events can create these types of jolts? It may be instructive to examine the system affects created by natural disasters. Whether the disaster is water, fire, wind, or earthquake, the potential for widespread system failure is real and can have devastating effects. The human and financial toll extracted by large natural disasters can sometimes seem overwhelming. According to the Federal Emergency Management Association (FEMA), the ten most costly natural disasters in the United States occurred between 1993 and 2001. They included five hurricanes and one tropical storm, two floods, and two earthquakes. The relief costs ranged from $621.2 million as a result of Hurricane Fran in 1996 to $7 billion caused by the Northridge earthquake in 1994. These are all single event costs and do not include an accounting for months long droughts and fires that can also cost in the billions of dollars. To what degree do well designed systems survive these sorts of environmental jolts?

It is particularly interesting to examine the effects felt across the Los Angeles basin as a result of the 1994 Northridge earthquake and particularly, the ability of the various businesses and organizations to recover as a result of this particular environmental jolt. In their 1998 paper, Dahlhamer and Tierney examined recovery outcomes from over 1,100 businesses damaged by the Northridge quake. They determined that business size, earthquake shaking intensity, disruption of business operations, and utilization of external post-disaster aid are all predictors of business recovery. As might have been expected, large businesses proved better at surviving disaster losses than did small business with more limited resources. One outcome that was unexpected in their research was the finding that receipt of post-disaster aid did not appear to provide much help in assisting businesses to recover and might in fact have had a negative effect given the additional debt this aid imposed on organizations. In their sample, 56% of the businesses were forced to close for a period of time after the earthquake. One of the more interesting outcomes of the Northridge earthquake is that the State of California in 1994 passed legislation that would require all acute-care hospitals in the state to retrofit, rebuild or close any non-compliant inpatient facility by 2008. Even more stringent seismic standards will be required by 2030. According to a report by the California Office of Statewide Planning and Development (OSHPD), 78% of hospitals in the state have at least one building that, unaltered, would be forced to close by 2008. The California Hospital Association estimates the cost of complying with this mandate to be $24 billion.

There have been other disasters that have challenged health care delivery organizations. A 1996 report in Modern Healthcare spoke to the concern that if hospital downsizing continued at the pace seen in the mid-1990s health care facilities would be unable to care for the sick and injured as a result of a natural or manmade disaster (MacPhearson). An editorial in the American Journal of Public Health in 1986 uses three environmental jolts as the context to discuss preparing for disasters (Merchant, 1986). Examining the effects of the eruption of Mount St. Helens, the toxic gas release at the Union Carbide plant in Bhopal, India in 1984, and the Mexico City earthquake in 1985 that killed over 7,000 persons, the author suggests five ways to plan for future disasters. These methods include using technology to forecast events, use of engineering to reduce risks, public education on potential hazards, coordinated emergency response, and a systematic assessment of the effects of a disaster to better prepare for future events. While preparation is stressed, nowhere in the AJPH editorial does the author speak to the organizational effects created by these disastrous events.

While five persons died and 17 were taken ill by the anthrax attacks of 2001 the potential for mass casualties in a widespread and well orchestrated bioterror attack would almost certainly overwhelm every hospital and health care provider in the affected area. Let us return to the scenario used to introduce this chapter. It is instructive to examine the sorts of environmental jolts that shook the hospital in question and ask what went wrong and what might have been done differently. The CEO was clearly burdened with both the volume and complexity of the change issues facing the hospital. Between the nurses labor action, merger talks, medical malpractice concerns, information system change over, and angry physicians, the pace of change and uncertainty clearly resulted in a turbulent environment. Each one of these variables is difficult enough on their own. Taken together, all the leadership skills of the CEO are put to the test as various parts of the administration and clinical staff have to deal with handling each of the items as well as the consequences of the respective decision.

There is one thing to keep in mind as the CEO ponders each of the decisions that must be made relative to each issue. None of these are particularly time critical and the opportunity exists to think about alternative options. The threatened labor action has almost certainly been building for some time and in all likelihood, there is a significant period of time to either blunt the unionization effort or prepare for life with a unionized workforce. The merger talks are just that. They are the opening round of a very long and protracted series of negotiations which will almost certainly take many twists and turns before a final outcome is achieved. The medical malpractice problem is endemic across the nation as physicians, hospitals, malpractice insurance companies, government entities, and others try to develop a coherent and workable solution. There is time available to develop options and possibly even develop a self-funded insurance option. The information system change over has been planned for some time and it is coincidence that this was the evening the switch was to occur. In the worst case, you could rely on the paper and pencil backups as your IS director suggested. Finally, the problem with the unhappy ED physicians is not going to get resolved overnight. Likely (as was the case with the nurses), there was a growing tension among the physicians because of real or perceived grievances. Developing a solution will require the time and efforts of a large number of stakeholders.

Common in each of these difficult problematic situations is that they reside in the world of the known and familiar. While they might result in headaches and overly long meetings with too many unhappy stakeholders, all can be understood and reasonable solutions can be proposed. Using a variety of tools and techniques, the problem is defined in the context of the particular organization. Relying on data from multiple sources coupled with the experience gathered from similar situations, the CEO breaks down each problem into manageable parts and works to devise the best solution possible. In every case, the executive has the luxury of time.

Now, what happens when time and the ability to ponder, consult, negotiate, and otherwise delay a decision is suddenly taken away? How does the organization (and internal decision-makers) respond when the environment is turned from turbulent to hyperturbulent in a heartbeat? In our scenario, those directly affected by the release of the toxin either in the theatre or from venting through the air conditioning system are quickly taken ill. In this case, those who are ambulatory enough to do so try to get to the hospital as quickly as possible. Without warning, the ED is overrun. As was noted, the disaster plan fails on two counts. First, it is stuck away on a back shelf and likely has not been updated or read in some time. Given the comments of the ED director, it is unlikely that the actions and activities called for in the plan have been practiced recently if at all. Secondly, the disasters planned for in the manual are common and well known. There is no provision for anything resembling the events that are currently unfolding. The inability to contact the physicians and nurses whose names are on the on-call list suggests that either they were all in the theatre when the gas was released (not a likely scenario) but more likely did not treat their pages as something serious that needed an immediate response.

In order to better understand the scenario, Gharajedaghi (1999) suggests that we think about the events taking place as something he calls a "mess" or stated another way, as a system of problems. The CEO simply can not deal with each of the events taking place that evening as independent events that can be understood and resolved apart from the others. Among the attributes of messes, Gharajedaghi observes that they occur as a natural consequence of the existing order and hence should be expected. Messes are seen as very resilient and have a way or regenerating themselves which is the characteristic that makes messes so difficult (if not impossible) to manage.

For most messes, opportunities exist to design solutions. Gharajedaghi (1999) recommends that an idealized redesign process be used as a means of defining purpose, function, structure, and critical processes. In the case of a bioterror or WMD attack, it is simply not possible to take the time and gather stakeholder buy-in as the CEO develops an entirely new way of dealing with the mess as the attack is in progress. While idealized redesign as it is currently constructed might not be feasible in the scenario built earlier, perhaps a parallel type of process is in order. Rather than wait for the mess to appear, might it be worthwhile to design a system based on the likelihood of a certain set of assumptions about the future? How could the processes and methods of consequence management be used to design a health care organization that is both resilient and sustainable even when faced with the unimaginable? In the final section of this chapter, we provide ideas and recommendations that organizations and health systems can use as they plan for a future where the WMD threat is not an abstract notion.

RECOMMENDATIONS AND REFLECTIONS – SHAPING THE BATTLEFIELD

As a general rule, uniformed professionals are usually reluctant to introduce phrases from the military vernacular to describe how civilian organizations should approach a problem. However, when it comes to developing a strategy that organizations and health systems can use to prepare for WMD events, the notion of "shaping the battlefield" is exactly the approach to consider. To begin with, it is important to recognize that the watershed events of 9/11 have forever changed our view of the battlefield as a *distant* place, one from which we can readily and easily maintain a physical and mental detachment. In the "era of the asymmetrical threat," the governmental, financial, industrial, and psychological centers of gravity that we can expect will be targeted for attack are the same cities, towns, businesses, and icons of history and culture that we live and work around each and every day of the year. Talk about an environmental jolt – we now live and work *on* the battlefield! If we are going to develop strategies to help mitigate the effects of asymmetrical events, shaping the battlefield before events occur is the proven way to gain advantage over our adversary and hopefully increase our collective chances of survival. How does our earlier discussion on the response of complex organizations to rapid and unexpected changes in their environment tie to the notion of being on the front lines of the new battlefield? What can health care organizations do to reframe the way they think about their role in this new environment?

First off, it is imperative that the collective medical community recognizes that they will be doing the heavy lifting if and when another event occurs. If we accept the definition of consequence management as the process, "to provide timely, appropriate, and effective support in order to save lives, relieve suffering, and assist in the mitigation of the further harm to an affected population" (Marghella,

2002), then no other community will bear a commensurate burden. In order to minimize the environmental impact on hospital-based organizations, CEO's and COO's must become actively engaged in developing executable medical plans and consequence management strategies – and their focus must be on the development of strategies that assume from the onset that extreme stress (jolts) will be placed upon their healthcare delivery systems.

It is very difficult to imagine casualty streams that run into the hundreds, if not the thousands or beyond. As a human species, our nearly reflexive reaction when considering events of such magnitude is to psychologically minimize the likelihood of such an event even occurring, or just not think about it at all. The fact that our main casualty production machine – the military – has not engaged in wartime operations that have produced significant casualty spikes (1000+ per single event) since the Second World War has only created the false impression that we are somehow invulnerable to events of strategic magnitude. The events of September 11 proved we are not.

In a traditional (symmetrical) view of the spectrum of conflict events, the progression from tactical to strategic impact is linear, and (usually) clearly demarcated to a specific type of operational environment (Fig. 2). Casualty production related to these events is (usually) also linear in progression. As participant force size (also known as a "population-at-risk" or "PAR") and operational intensity increased, we could reasonably expect that there would be a commensurate increase in casualty production related to the event.

In the new spectrum of conflict (the asymmetrical environment), chemical, biological, radiological, and high explosive (CBRNE) weapons of mass destruction

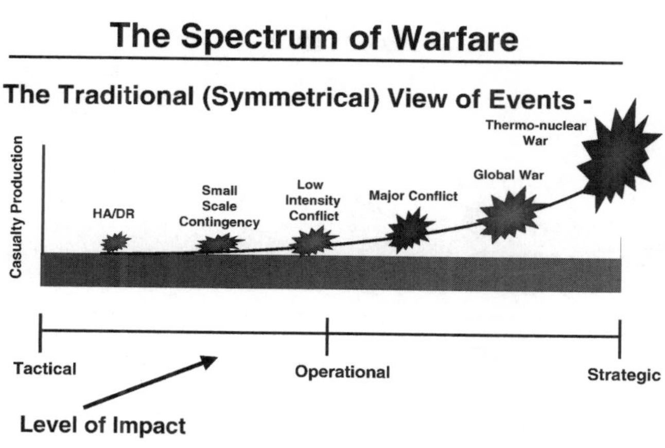

Fig. 2.

(WMD) can cross the linear spectrum, as the impact of the event in based on several variables:

- Type of Agent and Dose (size of weapon) used.
- Dispersion/Delivery Method.
- Concentration of the Population-at-Risk (PAR).
- Weather and Ambient Conditions During Dispersal.
- Overall Protective Posture of the PAR.

Figure 3 represents the asymmetrical perspective of the spectrum of conflict where biological agents are used. While it is possible to progress to strategic level casualty production based on the type of biological agent used, it is more difficult to "frame" or predict the outcome of the event due to the variables described above that can occur with these agents crossing the linear spectrum.

As with a thermonuclear event, it is a given that a high-end, strategic biological attack can put us at-risk of "busting" the collective inventory of resources the healthcare community can bring to bear, even at the national level. Barbish (2002) describes three levels of events in terms of casualty production:

- Level 1 – Up to 1000 Casualties.
- Level 2 – Up to 10,000 Casualties.
- Level 3 – Above 10,000 Casualties.

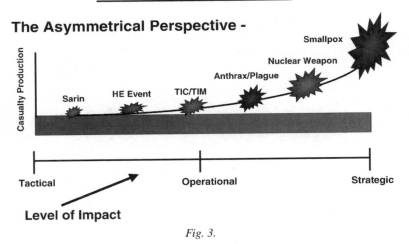

Fig. 3.

Realistically, a Level 3 event may be so overwhelming that it would preclude successful execution of even the most carefully considered consequence management (CM) plan (however, this should not be construed as limiting strategic level planners involved with organizations like the Departments of Health and Human Services and Defense from developing WMD CM strategies). At the healthcare organization level, the Levels 1 and 2 events merit the attention of the organizational leadership. Herein lies the challenge: as previously noted, in order to minimize the environmental impact of bioterrorism, healthcare organizations must develop strategies that can offer a reasonable expectation of success if required to be executed.

NETWORK-CENTRIC CONSEQUENCE MANAGEMENT

Let us return to the metaphor of the cellular telephone. This time, however, let's take a counter-intuitive or inverse perspective from the phone being a part of that larger communication's system we previously described. That description still applies, but before we can look at it and figure out how it all fits together at the macro-level, consider the phone from this new perspective as an analogy for a community healthcare network, and the components of that phone as the individual treatment facilities that support that community healthcare network.

The base assumption of a WMD event is that large-scale casualty streams associated with an event will over-burden and/or corrupt the resources of medical facilities that are geographically proximate to the affected population. Our recommended solution is to consider a network-based approach to mass casualty management, which relies on regional capabilities to support medical consequence management requirements. Figure 4 provides a graphic of this concept.

As the figure notes, a "concentric ring concept" of medical management using a networked facility approach precludes the resources of any one hospital, or even a "group" of healthcare assets, from collapse. As the health service support requirements increase concurrent with spikes in casualties related to the WMD event, the concentric rings of networked resources begin to take on the overflow requirements to support casualty management. Related to our cellular telephone analogy, the components (individual healthcare organizations) all come together as a network to produce the functioning capability needed to answer the call.

Remember, however, that all hospitals (as physical facilities) are already performing the mission of healthcare delivery prior to a BW event occurring. At any given time, roughly 90% of all hospital beds nation-wide are already filled. We labor under the notion that if a significant WMD event occurs and creates

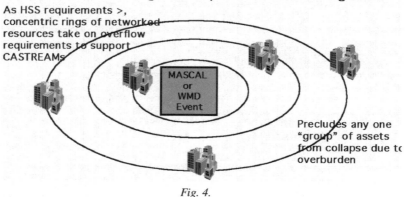

Medical Planning in Consequence Management

The "Concentric Ring" Concept of Medical Management:

As HSS requirements >, concentric rings of networked resources take on overflow requirements to support CASTREAMs

MASCAL or WMD Event

Precludes any one "group" of assets from collapse due to overburden

Fig. 4.

spiked casualty streams, our hospitals are simply going to "dump" their minimal care cases, open their doors, and be ready and able to treat potentially thousands of victims. This is just not so. If a hospital is already filled to near capacity, as soon as significant presentations begin to show up at the ER, the facility will virtually shut down. In this occurrence, we have suddenly shifted from the known to the unknown. Outside of urban areas where multiple health care organizations operate, it may not be possible to cease operations and just care for those already admitted to the facility.

The solution lies in the paradigm shift away from the hospital as the *locus in quo* for the medical management of WMD affected casualties to a reliance on the personnel and equipment resources of the hospital at a detached location where support can be provided (Fig. 4).

When a BW event occurs, the regional emergency response network would leverage large buildings of opportunity (such as gyms, schools, and malls) where bed-downs sites could be established. After processing through a triage area, casualties would be moved to these bed-down locations for management of their illness. Local hospitals would use their professional staff and equipment and supply capabilities to primarily resource the triage and mass casualty management location(s). Since a BW event would largely cause minimal care cases, patients at the bed-down sites would largely be managed through self and buddy aid, which could, for the most part, be community generated by the non-effected population.

Creating these networks should not be that difficult. In the wake of the 9/11 attacks on the United States, most healthcare organizations are acutely aware of the potential for significant environmental impact if they remain outside an inter-agency support arena. The real work that needs to be accomplished is defining areas of responsibility, coordinating resource management *prior to execution occurring*, and setting up the command and control and communications architecture that would be needed to manage a calamitous event. Without minimizing the potential enormity of a bioterrorism event, healthcare organizations could reduce the environmental impact simply by improving communications and sorting out roles and responsibilities. Trying to work through them as an event unfolds is a recipe for disaster. If there is any lesson to be learned from 9/11, it is that comprehensive medical planning is the key to successful consequence management.

In summary:

- Times have changed; planning for calamitous events is essential.
- Medical planning is a priority for consequence management related to complex emergencies.
- Observing the principles of deliberate planning prior to an event improves response in a crisis.
- Developing succinct medical plans for disasters (whether symmetrical or asymmetrical) ensures the improved provision of health service support.
- Developing "network centric" medical plans levels the playing field in emergency response situations.
- Inter-agency coordination is an important key to successfully mitigating crisis events.

CONCLUDING THOUGHTS

In the event of an asymmetrical attack involving one of several WMD, the rules for managing change even in hyper turbulent environments will likely be completely ineffective. The skills and experience for which health care executives have handling even the most difficult and contentious problem will be of little use when faced with a mass casualty event that overwhelms the entire infrastructure of the hospital or health system. Given the likelihood of another asymmetrical attack on this country much like what occurred on September 11, 2001, what should health care executives do in order to prepare for an event that can not be predicted with any degree of precision or certainty?

The most important suggestion that we can make is to begin to entertain two separate but interconnected thought processes. First is the perspective afforded by a systems orientation and complexity theory. Health care executives need to take seriously the realization that all parts of our health care system are interconnected and that changes in one part of the system can have a profound effect on other parts. We have a long history of trying to optimize the various parts of the system and then are dismayed to discover that despite this effort, the system fails to demonstrate significant or long lasting improvement. This system interconnectedness is taken to best advantage in Fig. 4 where a virtual health system is created when the primary locus of care is overwhelmed. Thinking "outside the box" is today considered a trite and useless phrase. However, it is exactly this type of systems thinking that will be required in the event of a major bioterror incident.

The second and final suggestion is the need to pay close attention to the concept of consequence management. In the language of emergency and disaster planners, a number of events have the potential to create a "disabling business interruption." For some businesses, shutting down their operations for even a short time has the potential of creating significant economic hardship as a result of high fixed costs and lack of revenue when goods or services are not produced. A case in point is the airline industry in the wake of the September 11 attacks. Shutting down commercial and private air traffic had a devastating effect on both the traveling public and on the financial health of the airlines. In late 2002, U.S. Airways filed for Chapter 11 bankruptcy protection and United Airlines chose the same fate, both coming as an aftermath of the events of September 11. While air travel is important, other local businesses are the lifeblood of a community and would certainly include utilities and hospitals. Imagine what would happen if utilities including telephone, water, and power were disrupted for a period of days or weeks. Think about the effects that a community would face if local hospitals were unable to treat acutely ill or injured patients. In an effort to mitigate the effects of this type of disabling business interruption, the development of consequence management strategies gives health care executives the tools and techniques needed to quickly recover and resume operations.

The communities of both health services researchers and health care practitioners has the potential to answer a number of interesting and important questions. Organizational resilience is critical if health care firms and entire communities are to survive the next terrorist attack in a way that allows the affected organization to recover in the shortest possible period of time. What are the factors that influence resilience and how can organizations learn the skills necessary to increase resilience? Another important area of investigation is the role that complex adaptive systems plays in helping understand the way that organizations deal with severe and unexpected change in their environments. Finally, while we did not specifically speak to the topic, leadership during times of crisis can be the distinguishing factor determining whether an organization survives the particular environmental jolt. What crisis management skills are found within leaders who have successfully led their organizations during the most difficult circumstances?

There can be no doubt that as a nation we feel vulnerable and exposed to the whims of persons who would do harm to us. As health care administrators, clinicians, policy makers, academics, or students, there is little that we can do to change the geopolitical forces that we face today. Given the experience of September 11, 2001, we are forced to confront a world defined by fear and uncertainty. Despite this, we have an obligation to everyone that depends on health care organizations to fulfill their mission to care for their communities. Failure to effectively plan for a WMD attack will have tragic consequences beyond the immediate loss of life. We owe it to our communities and everyone who depends on us to dedicate the time and resources needed in order to prepare for next unthinkable event.

REFERENCES

Barbish, D. (2002). Federal planning for consequence management. Presented at the 2002 American College of Contingency Planners (ACCP) Conference, Boston, MA (23 November).

Berwick, D. (1989). Continuous improvement as an ideal in health care. *New England Journal of Medicine, 320*(21), 1424–1425.

Dahlhamer, J., & Tierney, K. (1998). Rebounding from disruptive events: Business recovery following the Northridge earthquake. *Sociological Spectrum* (18), 121–141.

Edmonson, A., Bohmer, R., & Pisano, G. (2001). Speeding up team learning. *Harvard Business Review, 79*(9), 125–132.

Friedman, L., & Goes, J. (2000). The timing of medical technology acquisition: Strategic decision making in turbulent environments. *Journal of Healthcare Management, 45*(5), 321–335.

Gharajedaghi, J. (1999). *Systems thinking: Managing chaos and complexity.* Boston, MA: Butterworth-Heinemann.

Ginter, P., Swayne, L., & Duncan, W. J. (2002). *Strategic management of health care organizations* (4th ed.). Malden, MA: Blackwell.

Goes, J., Friedman, L., Seifert, N., & Buffa, J. (2000). A turbulent field: Theory, research, and practice on organizational change in health care. In: J. Blair, M. Fottler & G. Savage (Eds), *Advances in Health Care Management* (Vol. 1, pp. 143–180). New York, NY: Elsevier.

Gould, S. (1977). *Ever since Darwin: Reflections in natural history.* New York, NY: Penguin Books.

Grønhaug, K., & Falkenberg, J. (1989). Exploring strategy perceptions in changing environments. *Journal of Management Studies* (26), 349–359.

Kuhn, T. (1996). *Structure of scientific revolutions* (3rd ed.). Chicago, IL: University of Chicago Press.

Marghella, P. (2002). Medical planning considerations in consequence management. *Frontiers of Health Services Management, 19*(1), 15–23.

Merchant, J. (1986). Preparing for disaster. *American Journal of Public Health* (3), 233–235.

Meyer, A., Brooks, G., & Goes, J. (1990). Environmental jolts and industry revolutions: Organizational responses to discontinuous change. *Strategic Management Journal, 11*, 93–110.

Meyer, A., Goes, J., & Brooks, G. (1990). Organizations reacting to hyperturbulence. In: G. Huber & W. Glick (Eds), *Organizational Change and Redesign* (pp. 66–111). New York, NY: Oxford University Press.

Mintzberg, H. (1987). Crafting strategy. *Harvard Business Review, 65*(4).

Perrow, C. (1984). *Normal accidents*. New York, NY: Basic Books.

CHANGING ORGANIZATIONS FOR THEIR LIKELY MASS-CASUALTIES FUTURE

James W. Begun and H. Joanna Jiang

ABSTRACT

The threat of bioterrorism presents an opportunity for health care organizations to transform into more resilient, learning organizations. Rather than focusing solely on preparing for what is known or expected in a bioterrorist attack, organizations should strengthen their infrastructures to better manage surprises of all types. We advocate a combination of guidelines derived from conventional and complexity science perspectives on organizational change, including the need for leadership commitment, self-organization, culture change, and interorganizational connections.

Organizations in all sectors of industry and social service are better equipped than ever to learn about, prepare for, and respond to environmental events. Improved communications technology, in particular, has improved intelligence that organizations use about changes in the economy, the weather, competitors, and customers. Still, however, massive and unpredictable events can occur that challenge an organization's ability to be "ready for anything." Recently, a new class of such events, bioterrorism, has entered the organizational landscape.

In this chapter, we investigate how organizations can change in order to be better prepared for unforeseen, catastrophic events in general, and bioterrorism

Bioterrorism, Preparedness, Attack and Response
Advances in Health Care Management, Volume 4, 163–180
Copyright © 2004 by Elsevier Ltd.
All rights of reproduction in any form reserved
ISSN: 1474-8231/doi:10.1016/S1474-8231(04)04007-8

in particular. This is particularly critical for health care organizations, because they are community symbols of hope and service as well as the sites relied on by individuals for attention to the medical consequences of catastrophes. First, we classify bioterrorism relative to other environmental events. We describe the conventional methods of organizational change that could be applied to help organizations prepare for bioterrorism. Because of the nature of bioterrorism, however, conventional change mechanisms only go so far in preparing the organization. Newer concepts from complexity science assume a less knowable future and must be added to organizations' tools for preparing for bioterrorism. The chapter ends with a discussion of research needs that will improve our knowledge about changing organizations for their likely mass-casualties future.

MAJOR ENVIRONMENTAL EVENTS IN THE HEALTH CARE SECTOR

Over the last three decades, we have witnessed the health care environment becoming more complex and turbulent. Changes are frequent – with an increased volume and velocity – demanding that health care organizations constantly prepare for surprise and unanticipated consequences. To help understand the implications of bioterrorism for health care organizations, we begin with a review of more conventional environmental events and highlight the differences between bioterrorism and those events.

Figure 1 presents a three-dimensional framework for analyzing major environmental events in the health care sector. The first dimension relates to the source of the event – whether there is a small number or a large variety of contributing factors and whether these contributing factors can be identified. If the event derives from just a few identifiable factors, it is considered to be relatively *simple* compared to those events that are caused by a large number of different factors that are not easy to detect or have complex relationships connecting one and another. The second dimension classifies the speed of change in the environmental event. Some events unfold slowly, and others with suddenness and little foreshadowing. The third dimension, represented by the vertical axis, describes the predictability of the event in terms of the nature of its occurrence – time, place, speed, direction, scale, and impact. In other words, how well can we see the event coming and anticipate its trajectory of change and consequences?

Major environmental events include regulatory changes, workforce shortage, market transformation, technologic advancement, consumerism, natural disasters, epidemics, and terrorism. These events do not always share similar attributes when viewed from this framework. Regulatory changes have a relatively long history in

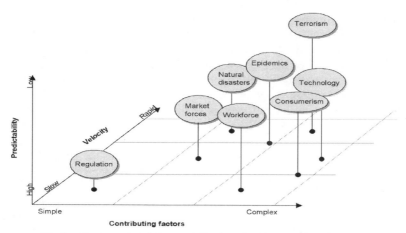

Fig. 1. Environmental Events Affecting Health Care Organizations.

the health care sector – from the enactment of Medicare and Medicaid programs in the 1960s to introduction of the Medicare inpatient prospective payment system in 1983 and the more recent implementation of the Medicare outpatient prospective payment system through the Balanced Budget Act of 1997. Regardless of whether the regulatory change comes from government or private accreditation entities, health care organizations typically have avenues to interact with the initiators of the change and provide input on the new regulation. In addition, regulatory changes tend to happen incrementally, which gives health care organizations time to prepare and respond. For example, the outpatient prospective patient system has been fully implemented over a period of years. Thus, compared with other environmental events, regulatory changes display simple contributing factors and high predictability.

Next on the scale is change in market forces, which has more highly interactive contributing factors and less predictable outcomes. The pace of change is more rapid, particularly in a local environment, than regulatory events. In health care, local market conditions have undergone substantial transformation in the 1990s due to multiple factors originating in the private sector. The growth of managed care significantly influences the type of services providers offer and the reimbursement mechanisms. The rise of the employers' role in health care purchasing generates an unprecedented pressure on providers to control cost and improve quality of care. Consolidation among providers further complicates the market landscape. It becomes increasingly difficult for health care organizations to predict the expectations of their stakeholders and the actions of their competitors.

Changes in workforce, technology, and the rise of consumerism present a higher level of uncertainty than changes in regulation and market forces. The workforce shortage has been an issue for decades in the United States – once for physicians and now for nursing and other allied health professionals. Multiple factors influence the demand and supply of the health care workforce, including population composition, economic conditions, regulation, and technology. Few predictions on the adequacy of various health professionals ever correspond to the reality that emerges. For instance, in early 1990s both the federal government and the Association of American Medical Colleges predicted an oversupply of specialist physicians by 2000. Now evidence has begun to show just the opposite – a severe shortage of specialists is predicted to occur (Cooper, 2002).

Technological advancement can be reflected in three main areas – genetics, medical equipment, and informatics. The development of new medical equipment allows more and more inpatient procedures to be performed in the outpatient settings. Genetic therapy is growing on the horizon, while information technology has already shown its impact by making clinical information available online within and across health care organizations and by giving consumers almost costless access to a vast amount of medical information. Given the scope and pace in current technology development, it is hard to fully foresee what type of new technology will emerge in what date and the implications for health care delivery.

Consumerism represents a relatively new movement in the health care sector. Its early signs can be traced back to efforts such as total quality management and patient-focused care. Consumerism has its roots in the American culture of individual choice and is catalyzed by factors such as backlash against managed care, economic prosperity, and Internet technology (Robinson, 2001). Consumers increasingly empower themselves with information obtained through the Internet and actively participate in medical decision-making. They also gradually assume a more important role in health care purchasing. Undoubtedly, health care organizations will face a whole set of new challenges posed by a consumer-driven system.

In comparison to all the conventional environmental events discussed above, terrorism displays the highest level of uncertainty in terms of lack of information beforehand, and sometimes even afterward, on identifiable source of the event, the time and place of happening, and the impact. First, "terrorism" means that the event is an unexpected attack from an anonymous source that could be a single entity or multiple groups. A bioterrorist attack against America is considered to be virtually inevitable after the September 11 event (Becker, 2002; Tierman, 2002), but it remains unknown as to when, where, and how such an event will occur. In particular, terrorist events may employ new technologies in producing various kinds of infectious agents and designing the route of exposure (Franz et al., 1997; Pringle, 1998), which adds to the difficulty of predicting and preventing these events.

Second, although terrorist events occur with little or no prior notification, how rapidly the resultant impact from the attack is manifested varies by the type of the agent (Franz et al., 1997; Macintyre et al., 2000). If a chemical weapon is used, mass casualties may present instantly and a large number of patients need immediate treatment. In contrast, most biological agents have various incubation periods and an attack will not be recognized until an increased number of patients present at their physicians' offices or the emergency departments with signs and symptoms caused by the infectious agent. By the time the agent and the route of exposure are identified, thousands of people could have been exposed to the infection. Health care facilities would need to provide both prophylactic medications and specific treatment. The radical impact caused by a terrorist event and the speed at which health care organizations are required to respond may far surpass those of any more conventional environmental event.

Third, in a terrorist attack, health care organizations become not only part of the rescue effort but possibly victims themselves, as revealed in the September 11, 2001 event (Becker, 2002). The financial and physical impact on a victimized health care facility and its staff could be enormous. For instance, the New York University Downtown Hospital, located just three blocks from the World Trade Center, was worst hit among all hospitals in the city. With the surrounding traffic disrupted and power disconnected, this 149-bed hospital had to provide medical care as well as food and shelter to the victims. During the subsequent months, patient volumes for surgeries and obstetrics dropped substantially because the physical destruction in the neighborhood created an access problem to the hospital. The hospital lost $2.6 million as a direct result of the event.

If involving the use of biological agents, terrorist events may resemble epidemics in some ways. For instance, in the most recent global epidemic of SARS (severe acute respiratory syndrome), mass infection occurred before the agent and route of contagion were identified. In treating the victims, some health care facilities failed to protect their staff and other patients from being infected, which unfortunately contributed to further spread of the disease. Nevertheless, most epidemics can be controlled once the virus is identified, interventions are implemented, and the public is educated. In contrast, terrorism adds another level of complexity, unpredictability, and fear because of the attacker's malicious intent, desire to surprise, and anonymity.

Terrorism is different from natural disasters, too. Although natural disasters also occur very rapidly, they do present measurable signs that allow the capture and tracking of their movement. Almost every health care facility has emergency plans for natural events like snowstorms, flooding, hurricanes, and tornados, and the past experience with these events can help alleviate the fear of any unanticipated consequences. In contrast, terrorism posts a completely new yet imminent threat to health care organizations.

In summary, health care organizations today are operating in a new kind of environment. They not only have to deal with continuous changes from regulation, the marketplace, technology, and consumer expectations, but are challenged to prepare for terrorist events that very likely will cause mass casualties and multiple surprises that simply are unknowable in advance.

ORGANIZATIONAL CHANGES PRESCRIBED BY CONVENTIONAL ORGANIZATIONAL THEORIES

Conventional organizational theories that have been employed to provide guidance for organizational changes in response to environmental events include: institutional theory, resource dependence, strategic contingency, structural contingency, transaction cost, agency theory, and population ecology. All these theories share a bias or common assumption influenced by Newtonian physics, that is, changes in the world are analyzable and predictable because of linear and deterministic relationships among discrete objects (Begun, 1994; Begun, Zimmerman & Dooley, 2003; McDaniel & Driebe, 2001). To prepare for change, organizations should analyze probabilities of different futures, set goals in light of those probable futures, and attempt to realize their goals. Based on facts and data, strategies and structures can be designed and implemented rationally to match the environment. When faced with actual environmental changes, organizations should implement control mechanisms to reduce any negative impacts. In addition, strategy and structure can be altered in concert with the actual environmental changes. Goes, Friedman, Seifert and Buffa (2000) categorize these conventional responses to first-order or incremental organization-level change as examples of adaptation (when the change is externally-driven) and teleology (when the change is voluntaristic).

Some recent studies have applied conventional organizational theories to examine hospital behaviors in response to changing market conditions. Using a strategic management perspective, Kumar, Subramanian and Strandholm (2002) suggested that hospitals adopt different strategies in matching their perception of environmental changes to enhance their financial performance. Efficiency-oriented strategies fit a relatively stable and certain environment while market-focused strategies work better with greater environmental uncertainty. Young, Parker and Charns (2001) employed the contingency theory to propose criteria in evaluating the effectiveness of provider integration within the context of local market conditions. Both studies were built on the assumption that as highly rational entities, organizations can adapt to changing environmental conditions through modifying their strategies or internal design. Conventional change theory leads to prescriptions for preparing and implementing change that can be quite valuable, given the time and resources to seriously implement them (Kotter, 1995).

While conventional theories have shown utility in understanding organizational changes in responses to certain environmental events such as changing regulatory and market conditions, they are not able to provide enough guidance for addressing terrorism. First, as discussed in the previous section, terrorist events are highly unpredictable and uncontrollable, complex, and rapid. These features clearly contradict the assumption of traditional theories that environmental changes are identifiable and predictable. Second, applying conventional theories to terrorism suggests a rational model of sophisticated and expensive planning and preparation that at some time is subject to diminishing returns and may even be counterproductive. Conventional change theory encourages health care organizations to attempt to discover all possible scenarios and develop response plans based on very limited information and substantial guesswork. The danger of doing this is that health care organizations would get trapped into a deception that they are in well control of the situation, severely undermining their ability to deal with any unexpected surprise. Finally, conventional organizational theories lead the organization to focus on key dimensions (e.g. efficiency, structural design) and ignore those factors that appear to be random. Such a perspective is particularly risky in handling terrorist events that are associated with a whole web of unknown factors. Health care organizations need to pay a great deal of attention to "random" details in both preparing for and responding to the event.

A COMPLEXITY SCIENCE PERSPECTIVE ON PREPARATION FOR UNPREDICTABLE CHANGE

To prepare organizations for surprise, conventional approaches to organizational change should be combined with newer approaches that focus on preparation for uncertainty and surprise (Anderson & McDaniel, 2000). These approaches call for radical and organization-wide change, classified as transformational in the schema developed by Goes et al. (2000). While these approaches derive from several sources, complexity science provides an umbrella that is consistent with the study of preparation for surprise. A complexity science perspective on terrorism allows us to effectively address those areas where conventional organizational theories fall short. Table 1 summarizes some key differences between conventional and complexity science perspectives.

Table 1. A Comparison of Conventional Theories and Complexity Science.

	Conventional Theories	Complexity Science
View of the environment	Independent of the organization	Coevolves with the organization
	Linear and deterministic relationships among environmental elements	Complex and nonlinear relationships among environmental elements
	Changes are predictable	Environment is relatively unknown
Organizational response	Develop control mechanisms to reduce dependence and uncertainty	Modify expectations to assimilate the changes and unexpected surprises
	Focus on strategy and structure	Focus on creation of meaning and adherence to value systems
	Manage through rules and algorithms	Manage through self-organization
	Ignore random changes	No change is random

Sources: Derived from Olson and Eoyang (2001); Begun, Zimmerman and Dooley (2003).

As noted earlier, conventional theories consider the environment to be independent of the organization and relatively knowable. A complexity-based approach notes that systems are intricately linked to environmental events (Kauffman, 1995) and that future environments are relatively unknowable. In the case of bioterrorism, members of the organization may have relatives or friends that are victimized. Members of the organization may even be instigators or supporters of terroristic acts. The actions or non-actions of the organization can affect the view of terrorists toward the organization and its community, or their propensity to attack the community. In these and many other ways, health care organizations shape and respond to the potential for bioterrorism.

Complexity science assumes that organizations, like other systems, face changes among interconnected agents, making the unfolding of change unpredictable and complex. This clearly is the case for surprises like terrorism. Organizations, like other systems, respond by modifying their expectations to assimilate surprise. Surprises fall outside of the original expectations and could cause unpleasant feelings including frustration, anxiety, and most deeply a loss of sensemaking (Weick, 1993). Understanding and quick acceptance of the reality is a key for organizational survival, as exemplified in the preparations Morgan Stanley, the giant investment bank, had made prior to September 11, 2001 (Coutu, 2002). Being the largest occupant in the World Trade Center, Morgan Stanley realized its vulnerability to terrorist attacks soon after the 1993 bombing in the Center. The company started a program preparing their employees to know exactly what to do in a catastrophe. On September 11, just one minute after the first plane hit the north tower, Morgan Stanley started evacuating about 2,700 employees in the south tower following their well-practiced protocol. The company lost only seven employees compared to other companies that were unable to do much to save their employees.

Organizational applications of complexity science suggest a need to focus on values and beliefs that form the foundation for expectations. When surprises hit, strong value systems shared by members of the organization can help them see meanings or make sense of the events (Coutu, 2002). The creation of meaning then shapes the frame for important decision making related to mission and strategy. This was demonstrated decades ago in Meyer's (1982) account of hospitals' responses to the environmental "jolt" of an unpredicted doctor's strike. For hospitals in New York City, even though the September 11 attack brought tremendous damage to downtown Manhattan and financial losses to many organizations, this event both drew upon and expanded a sense of connection among the hospitals as well as between the hospitals and the community. The community appreciated the value of its hospitals and the hospitals appreciated the hard work of their staff members – the spirit of thankfulness spawning across the city. As a hospital staff member reflected on the event, "we have survived, and we're stronger" (Becker, 2002). Health care organizations that build values of community, service, and caring into their internal cultures are more likely to be effective in time of community crisis.

Complexity science asserts self-organization to be driving force for organizational change. Self-organization refers to the process in which agents in the system (e.g. persons, groups, other subunits of the system) interact with one another in the absence of an overall system-wide blueprint (Stacey, 1999). System agents learn and adapt as the "surprise" unfolds. Informal networks are formed through self-organization so that different expertise can be rapidly pooled together in handling the crisis (Weick & Sutcliffe, 2001).

Finally, complexity science suggests organizations interpret external changes as non-random, and potentially meaningful, searching for anomaly before filtering or discarding input. Applied to bioterrorism, any "normal" clinical case may be the harbinger of a biologically-induced epidemic, if it is a piece in a larger pattern. This requires that organizations collect, or participate in the collection of, data that are examined for emergent patterns.

Weick and Sutcliffe (2001) label such organizations "mindful" ones. Mindful organizations are prepared internally to identify and deal with surprise. Table 2 outlines several internal characteristics of mindful organizations. Mindful organizations analyze their failures in order to learn. This characteristic may be of limited application to terrorism, since events are so rare. The other characteristics of mindful organizations are more relevant in the context of terrorism, however.

Table 2.   Characteristics of Mindful Organizations.

Characteristics	Description/Details
Preoccupied with failure	Pay attention to near misses as a signal of defection in other parts of the system Avoid overconfidence from past successes Accept and learn from failures Encourage self-analysis and knowledge sharing
Reluctant to simplify interpretations	View the world as complex, unstable, and unpredictable Challenge conventional wisdom Promote diversity in perspective and opinion Refine and complicate existing expectations, and invent new ones
Sensitive to operations	Treat operations as center of the system Systemwide sharing of real-time information on operations, across ranks, departments, and disciplines
Committed to resilience	Improvise to contain and bounce back from unexpected events Encourage open communications, informal networks, and fast real-time learning
Defer to expertise	See expertise and experience as more important than rank A hybrid of hierarchy and specialization decision structure to allow for authority migrating to the people with most expertise

Source: Compiled from Weick and Sutcliffe (2002).

Organizations must be in the habit of looking for complex rather than simple interpretations of events. Training clinicians to get extra information about diseases that potentially are caused by bioterrorism is an example. Organizations must be extra-sensitive to operations, so that perturbations – anything out of the ordinary – can be noticed as early as possible. Being sensitive to changes in disease patterns requires this kind of sensitivity, and requires data collection and monitoring efforts by the organization. An organization's commitment to and experience with resilience reminds members that the organization will face surprise, will on occasion fail, but will learn and ultimately will survive. It also gives members "permission" to be innovative when faced with unpredictable adversity. Lastly, in mindful organizations individuals defer to expertise. In times of crisis, as hospitals have known at least since the advent of the emergency room, a clear decision-making hierarchy, based on expertise, is necessary for selected categories of activities. In regard to bioterrorism, expertise needs to be identified and communicated well in advance of actual acts of terrorism.

In summary, the complexity science perspective contends that in the face of unpredictable, radical change, organizations need to rely on value systems in addition to operations manuals, and on learning in addition to rote practice. They can prepare for such events by purposefully pursuing an internal culture and values that are consistent with resilience and learning. How to move the organization from the conventional to the prepared is the subject of the next section.

CHANGING HEALTH CARE ORGANIZATIONS TO PREPARE FOR MASS-CASUALTIES

How does a "conventional" organization become an organization that is prepared for an unknown future? In this section, we explore several guidelines that are drawn from both conventional theories of organizational change and from the complexity science perspective. They are summarized in Table 3.

Table 3. Guidelines for Preparing Health Care Organizations for Bioterrorism.

Guideline	Rationale
Develop leadership commitment to organizational resilience and learning	Change doesn't occur without top commitment; change requires time and resources
Support conditions for self-organization Clear roles Training Seek out ideas Timely feedback Connections Communication Identify expertise	Plans will be incomplete in meeting bioterrorist events; anticipates complexity and unpredictability in responding to bioterrorism; distributes accountability and responsibility for response
Develop culture of resilience, learning, and social responsibility	Links bioterrorism to larger category of unpredictable events; avoids "flavor of the month" change
Connect and collaborate with the organizational field, including competitors	Competition is suspended in emergencies; bioterrorism response is interorganizational
Prepare individual organizational members and constituents for bioterrorism	Individuals need information and training; constituents need reasonable (vs. ideal) expectations
Contribute to and support governmental and professional association plans for preparedness	Enrolls organization as a participant in the prevention of bioterrorism; recognizes social responsibility and interconnectedness

Little can be accomplished without the commitment of a coalition of organizational leaders, whether they are the formal authorities among board and medical staff officeholders and the top management team, or others in the organization with significant power bases. A coalition of leaders needs to define bioterrorism as a real threat, responsibility, and opportunity, and convince others that the organization must devote time and resources and energy to change. Leaders can use bioterrorism as an example or metaphor for the new environment for the organization – bioterrorism can be presented as a case example of new and surprising events that the newly resilient organization must be prepared for. Preparing for bioterrorism is consistent with the social service mission of health care organizations, and again can be presented as an opportunity to extend that mission.

Absent leadership commitment, response to bioterrorism is likely to be perfunctory and superficial, and preparation may meet the letter of the law but not the spirit. Organizational members will do what they are asked but no more.

A new challenge for organizational change efforts in this era of "constant" change is the cynicism of organizational members about change, and their negative experiences with past changes. Overcoming cynicism may be less pressing a problem because bioterrorism is scary and has personal impact, but some organizational members may consider the threat to be overexaggerated and be wary of the need to be prepared. This skepticism should be recognized and appreciated. It reinforces the need to place organizational change for bioterrorism within a broader context of becoming more resilient as an organization. In the event that bioterrorist surprises do not materialize, others nearly as unpredictable, complex, and sudden will occur, rewarding organizations that are mindful and prepared for surprise. Thus, leaders can squarely counter the criticism that their particular organization may indeed never be directly affected by any given particular surprise.

A second organizational change strategy for preparation for bioterrorism is the encouragement of self-organization. To the extent that employees have not only the permission to, but the responsibility to find solutions to complex problems themselves, they are more likely to move decisively in the face of surprise. Conditions for self-organization include an understanding of "the big picture" by everyone in the organization, clear roles and expectation, training and development resources, seeking out different ideas, giving timely feedback, development of connections among individuals throughout the organization, and extensive communication from "headquarters" (Olson & Eoyang, 2001). Pockets of expertise in subjects matter relevant to bioterrorism need to be identified, communicated, and accessible.

Development of these self-organizing conditions enables confidence and responsibility for response to surprise to develop throughout the organization.

In addition to making the organization as a whole more powerful, top leadership is relieved of an overwhelming, cumulative sense of responsibility and account-ability that would follow from a totally centralized approach to preparedness. Heavy centralization of response also can lead to overload of input at one point in the organization, slower processing, and delayed or ineffective response as a result.

At a more cognitive level lies the culture of the organization – the shared values, beliefs, and norms – which must recognize and integrate the likelihood of bioterrorism and guide a response to it. Organizational leaders need to examine their organizations' missions and visions and values, and ask whether they accommodate and endorse a commitment to preparedness for bioterrorism. To be lasting and widespread, changes to mission and vision need to be widely discussed and debated. Such changes are an opportunity for leadership to educate members of the organization about terrorism, and enroll members of the organization in the "new" organizational culture of learning and resilience. Here, the fact that health care organizations will become a beacon of hope and a communications center for communities during a bioterrorism crisis should be discussed, requiring a re-commitment to social responsibility.

A fourth strategy for change is the cultivation of organizational connections. Connections become valuable in the search for solutions to complex problems. At the individual employee level, connections are a key to the ability to self-organize. At the organizational level, connections to other organizations are the key to efficient and effective rapid response. To the extent that connections are not only known but have been practiced, interorganizational innovation in times of crisis is more likely. Hospital CEOs Phillip Robinson (Robinson, undated) and Michael Covert (*Journal of Healthcare Management*, 2002) report that jurisdictional issues are critical to resolve quickly in crisis situations. Public health departments, emergency medical services, and emergency management agencies must coordinate communications, control, and resource distribution. Reasonable expectations should be set in advance with key stakeholders in the health care organization's community. Key communication channels with the media should be identified. In many cases, important connections in responding to bioterrorism may be those with marketplace competitors.

A fifth organizational change strategy is to prepare the individuals in and around the organization – employees, contractors, patients, etc. – for the likelihood of surprise. Recognizing that the first concern of organizational members in a bioterrorist event will be their own health and the health of their families, the health care organization will have emergency security procedures and will provide as much information as possible to allow members of the organization to help take care of themselves. Recognizing that patients and community leaders may hold unreasonable expectations for preparedness, health care organizations will encourage open communication of their limitations and their reliance on understanding and cooperation from their publics.

Response to completely unexpected surprise can be hampered by conventional centralized planning and preparation if the "plan" is viewed as the "answer." However, we view centralized planning as complementary, rather than contradictory, to the previously listed guidelines. Individuals and organizations are capable of both having plans and realizing the limitations of them. Therefore, the guidelines in Table 3 end with the need to be involved in conventional disaster response preparedness. Indeed, the most prepared organizations will be those with detailed, practiced protocols coupled with stark realization of the limits of the protocols in times of true surprise.

PREPARING FOR THE UNKNOWN: RESEARCH QUESTIONS

Research on preparing organizations for bioterrorism is constrained by the infrequency of cases of bioterrorism. In spite of this, research opportunities are plentiful and interesting, because the research employs new theory and new data, and the results are potentially of great practical significance, since thousands of lives are at stake.

First exists the need to learn from the experience of organizations that have proven to be resilient in the face of surprising events. Case studies of "high-reliability" organizations and "mindful" organizations are beginning to partially fill this void (Bigley & Roberts, 2001; Weick & Sutcliffe, 2001). The experiences of health care facilities in the Middle East that have treated bombing victims, hospitals that dealt with the Florida anthrax cases in 2002, as well as health care organizations in war zones, are ripe for case study. In the Middle East, terrorist events that originally were "surprising" may have transitioned into routine events, and lessons learned can help other organizations prepare more quickly and move through the transition more smoothly if bioterrorism were to proliferate. Lessons from preparations for natural disasters provide another source of relevant experience. At the individual level, organizations can certainly learn more about how to better prepare individuals for their role in the health care organization in the event of calamity.

Second, the development of instruments to monitor organizational preparedness is needed, beyond those that are checklist style and targeted at the operational details. The AHRQ tool, Bioterrorism Emergency Planning and Preparedness Questionnaire for Healthcare Facilities, for example, measures 42 items related to the capacity of hospitals to treat victims of bioterrorist attacks (AHRQ, 2002), in yes-no format. The tool covers training for hospital personnel, staff and bed capacity, availability of diagnostic and therapeutic resources, as well as policy and procedures for guiding emergency medical response to an overflow of patients. The Association for Professionals in Infection Control and Epidemiology (APIC, 2002) offers a mass casualty disaster plan checklist with similar features. Such instruments begin to measure preparation for the predictable, but instruments measuring the characteristics of mindful organizations (see Table 2) or a culture of resilience and learning in health care organizations are less common. Translating existing tools such as those developed by Weick and Sutcliffe (2001) for use in the health care sector would be helpful in addressing this gap.

Once instruments are tested for reliability and validity, longitudinal studies should investigate change in preparedness and its association with organizational change efforts as well as contextual variables. For example, not-for-profit health care organizations may more easily embrace bioterrorism preparedness as consistent with their missions, and system members may have better access to standardized preparedness routines. Larger institutions may exhibit more difficulty in developing the characteristics of resilient and mindful organizations due to bureaucratic inertia. Health care organizations experienced with natural disaster events may exhibit better preparedness. These and other hypotheses would facilitate targeted interventions.

Preliminary evidence predicts that organizations operating within, rather than at the edge of, "zones of safety" respond with more resiliency to unusual events, and organizations with more resources are better able to heed warning signs (Marcus & Nichols, 1999). The role of resource availability and allocation and use in preparedness could help drive policy decisions about the size of investment needed to assure preparedness. Studies of the financial and other costs of change programs and preparedness are critical as well, since health care organizations generally operate at low margins and have a variety of competing demands for new expenditures.

Research on bioterrorism preparation should take advantage of the current interest in patient safety and the prevention of medical errors, which often are "surprises" to organizations. Comparisons to and learnings from organizational efforts to instill patient safety cultures are of interest. On the research methods front, realistic computer simulations and agent-based modeling of systems with surprises and responses to them are becoming more feasible with increases in computing power and greater sophistication in the assumptions about the agents being simulated. Improvements in longitudinal study methods include the recognition that the object under study itself will change and emerge as the study evolves, requiring greater openness of researchers themselves to emergence in methods and surprise in findings (Meyer, Goes & Brooks, 1993).

Research on how organizations can effectively implement changes necessary to produce organizations "prepared" for a mass-casualty future is a necessary part of the process of adapting the social fabric to deal with new realities.

CONCLUSIONS

Phillip Robinson, CEO of JFK Medical Center, Atlantis, Florida, argues that hospitals are the front line in the fight against bioterrorism. No doubt he would include all health care organizations in that front line. Among lessons learned by his hospital's experience with anthrax cases were that the hospital needed better information about the contagiousness of biological agents, a better understanding of jurisdictional issues, and a rethinking of the design of the emergency department and other facilities. They needed to identify new off-site locations for crisis management and needed to modify their crisis plans (Robinson, undated). The key point is that the medical center exhibited resilience during the crisis, learned, and their learnings can and should be shared. The ability of health care organizations to behave as resilient, learning organizations could make the difference between a scenario of societal chaos, conflict, and confusion and a scenario of competence, caring, and hope the next time bioterrorism hits home.

The magnitude and difficulty of this shift from complacent to resilient organizations cannot be underestimated. Organizations in the health care sector have been slow to change in the face of "normal" environmental change. Christensen, Bohmer and Kenagy (2000, p. 1) conclude that "Health care may be the most entrenched, change-averse industry in the United States." Morrison (2000, pp. 199, 203) suggests that "Health care moves at glacial speed compared to most other industries," and that "Hospitals and physicians have organizational time clocks that are geared more to geological speed than to Internet speed." Those who believe in the obligation of health care organizations to do all that is possible and reasonable to protect their communities from bioterrorism can take advantage of new knowledge about the characteristics of resilient, learning, mindful organizations. They can envision for their constituents a radically new type of organization that looks forward with confidence rather than waiting for external pressures to force change. Health care researchers and practitioners can learn from and draw upon the occasion of bioterrorism to transform health care organizations and agendas for health care organization research.

REFERENCES

AHRQ (Agency for Healthcare Research and Quality) (2002). Bioterrorism emergency planning and preparedness questionnaire for healthcare facilities. http://www.ahrq.gov/about/ cpcr/bioterrtxt.htm. Accessed 12/14/02.

Anderson, R. A., & McDaniel, R. R., Jr. (2000). Managing health care organizations: Where professionalism meets complexity science. *Health Care Management Review, 24*(1), 7–16.

APIC (Association for Professionals in Infection Control and Epidemiology, Inc.) (2002). Mass casualty disaster plan checklist: A template for healthcare facilities. http://www.apic.org/ bioterror/checklist.doc. Accessed 12/12/02.

Becker, C. (2002). We have survived, and we're stronger. *Modern Healthcare, 32*(35), 22–26.

Begun, J. (1994). Chaos and complexity: Frontiers of organization science. *Journal of Management Inquiry, 3*, 329–335.

Begun, J., Zimmerman, B., & Dooley, K. (2003). Health care organizations as complex adaptive systems. In: S. M. Mick (Ed.), *Advances in Health Care Organization Theory* (pp. 253–288). San Francisco: Jossey-Bass.

Bigley, G. A., & Roberts, K. H. (2001). The incident command system: High-reliability organizing for complex and volatile task environments. *Academy of Management Journal, 44*, 1281–1299.

Christensen, C. M., Bohmer, R., & Kenagy, J. (2000). Will disruptive innovations cure health care? *Harvard Business Review, 78*(5), 102–112.

Cooper, R. A. (2002). There's a shortage of specialists: Is anyone listening? *Academic Medicine, 77*(8), 761–766.

Coutu, D. L. (2002). How resilience works. *Harvard Business Review, 80*(5), 46–55.

Franz, D. R., Jahrling, P. B., Friedlander, A. M., McClain, D. J., Hoover, D. L., Bryne, W. R., Pavlin, J. A., Christopher, G. W., & Eitzen, E. M., Jr. (1997). Clinical recognition and management of patients exposed to biological warfare agents. *Journal of American Medical Association, 278*(5), 399–411.

Goes, J. B., Friedman, L., Seifert, N., & Buffa, J. (2000). A turbulent field: Theory, research, and practice on organizational change in health care. *Advances in Health Care Management, 1*, 143–180.

Journal of Healthcare Management (2002). Interview with Michael H. Covert, FACHE, President, Washington Hospital Center, Washington, DC. *Journal of Healthcare Management, 47*(4), 212–215.

Kauffman, S. (1995). *At home in the universe.* New York: Oxford University Press.

Kotter, J. P. (1995). Leading change: Why transformation efforts fail. *Harvard Business Review* (March/April), 59–67.

Kumar, K., Subramanian, R., & Strandholm, K. (2002). Market and efficiency-based strategic responses to environmental changes in the health care industry. *Health Care Management Review, 27*(3), 21–31.

Macintyre, A. G., Christopher, G. W., Eitzen, E., Jr., Gum, R., Weir, S., DeAtley, C., Tonat, K., & Barbera, J. A. (2000). Weapons of mass destruction events with contaminated casualties: Effective planning for health care facilities. *Journal of American Medical Association, 283*(2), 242–249.

Marcus, A. A., & Nichols, M. L. (1999). On the edge: Heeding the warnings of unusual events. *Organization Science, 10*, 482–499.

McDaniel, R. R., Jr., & Driebe, D. J. (2001). Complexity science and health care management. *Advances in Health Care Management, 2*, 11–36.

Meyer, A. D. (1982). Adapting to environmental jolts. *Administrative Science Quarterly, 27*, 515–537.

Meyer, A. D., Goes, J. B., & Brooks, G. R. (1993). Organizations reacting to hyperturbulence. In: G. P. Huber & W. H. Glick (Eds), *Organizational Change and Redesign* (pp. 66–111). New York: Oxford University Press.

Morrison, I. (2000). *Health care in the new millennium: Vision, values and leadership*. San Francisco: Jossey-Bass.

Olson, E. E., & Eoyang, G. H. (2001). *Facilitating organization change: Lessons from complexity science*. San Francisco: Jossey-Bass/Pfeiffer.

Pringle, P. (1998). Terrorism: America's newest war game. *The Nation, 267*(15), 11–17.

Robinson, J. C. (2001). The end of managed care. *Journal of American Medical Association, 285*(20), 2622–2628.

Robinson, P. D. (undated). When disaster strikes: The role of the hospital. Slide presentation. JFK Medical Center, Atlantis, Florida.

Stacey, R. D. (1999). *Strategic management and organisational dynamics: The challenge of complexity* (3rd ed.). London: Trans-Atlantic.

Tierman, J. (2002). Hospitals create new models as they gird for bioterrorism. *Modern Healthcare, 32*(35), 8, 16.

Weick, K. E. (1993). The collapse of sensemaking in organizations: The Mann Gulch disaster. *Administrative Science Quarterly, 38*, 628–652.

Weick, K. E., & Sutcliffe, K. M. (2001). *Managing the unexpected*. San Francisco: Jossey-Bass.

Young, G. J., Parker, V. A., & Charns, M. P. (2001). Provider integration and local market conditions: A contingency theory perspective. *Health Care Management Review, 26*(2), 73–79.

PART III:
ORGANIZATIONS
RESPOND . . . OR NOT

MULTIPROVIDER SYSTEMS AS FIRST LINE RESPONDERS TO BIOTERRORISM EVENTS: CHALLENGES AND STRATEGIES

Myron D. Fottler, Kourtney Scharoun and
Reid M. Oetjen

ABSTRACT

The possibility of a bioterrorism event haunts all healthcare organizations. We believe a bioterrorism event is more likely in an urban area, and that urban multiprovider systems will be the "first line healthcare responders." Due to the lack of empirical research on this topic and firsthand experience with a bioterrorism event, this paper will provide the theoretical underpinnings to support the rationale for multiprovider systems as "first responders." This chapter outlines the nature and challenges of bioterrorism for a healthcare organization, the likely "state of the art" preparations for such events on the part of the four categories of healthcare organizations, and finally, the implications for the structuring of multiprovider systems to enhance their ability to plan for, and respond to bioterrorism events. Potential future research issues are also addressed.

The tiny crop dusting plane flies towards the stadium going virtually unnoticed. In an area so prone to planes flying overhead, a small plane such as this one seems commonplace, thus arousing no suspicion. As the pilot nears the stadium, beads of sweat form on his forehead. It

Bioterrorism, Preparedness, Attack and Response
Advances in Health Care Management, Volume 4, 183–209
© **2004 Published by Elsevier Ltd.**
ISSN: 1474-8231/doi:10.1016/S1474-8231(04)04008-X

is critical that this mission is carried out perfectly – one wrong move, and he could put himself in grave danger. He wrinkles his brow as he concentrates on the task at hand.

Thousands of feet below him, the spectators cheer as their team scores a touchdown. The road to the playoffs has been a long one, but these loyal fans have been there every step of the way. Nothing could keep these fans from seeing their hometown team in the playoffs for the first time in twenty years – not even repeated warnings of possible terrorist attacks.

As the merriment continues, the pilot reaches the stadium. As instructed, he quickly flips the switch, silently releasing millions of tiny anthrax spores. They float aimlessly towards the crowded stadium, and to areas of the city surrounding it. Unknowingly, the fans below inhale the spores into their bodies, seeing and feeling nothing. They continue cheering on their team, unaware that they have just been in contact with a potentially deadly disease-causing agent.

When the game ends, the fans file out of the stadium, excited at the prospect of more playoff games to come – their team could just go all the way this year. They excitedly discuss this prospect on the way home. Some go a few blocks to their homes, others to neighboring towns, and still other diehard fans to other parts of the country. They settle back into their lives, and conversation turns to the upcoming game next weekend. No one suspects they have been in contact with anthrax, nor does anyone show any symptoms out of the ordinary.

In the weeks to come, one by one the individuals start to fall ill with flu like symptoms. They shrug it off as "just the start of the flu season." No one suspects that it may be anthrax. By the time a physician makes a suspected diagnosis, thousands all over the country are showing signs of inhalation anthrax. Emergency departments, health departments, and doctor's offices all over the country are flooded with people, both the truly sick and those worried that they may have been exposed. Panic ensues nationwide as Americans deal with the first real widespread bioterrorist attack (as adapted from Osterholm & Schwartz, 2001).

The previous scenario, while fictional, touches close to home, especially in the wake of America's brush with anthrax in late 2001 and the attacks of September 11th. Like never before, Americans deal with terrorist warnings and the threat of an attack on a daily basis. Previously a problem that "only other countries dealt with," it has become a daily part of life in the United States and presents a host of problems for the healthcare community. Previous literature (Greenberg, Jurgens & Gracely, 2002; Helget & Smith, 2002; Treat, Williams, Furbee, Manley, Russell & Stamper, 2001; Wetter, Daniell & Treser, 2001) has focused primarily on the impact a bioterrorist attack would have on a hospital, yet none have assessed the potential impact on an organized delivery system.

In other disaster situations, the healthcare community has been able to respond in a manner that was based upon knowledge garnered from previous events. The issue with a bioterrorist attack is that, to date, there has been no widespread attack on American soil. The brush with anthrax in late 2001 provided a glimpse into the dark world of bioterrorism, but did not produce a widespread attack that would serve as a model for future disaster plans.

In the event of a bioterrorist attack, Americans will expect that all facets of the nation's healthcare system will continue to operate and serve the needs of the public (Barbera, Macintyre & DeAtley, 2001). However, questions abound as the healthcare industry struggles to deal with everyday patient flow. Will emergency departments be able to deal with the massive influx of patients seeking care? What role will each part of the healthcare community play in a bioterrorist attack – will doctor's offices, labs, and pharmacies be involved as individual entities, or as part of a multiprovider system? Or will it primarily be left to hospitals and the health departments to deal with? Is there one specific plan that every healthcare entity should adopt to be able to successfully deal with a bioterrorist attack and do these plans differ if the entity is part of a multiprovider system?

The answers, while currently unclear, will begin to flow from experience as more is discovered and prepared for in the war against bioterrorism. This chapter will serve to explore these, and other questions facing healthcare executives across the country, and provide some general guidelines that can be applied to multiprovider healthcare delivery systems. It is important to note that currently there are no empirical studies available for review that examine the relationship between bioterrorism preparedness and multiprovider system membership. Future research perhaps will help to definitively answer some of these questions. The present chapter will attempt to do so on a theoretical level, but as a result of the lack of empirical research and experience with bioterrorism events, does not tout that it provides any definitive answers that can be validated only through experience.

In the paper, we will define and describe bioterrorism, outline possible organizational responses to bioterrorism, define multiprovider healthcare systems, utilize theory to make linkages between multiprovider systems and bioterrorism response, compare four categories of healthcare providers in terms of their preparedness and response to bioterrorism, and conclude with strategies by which multiprovider systems can manage the process of preparing for and responding to a bioterrorist event. The chapter concludes with suggestions for future research.

BIOTERRORISM

Definition

Bioterrorism, as defined by Edlin, is described as, "the threat of mass destruction by weapons of biological origin such as bacteria, toxins, and viruses" (2001, p. 30). Building upon that definition in more expansive terms, the General Accounting Office (GAO) defined bioterrorism as, "the intentional use of any microorganism, virus, infectious substance, or biological product that may be engineered as a result of biotechnology, or any naturally occurring or bioengineered component of any such microorganism, virus, infectious substance, or biological product, to cause death, disease, or any other biological malfunction in a human, an animal, a plant, or another living organism in order to influence the conduct of the government or to intimidate or coerce a civilian population" (GAO/West, 2000). Regardless of the depth of the definition, the mere mention of the word drives fear into many as the reality of bioterrorism grows more plausible each day.

Key Characteristics

Unlike previous, more overt acts of terrorism, a bioterrorist attack tends to be of a more covert nature. It has the ability to be carried out silently, lying dormant for weeks at a time before truly surfacing. In fact, there are several key characteristics to consider when dealing with a bioterrorist attack that sets it apart from previous disaster scenarios. They are:

- "The onset of the incident may remain unknown for several days before symptoms appear;
- Even when symptoms appear, they may be distributed throughout the community's health system and not be recognized immediately by any one provider or practitioner;
- Once identified, the initial symptoms are likely to mirror those of the flu or the common cold so that the health system will have to care for both those infected and the 'worried well';
- Having gone undetected for several days or a week, some infectious agents may already be in their 'second wave' before the first wave of casualties is identified;
- Public confidence in government officials and health care authorities may be undermined by the initial uncertainty about the cause of and treatment for the outbreak;
- Health care authorities and hospitals may want to restrict those infected to a limited number of hospitals but the public may seek care from a wide range of practitioners and institutions; and
- Health care workers may be reluctant to place themselves or family members at increased risk by reporting to work" (American Hospital Association, 2000, p. 18).

Psychological Implications

In keeping with the unique characteristics of a bioterrorist attack, it is only natural that public panic should be considered when creating a disaster response plan. Although a large-scale bioterrorist attack has not yet occurred on U.S. soil, the brush with anthrax in late 2001 filled every American with a sense of fear and foreboding. The fear of the unknown gripped Americans as individuals flocked to doctor's offices and emergency departments – worried that they might have come in contact with anthrax. They demanded screening and testing for exposure to anthrax, and begged for prescriptions for antibiotics deemed effective for treating anthrax exposure. Alarmingly, as panicked as the American public was with this anthrax scare, a full-blown release of a biological or chemical agent could cause immeasurable levels of panic and mass hysteria.

Adding to the chaos, news headlines scream of war, and top government officials have stated, "it is highly likely that a terrorist group will launch or threaten a germ or chemical attack on U.S. soil in the next few years" (Bradley, 2000, p. 261). What was previously thought to happen only in other places of the world has come home to America's backyards and is seemingly here to stay.

The range of emotions and fears following a bioterrorist attack may include: horror, anger, panic, unrealistic concerns of infection, fear of contagion, paranoia, social isolation, and demoralization (APIC, 1999). In order to reduce some of these fears and minimize mass hysteria, clear lines of communication must be created and utilized (Simpson, 2002). If information is disseminated to the public in a lucid, easy to understand manner that explains both the risks and the logistics of the situation, levels of mass hysteria can be lessened (APIC, 1999).

Moreover, a 2002 study conducted by the Johns Hopkins School of Public Health found that people react to the information presented to them, and that public panic is rare and preventable. For example, the increase in gas mask sales post September 11th was not out of pure panic, but because the information presented to the public by the media portrayed the need for a gas mask as reasonable (Johns Hopkins, 2002).

However, public panic can occur, even if strides are taken to prevent it. A calm, well trained, and educated staff armed with clear information, is one of the greatest assets when calming the fears of patients and the general public. In addition, having psychologists on hand to deal with the gamut of emotions involved is an essential service to deal with the influx of worried individuals (APIC, 1999).

Challenges Bioterrorism Creates

As previously stated, there are numerous questions that surround the healthcare industry's level of preparedness for a bioterrorist attack. To be effective, bioterrorism plans will need to encompass a number of different areas and be a community-wide effort. Unfortunately, according to Donald Henderson, former director of the Johns Hopkins Center for Civilian Biodefense and current Chairman of the

National Advisory Council on Public Health Preparedness, "the major problem is that there is really no public health 'system' for dealing with infectious disease in this country, but, rather, a fragmented pattern of activities" (2001, p. 67). In the past, "the diverse initiatives taken by different agencies of government were not well-coordinated, even within the agencies themselves, and many have been designed with little comprehension of what is implied for the civilian population when a biological weapon is used" (Henderson, 2001, p. 66).

To be truly prepared, a comprehensive effort must be undertaken in concert with all those involved. This involves cooperation among medical and public health professionals, emergency management officials, the military, government, and law enforcement (Centers for Disease Control, 2000).

In addition, bioterrorism plans must address the critical areas of preparedness and prevention, detection and surveillance, diagnosis and characterization of biological and chemical agents, response, and communication (AHA, 2000; CDC, 2000; Johns Hopkins, 2001). Even if a hospital already has a disaster plan in effect, it is critical that a separate, detailed bioterrorism plan be added to the organization's planning component as a stand-alone policy (Evans, 2002). Although there is not one plan that will work for all healthcare facilities, it is critical to review those templates available and make changes based upon each facility's needs. This includes separate plans for each part of the organized delivery system that fit into an overall plan for the system itself.

Another challenge to establishing necessary community linkages involves legal issues. Currently, the Health Insurance Portability and Accountability Act of 1996 (HIPAA) privacy regulations and efforts to improve hospital disease surveillance capabilities conflict with one another. The HIPAA privacy regulations present roadblocks in the path of the healthcare community's efforts to share important health and demographic information. Lawmakers will have to explore ways in which multiprovider systems can circumvent current privacy regulations that do not allow hospital associations to share "protected health information" from one hospital to another hospital (Schulman, 2002).

ORGANIZATIONAL PREPAREDNESS AND RESPONSE TO BIOTERRORISM

A study conducted in March 2001 by Helget and Smith (2002) looked at the preparedness levels of hospitals, long term care facilities, and assisted living facilities in Nebraska from March through April 2001. Although 900 surveys were mailed out to eligible facilities, only 131 were completed for a response rate of 14.6% (Helget & Smith, 2002). It is important to note, that although the response rate was rather low, the researchers felt that they had a representative sample of relevant healthcare institutions in Nebraska (Helget & Smith, 2002). The findings suggested that only 49% of those surveyed believed that a bioterrorist attack was something that their community would encounter, although hospitals were more likely than other healthcare organizations to recognize bioterrorism as a potential threat (Helget & Smith, 2002). Additionally, a resounding 98% stated that they did not feel as if their organizations were adequately prepared for an attack of this nature (Helget & Smith, 2002). Contrary to their survey results, the researchers feel that the number of organizations that believe that bioterrorism cannot touch them will have significantly decreased and these facilities are presently better prepared (Helget & Smith, 2002).

In keeping with the findings of Helget and Smith, prior to September 11th, hospitals across the United States listed preparations for a bioterrorist attack at the bottom of the list of priorities (Costello, 2000; Edlin, 2001; Johnson, 2001). In fact, one physician in charge of planning and training hospitals for bioterrorism estimated in 2000 that "only 15% of hospitals have the equipment or training to properly decontaminate victims in the event of a bioterrorist attack" (Costello, 2000, p. 5). No one deemed it important enough to spend a great deal of time or energy on a problem that belonged only to other countries. The prevailing notion was that it was a problem dealt with overseas but that in the United States, it belonged only in a movie plotline (Johnson, 2001). According to Susan Pisano, the Vice President of communications for the American Association of Health Plans (AAHP), "a lot of emergency preparedness manuals and thinking prior to September 11th focused on natural disasters; however, after September 11th, these assumptions had to change" (Krizner, 2002, p. 28).

Although no studies have been published demonstrating increased levels of preparedness since September 11th, the literature suggests through small-scale surveys that this attitude has changed (Johnson, 2001). As early as one month prior to September 11th, experts believed that American Hospitals were unprepared for a bioterrorist attack. However, one month after the attacks, hospitals are showing signs of progress in the quest to plan for bioterrorism (Johnson, 2001).

The response to the September 11th terrorist attacks at the World Trade Center and the Pentagon by Emergency Department (ED) physicians provided several important, firsthand lessons for disaster preparedness. Coincidentally, a group of 33 ED physicians were meeting two miles from the World Trade Center at the time of the attack and responded. From this response, they learned that communication from the disaster site was difficult, a system to verify credentials of volunteers was critical, need for identification of volunteers and their role was essential, field triage occurred infrequently, and victims presented at the closest hospitals, rather than the most appropriate hospital. The most important lesson learned was that all hospitals must be prepared to handle the trauma of a disaster and must establish a template of an effective plan and adapt it to the needs of the particular situation (ED Management, 2001).

Perhaps the greatest help in this quest has been in the form of monetary assistance. For example, in February 2002, the Department of Health and Human Services (DHHS) announced $20 million in funding for a nationwide network of Centers for Health Preparedness (Krizner, 2002). Additionally, DHHS asked for $518 million to help prepare the county's hospitals for a bioterrorist attack for the 2003 fiscal year (Krizner, 2002). These additional funds, coupled with a greater awareness have led to better, more focused preparations for hospitals across the United States.

Multiprovider Systems

No healthcare system in the world has undergone as much structural change as the United States has over the past three decades. It has been suggested that the extent and the swiftness of structural change in U.S. hospitals are unprecedented in postindustrial society. Some have characterized this change as fundamental and perhaps revolutionary. Nowhere is this more evident than in the transition to multiprovider healthcare delivery systems. The previous cottage industry of individual, freestanding hospitals has become a complex web of systems, alliances, and networks. Expectations of how managed care has and will reshape the industry and how organizations will ultimately respond have influenced many of the recent changes (Fottler & Malvey, 2003).

The development of hospital systems in the United States initially encompassed the horizontal integration of facilities and resulted in the creation of multi-hospital systems that provided similar acute care services in multiple locations. Later, system capability expanded through vertical integration and diversification into activities that may or may not have been related to a hospital's inpatient acute care business. More recently, expansion has reflected "virtual" integration that involves relationships based on contracts.

This system development reflects the transformation of multiprovider systems, from providers of acute care to providers that are capable of addressing a continuum of healthcare needs. Multiprovider systems have become dominant in healthcare, with more than 50% of hospitals belonging to systems and the remainder involved in some other type of collaborative relationship (Shortell et al., 1996). For the purposes of this chapter, a multiprovider system is defined as two or more hospitals or other provider organizations resulting from horizontal, vertical, or virtual integration (Fottler & Malvey, 2003).

THEORETICAL UNDERPINNINGS

Scholars have recognized that multiproviders organizations (sometimes called alliances) have both advantages and disadvantages (Zajac & D'Aunno, 1997). They believe that any assessment of the risk (i.e. disadvantages) should be balanced with an assessment of the expected benefits including improved financial performance, innovation, organized learning, and the opportunity costs of "going at it alone." Participation in a multiprovider system may appear risky or costly when the baseline comparison is not made explicit. The alternative may be to not serve a particular market at all and to fail to meet important community needs.

The fundamental question is: Which is riskier – going at it alone, doing nothing, or participating in a multiprovider system? (Zajac & D'Aunno, 1997, p. 331). Multiprovider systems are one method of responding to a series of "environmental jolts" which have impacted the healthcare industry in recent years (Meyer, 1982). These are relatively abrupt, major, and often qualitative changes in the environment that threaten organizational survival (like a bioterrorism event). Multiprovider systems reflect the perception and reality that it is better to face life's uncertainties with partners than to go it alone.

One potential benefit of multiprovider systems is the formation of a "trading alliance" within the system, which brings together organizations that contribute different (often complementary) resources needed to enhance a given objective (Zajac & D'Aunno, 1997). System members enhance their own innovation and learning by being part of a system. Finally, that member might also enhance their own bargaining power, reduce uncertainty, and share risk by being part of a system.

Success in multiprovider systems is more likely if the individual units provide complementary resources, understand one another's goals and objectives, have aligned incentives, avoid common ownership, have a designated management team focused on the success of the system, have conflict and problem resolution mechanisms, change as member organization's needs change, and minimize bureaucratic impediments (Zajac & D'Aunno, 1997).

Although a paucity of theory exists to directly explain the relationship between multiprovider systems and bioterrorism preparedness, linkages can be made through the use of Porter's Five Forces and Hoskisson, Hitt, Wan and Yiu's (1999) structure-conduct-performance model.

Hoskisson et al.'s (1999) structure-conduct-performance model suggests that an organization's performance is "related to the strength of forces that define the structure of the industry environment" (p. 417). Utilizing this model, the relationship between the structures of the healthcare organization (i.e. whether or not it is part of a multiprovider system) can be used to predict the level of preparedness it is capable of in the case of a bioterrorist attack. In fact, many suggest that the current consolidation of the healthcare market into multiprovider systems is a way to increase efficiency and improve their strategic positions in the healthcare market, thus suggesting that those that are a part of a multiprovider system will be better prepared to deal with a bioterrorist attack. In fact, Douglass and Ryman's (2003) study on the competitive advantages of multiprovider systems indicates that such entities are able to establish competitive advantages and mitigate some of the negative aspects of the healthcare market.

Porter (1991) identified two key factors of an organization's success which are important factors in determining an organization's ability to respond to crises, such as bioterrorism – the organization's external environment and its core competencies. Utilizing Porter's (1980) framework for analyzing the external environment provides a means for understanding the competitive dynamics of a particular healthcare market. Porter (1980) suggests that the level of competitive intensity is the most critical factor when evaluating an organization's environment. The competitive intensity is a function of the threat of new entrants to the market, the level of rivalry among existing organizations, the threat of substitute products and services, and the bargaining power of buyers and customers. The strength and impact of these five forces must be carefully considered in order to determine the potency and viability of an individual organization. It is posited that the stronger an organization is in these five measures of competitive intensity, the more equipped it is to adapt to environmental jolts and strategically prepare to manage crises.

Porter (1991) also points to an organization's core competencies as another dimension of success. Porter draws upon a resource-based as another source of competitive advantage. The connection between resources and activity is fundamental to the viability of an organization. Resources can originate from performing activities over time, acquiring them from the outside, or a combination of the two. Resources can be classified into three categories: (1) activities, such as its medical staff; (2) skills, organizational routines, or other assets attached to particular activities or groups of interrelated activities; and (3) external assets such as its reputation and relationships. Because multiprovider systems typically have access to more resources than individual hospitals, they may have an increased capacity and capability to prepare for and respond to a bioterrorist attack.

REVIEW OF MULTIPROVIDER
SYSTEMS LITERATURE

The literature on multiprovider systems has been recently reviewed by Fottler and Malvey (2003). Their review found several correlates of successful systems. One study examined the 17% of systems, which report operating margins of 4% more (Bilynsky, 2002). That survey found that these best performing systems:

- Focus on core competencies.
- Focus on quality of clinical outcomes and services (not size per se).
- Have not become complacent.
- Focus on execution of details.
- Focus on quality of physician integration (not quantity).
- Reduce duplication of services.
- Control future growth.

Another study conducted by Arthur Andersen based on interviews in seven systems concludes that: (1) communications are vital to the success of multiprovider systems; (2) sufficient staff, money, time, and energy must be devoted to planning, preparation, and training for systems integration; and (3) systems must research and understand community needs (Egger, 1999).

Fottler and Malvey (2003) also note some additional attributes of successful systems:

- Development of a shared system culture.
- Development of a formal strategic plan for the system.
- Development, implementation, and communication of explicit measures of quality of care, patient satisfaction, efficiency, and community benefit.
- Development of a simple, lean, flat, responsive, and customer-driven organizational structure.
- Selectivity in adding new system members.
- Alignment of incentives.
- Development of incentive structures to enhance system performances.

The literature (Greenberg, Jurgens & Gracely, 2002; Helget & Smith, 2002; Treat et al., 2001; Wetter, Daniell & Treser, 2001) is in agreement that the internal characteristics of the hospital, large size and urban location, tend to "predict" how well the hospital is prepared to deal with a bioterrorist attack, and how well the hospital can adapt to a situation as conditions warrant. Although none specifically addressed multiprovider organizations, one could predict an association between higher levels of preparedness with those in a multiprovider organization, due to the amount of resources available to the organization and the strong backbone of support they are afforded as a result of their system membership and relationships to other system providers. Moreover, those in an urban locations are likely to be larger in terms of the number of affiliated providers, have more potential "backup" resources (i.e. testing labs and hospital beds), and more likely to experience a bioterrorism event; thus, they are more apt to be prepared.

For example, "In general, respondents from urban hospitals reported higher levels of awareness in case of a bioterrorist attack equal to or higher than those reported by respondents from rural hospitals, and respondents from larger urban hospitals reported the greatest awareness" (Wetter, Daniell & Treser, 2001, p. 712). Similarly, a statistically significant correlation ($p < 0.01$) was identified between higher patient volume in the emergency department and likelihood of having a written plan in place to deal with victims of a biological or chemical attack (Greenberg, Jurgens & Gracely, 2002).

Another study, conducted by Treat, Williams, Furbee, Manley, Russell and Stamper (2001) in the Federal Emergency Management Agency's (FEMA) region III, attempted to assess the levels of preparedness at hospitals within this region. This study pointed out a strong association between preparedness and location of the hospital, with particular focus on the association between higher levels of preparedness, albeit perceived in some instances, and those in urban locations.

Figure 1 shows the categorizations of healthcare providers based on their membership in multiprovider systems and urban location. We believe that urban, multiprovider systems will be "first-line responders" to bioterrorism events with urban freestanding providers as "key backup." Rural areas are less likely to be bioterrorism targets and less likely to be primary or secondary responders.

Type of Provider

Location of Provider	Freestanding	Multiprovider System
Urban	"Key Backup"	"First-Line Healthcare Responders"
Rural	"Unlikely Responders"	"Secondary Backup"

Fig. 1. A Typology of Healthcare Organization Preparedness and Potential Response to Bioterrorism Based on Membership in a Multiprovider System and an Urban Location.

First responder traditionally refers to law enforcement personnel, fire and rescue, and emergency medical services (EMS) that provide prehospital care for victims of illness and accidents (CFSI, 2002; Kreiter, 1994). However, in the case of Bioterrorism, first responders involve public health, hospitals, physicians, and community emergency management officials along with the "traditional" first responders mentioned previously (Schulman, 2002).

MANAGEMENT STRATEGIES FOR RESPONDING TO BIOTERRORISM

Reporting Requirements

As the first responders to a confirmed or suspected bioterrorist attack, the multiprovider system bears the responsibility of reporting the event. Initially, the emergency department must request that the emergency response system be activated to respond to the attack, thus alerting the proper agencies, and informing them of the nature and logistics of the bioterrorist event (APIC, 1999). It is this initial reporting that will activate the emergency response system and ensure cooperation amongst local, state, and federal officials to properly deal with the bioterrorist event. Figure 2 outlines the notification procedures of a confirmed or suspected bioterrorist event.

In addition, this initial reporting will allow for the divisions of the organized delivery system to put their disaster plans into effect and begin to address the situation on an as needed basis. The activities involved at this stage may be as minimal as putting the organization on a standby alert for further action, to actually enacting the disaster plan and treating potentially infected patients.

Safety Precautions

In any healthcare setting universal precautions must be utilized to ensure the safety of the staff as well as the patients. However, a bioterrorist event often yields new problems and the implementation of a host of new protocols to ensure the safety of all those involved. As a general rule, standard precautions such as proper hand washing, utilization of gloves, masks, eye or face shields, and gowns must be implemented to prevent direct contact with any bodily fluids and mucous membranes (APIC, 1999).

Even when universal precautions are used, it is extremely important to ensure that hand washing occurs after every patient contact. This ensures that bodily

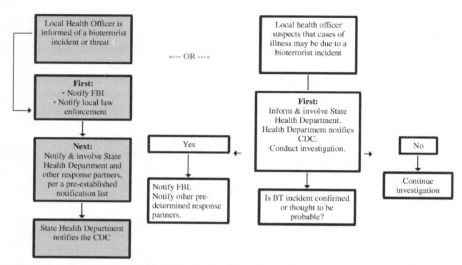

Fig. 2.   Interim Recommended Notification Procedures for Local and State Public Health
Department Leaders in the Event of a Bioterrorist Incident (CDC, 2002).

fluids are not transmitted to others even if gloves are used as a barrier (APIC, 1999). Additionally, the most effective way to prevent transference of germs or microorganisms is to use an antimicrobial soap when washing one's hands.

Furthermore, gloves are to be worn at all times when one is in contact with a patient. Universal precautions are critical in preventing the spread of bacteria or other agents. As a rule, gloves must be changed between patient contacts, and even changed during procedures on the same patient if there is contact with a contaminated material (APIC, 1999).

Masks, face, and eye shields also provide a certain degree of safety for healthcare workers when working on a patient. These items should be utilized when performing patient procedures that may cause any splashing of a bodily fluid such as blood, excretions, or secretions (APIC, 1999). It is critical to utilize the proper safety equipment based upon the type of agent that is suspected.

With regards to patient placement and transportation, it is necessary to know what bioterrorist agent one is dealing with before making decisions on how to transport and place patients. Erring on the side of caution is highly recommended if the agent has yet to be identified, and patients should be quarantined until a positive identification can be made. In this instance, and even when the agent is known, it will become necessary to isolate a section of the hospital to care for these patients without potentially contaminating those that are unaffected (APIC, 1999).

Disease specific requirements must be followed to ensure the safety of all those involved (Center for the Study of Bioterrorism, 2001). For example, because anthrax cannot be spread from person to person, it is not necessary to quarantine off patients and keep them separated from others (APIC, 1999). Conversely, small pox patients need to be kept away from others, as they are highly contagious (Center for the Study of Bioterrorism, 2001).

Sterilization of Equipment

Even though sterilization of equipment occurs regularly within a hospital, a bioterrorist event may yield further scrutiny of these procedures, and require additional precautions. Additional precautions include:

- Provide approved germicidal cleaning supplies to properly cleanse equipment and other materials that may have been in contact with any contaminant.
- Sterilize of instruments routinely to ensure the removal of any contaminants or bodily fluids.
- Disinfect, and clean environmental surfaces such as bed rails, beds, bedside equipment, and other equipment must be completed.
- Handle any contaminated equipment with care to reduce any chance of further contaminating others or other equipment.
- Ensure that reusable equipment is not reused before proper disinfection and sterilization.
- Handle patient linens in accordance with standard precautions to ensure the proper decontamination before reuse.
- Sort and discard contaminated waste in accordance with local, state, and federal regulations (APIC, 1999).

Staff Roles

"Since the employees of hospitals will often represent the 'first line of defense' in a bioterrorist attack, they must first and foremost receive adequate training to not only deal with the situation, but to recognize the symptoms of biological and chemical agents to ensure proper diagnosis" (Scharoun, van Caulil & Liberman, 2002, p. 86). There are numerous training programs available to educate staff on the specifics of biological and chemical agents. Since no one program will fit each healthcare facilities' needs, it is necessary to assess the facility's needs and schedule education and training as conditions warrant.

In keeping with the additional requirements of bioterrorism planning, it is critical that a chain of command be set up as part of the bioterrorism readiness plan. The idea behind this concept is to designate one person as the chief operating officer (CEO) of the entire operation, and have other roles and duties designated for all other staff in the healthcare facility ("ED's disaster," 2002). This will enable a clear chain of command in addition to well-delineated duties for all staff members, thus reducing potential chaos.

A hospital in Waterbury Connecticut has utilized this concept and has implemented an incident command system for use in the event of a bioterrorist attack ("ED's disaster," 2002). The essential elements of their command system include:

- An incident command field – 1st on the scene to evaluate and declare disaster if necessary; consists of the fire chief, EMS command, police command, primary and secondary triage officers, treatment officer, etc.
- An incident command hospital – consists of the director of emergency services, nursing supervisor, EMS coordinator and security management; only area to communicate directly with the incident command field.
- Personnel pool – incoming non-clinical staff report to CT scan corridor for further assignment.
- Main lobby – clinical staff report here for further assignment.
- Media – are guided to the hospital day care center where contact from the hospital will brief them on a regular basis.
- Security – facilitate lockdown if necessary and manage incoming/outgoing patient flow.
- Other areas of the hospital – designated for routine care vs. suspected exposure to a biological or chemical agent.
- Triage – personnel will assess and determine to which area of the hospital the patient should go ("ED's disaster," 2002).

Utilizing this system, the hospital feels that it will be able to handle the influx of patients, while still keeping the other non-contaminated patients safe from exposure and further harm.

Moreover, it is critical to ensure that an adequate number of staff will be on hand to deal with the influx of those seeking care. The nation's emergency departments are already overcrowded, and are struggling to deal with everyday emergencies. "While the public and the political communities assume that healthcare systems are adequately preparing for terrorism incidents that would generate catastrophic casualty loads, the medical community is struggling just to maintain its everyday capacity" (Barbera, Macintyre & DeAtley, 2001, p. 1). "While nearly 39 million people were uninsured for the entire year in 2000, it is estimated that approximately 45 millions people will have no health insurance by the end of 2002" due to the economic events following the September 11th tragedies (Miller, 2001, p. i). This will only increase the influx of patients utilizing the emergency departments for routine medical care. If a bioterrorist attack is added to the mix, the already overworked staff will have a huge increase in the number of patients to care for with the same number of staff on hand, thus collapsing the nation's healthcare "safety-net."

To alleviate this problem, the American Hospital Association (AHA) devised some guidelines to remove a portion of the burden from the current staff. In their August 2000 report, the AHA suggested that a reserve staff list be developed to deal with the added influx of patients. These reserve employees (i.e. retired employees with clinical backgrounds, or those that may be in administrative roles, but have a clinical background) can be called upon to help augment the staff and ensure that an adequate number of staff is on at all times (AHA, 2000). In addition, multiprovider systems can utilize staff from other non-affected entities to help deal with the increased patient load.

Communication

As previously discussed, communication is a crucial issue in creating a bioterrorism response plan. "Planned and structured arrangements for communication throughout the incident and during its response are critical components of hospital and community preparedness" (AHA, 2000). The media will be a critical actor in this step of the plan, and can help to get information out to the public (Johns Hopkins, 2002). In addition, the healthcare facility should have a designated spokesperson that will be solely responsible for addressing the media on behalf of the system (AHA, 2000). This will allow for uniform presentation of information to be presented, without conflicting stories being circulated to the general public.

Internally, communication is also critical. A system must be activated to communicate not only within the healthcare facility, but to other parts of the system as well. In addition, a backup system must be planned for to ensure that if the primary system fails, uniform communication will continue (AHA, 2000). One centralized means of communication will ensure that everyone knows where to go for information, and will allow for a uniform means of communication.

Hospitals perform an essential role as part of America's vital healthcare infrastructure – one that is destined to increase in the event of a bioterrorist incident. Hospitals can use their existing EMS, trauma coordination, and other relationships as a framework upon which to build expanded relationships for mass casualty readiness. Hospitals must adopt community-wide perspectives and involve other community partners, such as other multiprovider organizations, in order to increase the ability to respond to mass casualty incidents. To accomplish this task, multiprovider systems must communicate with one another (Schulman, 2002).

Hospitals need to be proactive and establish open and ongoing relationships with the local health departments. Bioterrorism incidents, in particular, require community-wide surveillance and control efforts in order to recognize possible incidents. Communication is essential to identifying apparently isolated symptoms into a recognizable pattern that alerts the community's health care and the public health system about the potential emergency situations. Developing a communication network will enable these organizations to initiate more appropriate public health interventions to combat a bioterrorism incident (Schulman, 2002).

Multiprovider systems will need to create linkages between their data reporting systems to provide a community-wide assessment of health needs and health care resources. Systems that share a common architecture and that have the ability to integrate real-time data from other institutional operations will provide the best means to matching community needs to available resources. These relationships must be developed prior to responding to a large-scale disaster; otherwise they will not be able to meet the increased demand on all of the community's health resources simultaneously. During crisis situations, there is not enough time or available staff to survey hospitals and other facilities in order to inventory capabilities. The ability to rapidly and accurately assess the situation and share such information will be critical to identifying and responding to an unusual pattern of symptoms that could indicate that a bioterrorist attack has occurred (Schulman, 2002).

In the event of a disaster, communities will need the ability to assess in a rapid and accurate way what health care resources are available for response. Developing linkages that improve the capacity of multiprovider systems, public health departments, and clinicians to engage in disease surveillance will be critical in determining if a cluster of disease may be related to the intentional release of a biological or chemical agent and in expediting an effective response. The ability to continue to share such information could be critical in identifying an unusual pattern of symptoms that could indicate that a bioterrorist attack has occurred (Schulman, 2002).

COMPARISON OF THE FOUR CATEGORIES OF HEALTHCARE PROVIDERS

Table 1 shows our summary of existing theory and our judgments concerning the current "state-of-the-art" preparedness and prevention for a bioterrorist event

Table 1. A Comparison of Four Categories of Healthcare Providers in Terms of Preparedness and Prevention of Bioterrorism.

Preparedness and Prevention Area	Category			
	Rural Freestanding	Urban Freestanding	Rural Multiprovider	Urban Multiprovider
Up-to-date disaster response plan with appropriate responses/actions	U	SL	L	L
Established hospital emergency incident command system	L	L	L	L
Plan widely available throughout facility/network	U	SL	SL	SL
Multi-disciplinary disaster planning committee	U	SL	SL	L
Established specific collaborative relationships and linkages with local EMS, EMA, and local health department	U	SL	L	L
Established linkages with other hospitals, poison control centers, and physician offices	U	SL	L	L
Development of "surge capacity" capability	U	U	SL	L
Identification of key position holders in the event of a bioterrorism event	U	SL	SL	L
Emergency call-up staffing plan supported by communication/transportation strategies	SL	SL	L	L
Enhanced bioterrorism disaster plan formulating and continuous bioterrorism related training for healthcare professionals	U	SL	SL	L
Adequate security with regards to entry, exit, and vehicular traffic points	U	SL	U	SL
Disaster plan specifies where multiple casualties can be received, identified, triaged, registered, treated in designated areas, admitted, transferred, and transported	U	SL	SL	SL
Adequate supplies of staff, medical equipment, supplies, and pharmaceuticals with designated backup suppliers outside region	U	SL	SL	L
Ongoing, mandatory disaster exercise programs with formal critiques	U	SL	SL	SL
Inter-organizational joint disaster training seminars	U	U	U	U

U = Unlikely; SL = Somewhat Likely; L = Likely.

among the four categories of healthcare providers identified in Fig. 1: freestanding rural, freestanding urban, multiprovider rural, and multiprovider urban. The latter category, which we characterized as "first-line responders," is most likely to be at the forefront of responding to a bioterrorism event. It should be noted that our judgments for this and upcoming tables are based on observation and previous research indicating large urban hospitals are better prepared for bioterrorism events than others (Greenberg, Jurgens & Gracely, 2002; Helget & Smith, 2002; Treat et al., 2001; Wetter, Daniell & Treser, 2001).

The four categories of healthcare providers can be better understood by first defining the terms multiprovider and freestanding. The term multiprovider indicates that two or more hospitals have combined their resources and expertise, either informally through a working agreement or more formally with a coordinated management structure, in order to avoid duplication of services. The multiprovider system provides patients with the full spectrum of services and allows each individual healthcare organization to benefit from the economies of scale and combined resources (Raffel & Barsukiewicz, 1989). The term freestanding indicates that the healthcare organization does not have a formal or informal affiliation with another healthcare organization, but rather, is a separate entity.

The area in which the hospital is located determines whether or not it is classified as urban or rural. According to the 2000 U.S. Census, an urban area, or "cluster" is defined as "a densely settled area that has a census population of 2,500 to 49,999" (www.census.gov, 2002). Conversely, a rural area is defined as, "all territory, population, and housing units located outside of urbanized areas and urban clusters" (www.census.gov, 2002).

Table 1 indicates that urban multiprovider systems are more likely than others to be better prepared along a number of dimensions than are their rural or freestanding counterparts. This is particularly evident in terms of an up to date disaster plan with specific outcomes, linkages with other providers to provide "surge capacity" backup resources, and continuous bioterrorism exercises and training for all personnel.

Detection, Surveillance, and Diagnosis

Table 2 shows the same four categories of healthcare providers in terms of their ability (alone or in concert with others) to detect, survey, and diagnose a bioterrorism event. Again, the urban multiprovider systems appear to be better prepared to perform these functions then do their freestanding or rural counterparts.

In particular, they appear to be better positioned to evaluate, track, and notify infection control in the event of a bioterrorism event. This is due to their adequate

Table 2. A Comparison of Freestanding *vs.* Multi-Provider Delivery Systems/Networks in Terms of Bioterrorism Detection, Surveillance, and Diagnosis.

Preparedness and Prevention Area	Category			
	Rural Freestanding	Urban Freestanding	Rural Multiprovider	Urban Multiprovider
Detection and surveillance				
An established process to evaluate, track, and notify infection control of deviations from baseline data 24 hours a day, 7 days a week	U	SL	SL	L
Specified number and location of isolation or protective environment rooms	U	SL	SL	L
Adequate epidemiological capacity to detect and respond to biological attacks	U	SL	U	SL
Appropriate personnel aware of importance of reporting unusual disease presentations.	U	SL	SL	L
Partnership of hospital ER with other healthcare facilities for routine surveillance	U	SL	SL	L
Dedicated staff, phones, and fax to support rapid reporting	U	SL	U	SL
Diagnosis				
Specification of the circumsatnces under which the disater plan will be activated	U	SL	SL	SL
Activation steps established and roles outlined at each stage (i.e. alert, stand by, call out, stand down)	U	U	U	U
An established chain of command and processes for notifying internal and external stakeholders of bioterrorism event	U	SL	SL	L
Ability to perform in house diagnostic tests	U	U	U	SL
Abity to become part of network to do initial testing	SL	U	SL	L

U = Unlikely; SL = Somewhat Likely; L = Likely.

epidemiological capacity relative to others and their partnerships with other key players. They are also more likely to have established a chain of command and a process for notifying internal and external stakeholders in the event of a bioterrorism event. However, we do not believe that any of the four categories of healthcare providers have adequately addressed the issue of specified roles and relationships for each stage of a bioterrorism event (i.e. alert, standby, call out, and stand down).

The first stage, Alert, occurs when information is received on a possible bioterrorism situation which could escalate, or which may require the coordination of resources and support. The second stage, Standby, occurs upon the receipt of information that disaster situation is imminent and may require deployment of personnel and resources. The third stage, Call Out, is activated when a major incident/disaster has occurred which requires the deployment of personnel and resources. Stand Down occurs when it is determined that the incident has been resolved and emergency operations are no longer required (NSW Health, 2003).

Table 3 compares the four categories of healthcare providers in terms of their ability to respond and communicate during a bioterrorism event. The major response advantages of urban, multiprovider systems are their ability to respond to a major influx of patients both internally and through partnerships with others; dedicated areas for decontamination; and established mechanisms to deal with non-emergency patients, visitors, and volunteers. They are more likely to be part of established communication networks with key external stakeholders and have backup systems if any of these fail.

DISCUSSION

The problem facing the U.S. in the event of a bioterrorism event is that hospital "surge capacity" and specialized medical capability across the country has never been more restricted (Barbera, Macintyre & DeAtley, 2001). The medical community is struggling just to maintain its everyday capacity. Surge capacity refers to the healthcare organizations' ability to provide excess medical capability and improved response in the event of a crisis, such as a bioterrorism event (NGA, 2002).

Yet while hospital and physician charges have been constrained, no similar external controls have been imposed on their own costs of doing business (i.e. supplies, building, energy, personnel). Adequate preparedness for bioterrorism is expensive and time consuming. Limited time and attention can be devoted to emergency preparedness activities, and it can be overly taxing for individual institutions

Table 3. Provider Delivery Systems/Networks in Terms of Response to and Communication Regarding Bioterrorism Events.

Preparedness and Prevention Area	Category			
	Rural Freestanding	Urban Freestanding	Rural Multiprovider	Urban Multiprovider
Response				
Interval plan to respond to external disaster indicates how facility will respond to abnormally large (>10%) influx of patients	U	U	U	SL
Individual department plans to provide services 24 hours a day	U	U	U	U
Dedicated areas for contaminated patients and decontamination	U	SL	U	L
Established internal and external traffic flow and control plan	U	U	U	SL
Established mechanisms to deal with volunteers and visitors, and non-emergency patients during bioterrorism events	U	U	U	SL
Established procedures for relocation of patients under agreements with other healthcare facilities	U	U	U	L
Established post-disaster recovery plan	U	U	U	SL
Established secure command and security systems	U	U	U	SL
Continually updated website	U	U	U	SL
Ability to increase capacity in the event of mass casualties	U	SL	U	L
Communication				
Established internal and external means of communication during bioterrorism event with backups for both	U	U	U	SL
Established external communication network includes local EMS agency, state health departments, emergency management, and media	U	U	U	SL
Established plan for continuation of operations in the event of communication failure	U	SL	U	SL
Communication and transportation strategies that support emergency plan	U	U	U	SL

U = Unlikely; SL = Somewhat Likely; L = Likely.

to participate in even low-key, preplanned exercises (DCHA Hospital Mutual Aid System, 1999). Yet no formal appropriations for hospitals and multiprovider systems have been made, even though appropriations have been made to enhance the training and equipment of various communities "first responders" (i.e. fire fighters and police).

While public policy initiatives should be continue to be developed to provide funding for bioterrorism prevention and response, we believe these resources also need to be focused on multiprovider systems in urban areas, which we define as "first-line healthcare responders." They are the healthcare institutions to which the ambulances will bring the patient victims of a bioterrorism event after the EMS, police, firefighters, and other first-line responders have done their jobs.

Table 4. Bioterrorism Challenges and Strategies for Managing Challenges in Urban, Multi-provider Delivery Systems (Healthcare First-Line Responders).

Challenges	Strategies and Tactics
Preparedness and prevention	
Disaster plan preparations	Development and communication of disaster plan with actionable steps for all personnel
Disaster plan training	Training/disaster exercise drills for all personnel
Collaborative agreements	Development of collaborative agreements with external bioterrorism stakeholders (i.e. health dept, EMS, EMA, MD offices)
Reserve capacity	Development of "surge capacity" capability in terms of staff, bed capacity, pharmaceuticals, etc.
Detection and surveillance	
Routine surveillance	Process to detect "deviations" from baseline data internally and/or thru partnerships
Protective environments	Specified protective environment rooms
Diagnosis	
Testing	Capacity to do in-house diagnostic testing or participate in testing network
Response	
Capacity expansion	Plan to internally or collaboratively expand capacity to respond to bioterrorism event
Dedicated areas	Dedicated areas for contaminated patients
Command and Control	Secure command, control, and security systems
Patient transfer	Collaborative agreements for patient transfer
Communication	
Internal communication	Disaster plan with internal communication and transportation
External communication	Established external communication network

Table 4 outlines some of the challenges and strategies these first-line healthcare providers will need to address in preparation for and response to a bioterrorism event. The key is to develop and practice a plan, which includes specific activation protocols. A key lesson learned from the September 11th attack was that all hospitals should establish a template of an effective plan and adapt it to the needs of the particular situation (ED Management, 2001). Table 4 provides an outline of key strategies and tactics available to all first-line responders.

A major managerial challenge is to build and sustain successful systems to prepare for a bioterrorism event and make that goal an explicit part of their strategic planning process. What we know now is that system success is related to a number of strategic, structural, and staffing variables discussed earlier in the paper. Further study is required to determine whether these same variables (i.e. system resources, etc.) are related to system success in preparing for and responding to a bioterrorism event.

Future research should address the issue of funding for hospitals' preparedness for bioterrorism, local planning to identify first healthcare responders in a given community, identification of gaps in the preparation and response to a bioterrorism event based on staged exercises, and analysis of communication and coordinated problems during some of the staged events. Empirical determination of whether the perceived relationships identified in Tables 1–3 are valid is also in order. We would hope to never do empirical research on an actual bioterrorism event. However, that hope is probably idealistic. Empirical research on an actual event will enhance the ability of multiprovider to prepare and respond to similar future events.

REFERENCES

American Hospital Association (2000). Hospital preparedness for mass casualties: Final report August 2000. Summary of an Invitational Forum Convened March 8–9, 2000. Chicago, IL: American Hospital Association.

Association for Professionals in Infection Control and Epidemiology (1999, April 13). Bioterrorism readiness plan: A template for healthcare facilities.

Barbera, J., Macintyre, A., & DeAtley, C. (2001, October). Ambulances to nowhere: America's critical shortfall in medical preparedness for catastrophic terrorism. BCSIA Discussion Paper 2001–15, EDSP Discussion Paper ESDP-2001-07. John F. Kennedy School of Government, Harvard University.

Bilynsky, U. (2002). Integrated best performers: Seven habits of successful healthcare systems. *Health Care Strategic Management, 20*(1), 12–14.

Bradley, R. N. (2000, July–September). Health care facility preparation for weapons of mass destruction. *Prehospital Emergency Care, 4*(2), 261–269.

Centers for Disease Control and Prevention (2000, April). Biological and chemical terrorism: Strategic plan for preparedness and response. *MMWR, 49*, 1–14.

Center for Study of Bioterrorism (2001, October 21). Bioterrorism Infection Control: Guidelines for Patient Management. Retrieved from the Saint Louis University Center for Study of Bioterrorism: http://www.slu.edu/colleges/sph/csbei/bioterrorism/quick/Isolation.PDF.

Congressional Fire Service Institute (CFSI) (2002). Fire service position paper on the proposed department of homeland security. Retrieved April 9, 2003, from http://www.nvfc.org/pdf/position_paper_on_dhs.pdf.

Costello, M. A. (2000, July). Preparing for bio, chemical terrorism. *American Hospital Association News, 36*(28), 5.

DCMA Hospital Mutual Aid System (1999, November 11). Hospital tornado drill outliers. Washington, DC: District of Columbia Hospital Association.

Douglass, T. J., & Ryman, J. A. (2003). Understanding competitive advantage in the general hospital industry: Evaluating strategic competencies. *Strategic Management Journal, 24*, 333–347.

Edlin, M. (2001, April). Barefoot in a deadly garden. *Managed Healthcare Executive, 11*(4), 30–33.

ED's disaster plan uses incident command system (2002, March). *ED Management, 14*(3), 31.

Egger, E. (1999). Integration the right strategy despite health executives increasing apprehension. *Health Care Strategic Management, 17*(6), 10–11.

Evans, G. (2002, June). Stanford sets the standard for bioterrorism planning; a separate piece: Standalone plan advised. *ED Management, 14*(6), S4.

Fottler, M. D., & Malvey, D. (2003). Multiprovider systems. In: L. Wolper (Ed.), *Health Care Administration: Planning, Implementing and Managing Organized Delivery Systems*. Philadelphia, PA: Lippincott.

General Accounting Office and 18 U.S. C.A § 178 (West 2000). Definition of bioterrorism.

Greenberg, M., Jurgens, S., & Gracely, E. (2002, April). Emergency department preparedness for the evaluation and treatment of victims of biological or chemical terrorist attack. *The Journal of Emergency Medicine, 22*(3), 273–278.

Helget, V., & Smith, P. W. (2002, February). Bioterrorism preparedness: A survey of Nebraska health care institutions. *The American Journal of Infection Control, 30*(1), 46–48.

Henderson, D. (2001, September). The threat of bioterrorism and the spread of infectious diseases. Testimony before the U.S. Senate Foreign relations Committee. 107th Congress, 1st session, 62–69.

Hoskisson, R., Hitt, M., Wan, W., & Yiu, D. (1999). Theory and research in strategic management: Swings of a pendulum. *Journal of Management, 25*, 417–456.

Johns Hopkins Center for Civilian Biodefense Studies (2001, October 8). Enhancing bioterrorism preparedness and response post-September 11th: Interim Actions for the Medical and Public Health Community. Retrieved from the Center for Civilian Biodefense at Johns Hopkins website: http://www.hopkins-biodefense.org.

Johnson, D. E. (2001, October). Hospitals and country are "woefully unprepared" for bioterrorist attacks. *Health Care Strategic Management, 19*(10), 14–20.

Kreiter, T. (1994, May–June). Vital response: A first responder training manual. *Saturday Evening Post, 266*(3). Retrieved April 9, 2003, from http://galenet.galegroup.com/.

Krizner, K. (2002, May). The enemy within: Bioterror forces new contingency plans to better coordinate resources and information to contain and minimize an outbreak. *Managed Healthcare Executive, 12*(5), 28–32.

Meyer, A. (1982). Adapting to environmental jolts. *Administrative Science Quarterly, 27*, 515–537.

Miller, J. (2001). A perfect storm: The confluence of forces affecting health care coverage. Retrieved September 29, 2002 from http://www.nchc.org. *National Coalition on Health Care*, Washington, DC.

National Governors Association Center for Best Practices (2002). *Issue brief: States homeland security priorities.* Retrieved April 9, 2003, from http://www.nga.org/cda/files/081202HSPRIORITIES.pdf.

NSW Health (2003). *Recognized stages of activation.* Retrieved April 10, from http://www.asnsw.health.nsw.gov.au/publichealth/cdop/cdu/hplan/hplan3.html.

Osterholm, M., & Schwartz, J. (2001). *Living terrors.* New York: Delacorte Press.

Porter, M. E. (1980). *Competitive strategy: Techniques for analyzing industries and competitors* (pp. 3–33). New York: Free Press.

Porter, M. E. (1991). Towards a dynamic theory of strategy. *Strategic Management Journal* (Winter Special Issue) (12), 95–117.

Scharoun, K., Van Caulil, K., & Liberman, A. (2002, September). Bioterrorism V. health security: Crafting a plan of preparedness. *The Health Care Manager, 21*(1), 74–92.

Schulman, R. (2002). Statement of the American Hospital Association before the National Committee on Vital and Health Statistics Panel on national preparedness and a national health information infrastructure. Retrieved April 9, 2003, from http://ncvhs.hhs.gov/020226p1.htm.

Simpson, R. (2002, March). Our first line of defense against bioterrorism, part 1. *Nursing Management, 33*(3), 10–13.

Treat, K. N., Williams, J. M., Furbee, P. M., Manley, W. G., Russell, F. K., & Stamper, C. D. (2001, November). Hospital preparedness for weapons of mass destruction incidents: An initial assessment. *The Annals of Emergency Medicine, 38*(5), 562–565.

Wetter, D. C., Daniell, W. E., & Treser, C. D. (2001, May). Hospital preparedness for victims of chemical or biological terrorism. *The American Journal of Public Health, 91*(5), 710–716.

Zajac, E. J., & D'Aunno, T. A. (1997). Managing strategic alliances. In: S. M. Shortell & A. D. Kaluzny (Eds), *Essentials of Health Care Management* (pp. 328–354). Albany, NY: Delmar Publishing.

BIOTERRORISM VISITS THE PHYSICIAN'S OFFICE

Lawrence F. Wolper, David N. Gans and
Thomas P. Peterson

ABSTRACT

As a key component of the American health care system, the physician office could be the front line in a bioterrorist attack. Nationally and locally, the primary focus on this subject appears to be from a hospital preparedness and public health agency perspective, with little attention devoted to primary physician providers in their own offices, and those specialists to whom patients may be referred. While unrelated to bioterrorism, the recent SARS outbreak also brings to the forefront the need for physicians offices to be able to clinically, operationally, and managerially respond to illnesses that mirror the symptoms of known illnesses, but may be more virulent new organisms or hybrids of existing organisms. If the face of bioterrorism is subtle and slow in its presentation, physicians, in their own offices, could be the first providers of care. Will they be prepared, or will they be among the first fatalities in a bioterrorist attack?

MONDAY MORNING – WINTER, ANY YEAR

It is a typical Monday in the office of Best Internal Medicine Consultants, P.C. At 8:15 a.m. the office staff of thirty clinical, administrative, and billing staff begin to arrive. Not unlike most mornings in the office, there is casual discussion about

Bioterrorism, Preparedness, Attack and Response
Advances in Health Care Management, Volume 4, 211–258
© 2004 Published by Elsevier Ltd.
ISSN: 1474-8231/doi:10.1016/S1474-8231(04)04009-1

social and other subjects of interest. Most staff anticipate a busy day with four physicians in the office. One hundred patients have already scheduled appointments between the hours of 8:30 a.m. and 8:00 p.m. Instinctively, the staff knows that many patients will call for an appointment this morning as a result of feeling ill over the weekend. Call-ins occur every day, but after weekends and holidays they are more frequent, stretching the limits of the office staff and physicians beyond the already high stress levels inherent in many physician offices. An aging population with more illnesses, managed care companies that inhibit the ability of the primary care physician's referral to specialists, slow reimbursement from insurers, and the burden of paperwork have made the practice of medicine more difficult and stressful.

With the first patients scheduled to arrive at 8:30 a.m., the Front Desk receptionist turns on the computer to check the day's schedule of patients. Medical records for the day's patient visits have been "pulled" and are at the Front Desk. As is common, the first patients are arriving earlier than their appointment time, precluding the Front Desk receptionist the opportunity to check with the practices answering service to determine if non-emergency patients who called during the weekend need to be called back promptly. The answering service would have directly called the physician on weekend call if there were an emergency. Arriving patients come to the Reception Desk, sign in with the pen and pad provided by the practice, and engage in casual conversation with the receptionist. The office already is behind schedule and certain functions, such as calling the answering service for weekend messages, must be delayed. The doctors, having completed their "rounds" in the hospital, arrive at 8:45. The first patients have been escorted to the exam rooms. On any given day, about 10% of the patients come for an annual physical, and are requested to provide a urine sample before proceeding to an exam room. Within the first half hour, there are a dozen patients in the waiting room, with another six patients in exam rooms being seen by technicians who are measuring blood pressures, taking patient histories, and drawing blood for lab work. Some patients, after being seen by the physician, will receive X-ray studies.

With the initial rush over, the receptionist has time to call the answering service. As this is occurring, patients are calling to see if they can come to see a doctor even though they do not have an appointment. A new staff member, with limited experience, joins the receptionist to assist in answering calls. The practice does not utilize a call center telephone system, so calls from patients queue up, causing a ten minute wait before a patient is able to speak to a staff member.

During most telephone conversations, the patient will identify their symptoms in general terms. Normally there is no additional telephone triaging by the receptionist to determine more specific symptoms to assess if the patient needs to be seen sooner by a physician. While some physician offices, usually specialists, utilize a triaging questionnaire to assess whether a patient really needs to be seen on the day of the call, most offices, particularly those providing primary care services, generally do not use a triage questionnaire.

It is February, one of the heavy months for flu. It also is the time of year in which colds and other bacterial or viral ailments are more prevalent. Therefore, the number of calls for same day appointments has increased. One of the receptionists receives a call from a long standing patient, John Whittington, a man of about 42 years of age, who complains that he has a fever of about 102 degrees, aching joints, and has vomited twice. He also has advised the receptionist that his wife and two teenage children are suffering similar symptoms. They have had the symptoms for about three days, having first become ill on Friday. They elected to wait through the weekend before calling because they were certain that they had the flu or colds, and felt the illness would improve without physician intervention. Because of their age and good health, neither he nor his wife elected to receive flu vaccinations.

Although the practice is backlogged, as a result of their symptoms and the fact that they are existing patients, the Whittingtons are asked to come in at noon. They are advised that the wait will be longer than the usual 75 minute waiting time.

Patient Whittington and family arrive shortly before noon and sign-in at the Front Desk. The father shares some niceties with the Front Desk receptionist whom he has known for a number of years. They are not asked for any demographic or insurance information as they have all been seen within the past year. The policy of the practice is to update patient information such as address, telephone numbers, next of kin, and insurance coverage only if the patient has not been seen for over one year. Although the Whittingtons moved four months ago, and have new telephone numbers, the receptionist makes no inquiry. The receptionist does not obtain any further information about the symptoms of the family members, however, she does inquire if the insurance coverage is the same.

During the conversation, Mr. Whittington mentions that he has been unusually tired for the past three days. He attributed his fatigue to the fact that he has traveled extensively out of the country, and does not sleep well in hotels. On his last business trip to Bali and Indonesia, he brought his entire family where they enjoyed staying in a luxury hotel suite.

Patient Whittington selects a few magazines to read, and sits down in the waiting room with 18 other patients, including his family. The waiting room is small and the seating configurations consist of bench-like seats that keep patients in close proximity to one another. John Whittington and his wife read magazines and, although feeling poorly, converse casually with other patients over the next hour.

After an hour wait, a practice technician approaches the family and asks them to follow her to the exam room area. The technician walks, shoulder to shoulder with the Whittington's. When they arrive at the exam rooms, she asks that they remove their outer clothing and put on cloth gowns. They are all advised that the doctor will see them in about 15 minutes. Realizing the additional wait, Mr. Whittington asks to use the bathroom. He is directed to the lavatory that is also used by the office staff, other patients, and physicians. Prior to turning the bathroom doorknob to enter, he suddenly feels the urge to cough, and does so while covering his mouth and nose with his hands. He then opens the lavatory door. The doorknob, flushing lever on the toilet, the faucet at the sink, and handle on the paper towel dispenser all come in contact with his hands. These surfaces are not routinely cleaned or disinfected during the day. The patient rinses his hands briefly without soap.

Mr. Whittington returns to the exam room. The physician arrives, takes a brief history of his symptoms, and gives him a physical. Eventually, the same is done for the other family members. The patient fails to mention that he has been traveling recently with his family overseas. In turn, he is not asked whether he has been abroad, or has been exposed to any unusual settings. Historically, the practice has had little basis for this type of questioning, unless an unusual event like Rocky Mountain spotted fever is suspected in the area. Because he was seen within the last year, the exam is not comprehensive. Throat and nose are examined, chest is listened to, glands and abdomen are palpated, blood pressure and temperature taken. The chest sounds clear, there is no upper respiratory congestion, but there is a slight red throat and his current temperature is 103 degrees. To rule out bacterial infection a nurse takes a throat swab for a culture. A phlebotomist draws blood samples. During the exam, the physician is in close proximity to the patient. The Doctor suggests that the Whittingtons probably all have the flu. It is suggested that they drink fluids, and take Tylenol for the aches and fever. The Doctor advises the parents that they will be notified if any of the tests come back positive. If the throat cultures define a Streptococcus infection they all will be put on antibiotics. It will take about two to three days for the test results. The doctor leaves the room to see his next patient. The doctor does not wash or disinfect his hands before seeing the next patient.

The technician escorts the Whittingtons to the front desk where the cashier is given the required co-payments, and issues a receipt. The cashier enters the charge into the computer for the visit, as well as a diagnosis code of Influenza.

The Whittington family goes home. Two days later, John Whittington calls the office complaining to the receptionist that they are all feeling worse. He identifies that he also has noticed small rashes on their tongues and in their mouths. The receptionist suggests that they all come back to the office for a re-check. Because the receptionist is untrained regarding the symptoms from exposure to bioagents, she has responded to the patients reporting they were feeling worse, rather than any new symptoms expressed by the patient. Therefore, neither the nurse nor the physician is advised by the receptionist of possible changes in symptoms. Coming back to the office, the Whittingtons again wait, are again exposed, in close quarters over a prolonged period of time, to other patients in the waiting room, and to office staff. On this visit, a different physician sees the family. As a result, the new physician is unfamiliar with the situation, briefly reads the recent medical record entries, and makes further inquiries of the family during the examination. The new complaints of rash-like red spots on the tongue and mouth are of concern. Upon observation, the physician recognized that the Whittingtons have symptoms consistent with smallpox, and that it is in the early rash phase of the disease – the most contagious stage. On this and the prior visit, the staff did not wear surgical masks, nor were they gloved except for the phlebotomist.

Through inquiry, the physicians in the group determine that during the "Initial Symptom Stage" and the "Early Rash" phase of smallpox, the Whittington family has been in direct and prolonged close contact with approximately 39 friends and relatives, as well as four physicians, members of the physicians' office staff and waiting patients. Although patients are most contagious during the "Early Rash Phase," they also can be contagious during the "Initial Symptom Phase." One of the physicians notifies the appropriate local health department and the Centers for Disease Control and Prevention (CDC).

It was later confirmed by the hotel in Bali that the Whittingtons were exposed to smallpox. They all stayed in a hotel room where one or more bioterrorist employees that worked at the hotel on a part-time basis intentionally contaminated the bedding, towels and robes in many rooms of this hotel and other hotels in the area.

After reporting the situation to the CDC, the physicians meet with key office staff. Although the probabilities were small that such an event would occur in their practice, they now realize they have, in fact, been exposed to the outcome of a bioterrorist event. As a result, some of the staff and their patients may be among the victims. They conclude that, not unlike the first hospital to receive anthrax patients in Florida in 2001, they were unprepared for this situation (Jenks & LaPoll, 2001). They also conclude that there are a substantial number of operational, managerial, and policy/procedural issues that require action in the practice, including policies and procedures for cooperating with local and federal agencies. Among the many areas requiring change, they determine that they have to assess which of their physicians, patients and staff may have been exposed to the Whittingtons during their two visits to the office, and who came in contact indirectly with infected body fluids, or contaminated objects/surfaces in the office or lavatory? Further, they conclude that they need to create a mechanism so that in the future they can determine whether an unusually high number of suspected flu cases seen in the office are in fact patients exposed to a bioagent? Among the many other considerations, they need to create and implement disinfection policies and procedures.

This theoretical event, reinforced by the recent SARS epidemic, demonstrates the range of clinical and operational questions that contemporary physician's offices should address, and are the subject of this chapter.

THE OPERATING CHARACTERISTICS OF THE CONTEMPORARY PHYSICIAN'S OFFICE

The previous hypothetical situation could have occurred in many, if not most physicians' offices around the country. Therefore, the purpose of this chapter is to:

- Overview the manner in which most physician offices currently operate with regard to policies, written administrative procedures, allocation of responsibility, and authority.
- To overview the range of bioagents and the related symptoms about which physician offices should be alert.
- To recommend a range of operational and procedural changes that would enhance the ability of a physician's office to handle the outcomes of exposure to patients that exhibit symptoms that are consistent with bioagents.

The Organization of a Physician's Practice

There is no way to generalize about the manner in which physicians offices operate or are organized. The practice of medicine ranges from solo physician practice settings to large physician organizations consisting of hundreds of physicians practicing in multiple offices. Most physician practices are more likely to fit into the three to five physician range.

This chapter will focus on physician practices in which the physicians are owners and/or employees of the practices, rather than employees of a medical school or hospital based practices. Of the approximately 675,000 U.S. physicians involved in patient care, over 75% practice in independent solo practitioner or group practice settings (Pasko & Smart, 2003). In a large-scale bioterrorist event, it is less likely that the independent physician practice will be the first-line of medical intervention. Hospital emergency departments and freestanding emergency clinics will absorb the initial impact. On the other hand, in a bioterrorist situation that is more "subtle," such as discussed in the hypothetical example at the beginning of the chapter, many smaller independent practices may become the first stop for patients that are feeling the initial effects of a bioterror event. The reality of the 2001 Brentwood anthrax attack demonstrated that victims will go to the physician practice as the first stop. The first two patients seen were not recognized as having anthrax, but were diagnosed with "the flu." By the time the victims sought additional help, it was too late for effective treatment (Borio et al., 2001).

The probability and manner in which a bioterrorist attack may unfold cannot be predicted, and, as such, all private physician practices, particularly those providing primary care, such as Family Practice, Pediatrics, Internal Medicine, and General Practice, should become aware of what changes are required to respond to these new and serious threats.

Notwithstanding size, almost all physician practices consist of operational and organizational elements that are similar. All practices need to make appointments, register patients to collect pertinent demographic and billing data, bill and collect for services rendered, and capture medical information. Before exploring what needs to change within a physicians practice, it is important to understand how most practices operate under normal circumstances.

The following will be discussed:

- Responsibility and Authority
- Appointment Scheduling and Triaging
- Patient Registration
- The Clinical Experience
- The Medical Record
- Patient Check-Out

Responsibility and Authority

Management is defined as working with, and delegating tasks to others in an organization to achieve the objectives of the organization. Physicians are comfortable with delegating tasks, responsibility and authority to other physicians and to other trained clinical personnel such as nurses and technicians. However, they are often less comfortable delegating responsibility and authority to lay individuals. Characteristic of most physician practices is that personnel who perform many of the office functions such as appointment scheduling, registration, billing and collections, including many roles of a clinical nature, do not have advanced educations or college experience. Contrary to long standing management principles, in many practices there is delegation of responsibility, but without authority, or not enough authority to be commensurate with the level of responsibility. A typical practice office manager will have a broad range of management responsibilities, but limited range of authority. Physician practices, unless very large in size, are closely held businesses that operate similar to partnerships, regardless of the manner in which they are legally structured. Therefore, physicians in private practice make many decisions by consensus, and leave little decision-making authority to lay individuals.

As demonstrated in the hypothetical situation presented earlier, often compounded by the fact that the physician with business responsibility for the practice may not be in the office every day, it is important that key personnel have the responsibility and authority to act effectively, decisively, and, if necessary, independently. In addition, it is important that a practice have written policies and procedures for all major operations occurring in the office, including preparedness for the outcomes of bioterrorism. In many physician offices this will require not only operational changes, but also modifications of practice culture. This is a major change in mindset, not only for practice staff, but also patients.

Appointment Scheduling and Triage

Appointment Scheduling
This function is the first point of contact between a new or existing patient, and a physician practice. There are basically two methods used to schedule appointments. The oldest method is the appointment book, which is still prevalent in smaller practices. Most larger physician practices utilize computerized billing systems, the majority of which also have computerized patient scheduling (appointment scheduling) "modules."

During the appointment scheduling process, demographic information often is taken if the patient is new to the practice. Typically the information that is solicited is broad in nature including an address or telephone numbers, and is obtained to create a "patient record and identification number" in the computer system. When the patient arrives at the office, more detailed information is gathered.

Triaging
Triaging refers to an analytical process in which the severity of a patients problems or symptoms are assessed, and a determination made as to the patient's need to be seen immediately by a physician. Triaging is normally accomplished by physicians or nurses through direct contact with the patient, at which time a health history and brief physical examination is conducted to determine the patient's condition. If a patient's condition is considered life threatening, the patient is deemed emergent and normally referred directly to a hospital emergency department.

In a physician office, the appointment scheduler may, through inquiry, solicit information from a patient or referring physicians office that will enable them to determine if the patient needs to be seen in the office today, referred to a hospital emergency department, or scheduled for another day. Often, the appointment scheduler does not routinely inquire of the patient if specific symptoms are present, rather, the patient usually volunteers their symptoms.

In the hypothetical example given, the patient volunteered that he and his family had fevers. A high fever, in and of itself, does not indicate active smallpox. Given the time of year, it is more likely the appointment scheduler would assume the patients had "the flu," or another routine cause of fever. The scheduler did not inquire further about other symptoms, duration of the symptoms, or recent travel of the patients.

Patient Registration

In most physician offices, registration occurs when the patient arrives. A new patient to the practice, typically receives a medical history form to complete, as well as other consent forms that give the physician practice permission to complete necessary billing procedures and access necessary medical information within the confines of the Health Insurance Portability and Accountability Act of 1996 (HIPAA). Many practices adopt the practice of having a staff member, such as a nurse or technician complete the form while "taking a history" from the patient.

The accuracy of this information is a function of the data provided by the patient as entered into the practice data system. In the case of existing patients, physician offices have varying policies regarding "re-registering" an existing patient. Some offices will attempt to update information every time a patient revisits the office. Other practices will seek to update information every six months to a year. Common errors that make it difficult to find a patient's record in the practice data system are misspellings of name as well as incorrect addresses, social security numbers, or insurance information.

If a physician practice needs to contact patients who were in the office on a specific day, the practice with a contemporary computer system could run a report that would identify those patients with their addresses and telephone numbers. However, if a computerized practice has not updated their patient information system with accurate data, it would make it more difficult and time consuming for the practice and/or governmental agency to notify patients that may have been exposed to a bioagent induced disease. The time lag that results from incorrect patient address and telephone information could preclude an individual from obtaining appropriate treatment or vaccines within the prescribed period for successful treatment of the disease agent.

In the minority number of practices that are not computerized, a very laborious process would be required to obtain simple information such as an address or telephone number. In a worst-case scenario, the appointment book would provide information about the patients that were seen in the office, with the staff researching each patient's medical record to create a list of addresses and telephone numbers.

The Clinical Experience

In the hypothetical situation described earlier, the patient with an active case of smallpox contaminated the waiting room, exam room area, and lavatory used by office staff, physicians, and other patients. Under normal operating circumstances, areas in the practice where patients are treated should be cleaned and disinfected regularly throughout the day by the staff, and comprehensively at the end of each day by trained personnel. Of high importance are lavatories and areas where patient blood and body fluids are handled. With appropriate signage in lavatories, patients should be instructed how to properly wash their hands, and be provided with appropriate soaps and disinfectants. Patients should be instructed how to dispose of urine sample cups. The staff assigned to quality assurance and infection control should approve and monitor the methods for collecting urine, containing spillage and minimizing contamination.

Office staff and physicians should be instructed and re-instructed about the proper methods for washing and disinfecting their hands. Interestingly, the failure to wash hands is a practice that has been noted in recent professional literature as becoming more common (Landers, 2002). In a recent report of 34 hand-washing studies, physicians, nurses and other health care workers cleaned up only 40% of the time. The shortcoming was associated with heavy workloads and hectic schedules. As a result the CDC issued new guidelines for hand washing in hospitals, clinics and physician offices. It was suggested that the CDC estimates that 2 million patients in the United States get infections because of being hospitalized, and about 90,000 die as a result (Boyce & Pittet, 2002).

In order to minimize the impact of infecting office patients and staff with life threatening bioagent disease brought into the office by unsuspecting patients, hand-washing procedures, and appropriate glove and mask procedures should be put in place and monitored to ensure compliance.

The Medical Record

In hospitals, clinics, and physicians offices the medical record is the central depository for all relevant patient medical information. The medical record is a file or electronic record in which all important patient documents are retained. This includes laboratory, radiology and pathology reports, results of other diagnostic tests, physician and nurse notes, and records of medications prescribed for the patient. Also contained in the record is the patient medical and social history, and notations from each office visit. To this day, in some physician practices, medical record information is recorded on index cards. If such a practice needed to develop information about the incidence of certain diseases within prescribed periods of time, it would be required to manually access and synthesize the information from all of its medical records. Comparing disease incidence rates and patterns within the same periods of time, but from one year to another, would be an even more difficult task.

While some physician practices use electronic medical records to gather and retain the medical information in a computerized or digital format, most do not. Digital formats make inquiries about disease patterns within periods of time for specific patient or age groups an easier task. For example, Kaiser-Permanente's state-of-the-art data retrieval systems enabled them to rapidly achieve direct contact with all patients in the Brentwood area for anthrax prophylaxis treatment (Simon, 2001).

Patient Checkout

After the physician has seen the patient, they are directed back to the reception area to a separate checkout counter. The patient's medical record and encounter form are brought to the checkout counter by the patient or a staff member. The encounter form is used for the physician or staff member to identify what services were rendered to the patient and the diagnosis(es) that were determined. This information is used to generate the claim forms for submission to insurance companies and billing statements for patients.

Procedure or services rendered, and diagnoses information is entered into the computer using standardized codes. If a practice wishes to run reports to detect trends in diagnoses, an inquiry is run on the computer listing all patients seen in the office within a specific period of time with a specific diagnosis. The report could be run for the same time periods for two or three years to detect if the incidence rate of a particular diagnosis had changed. If a significant increase has occurred, barring any other explanation, it might be an indicator that patients seen in the office during the time period in question have been exposed to a bioterror induced disease.

PREPARING THE PHYSICIAN'S OFFICE TO RESPOND TO BIOTERRORISM

As currently structured, and as illustrated earlier, physician practices are ill equipped to recognize and respond to bioterrorist events. Without significant planning, alteration in current operating policies, and on-going training of medical and office employees, the physician practice may be poised to be the first fatality in a bioterrorist attack. The remaining sections of this chapter provide guidance to physicians and practice staff regarding the preparation of their offices to effectively respond to the results of bioterrorism.

Defining the Agents of Biological Warfare – Clinical Preparedness

The risk that biological agents will be employed against the citizens of the United States is, in all likelihood, very low. However, the consequences of bioterrorism are so great that it is imperative that physicians, key physician office staff, and health care organizations, be prepared to diagnose and treat patients exposed to biological agents or toxins. Discovery that a biological agent or toxin has been employed will most likely occur when an infected patient visits a hospital emergency room, urgent care facility or doctor's office seeking care. Biological agents may be the ideal weapons for an attack on an unsuspecting public. If used in a covert manner, many persons will be unknowingly exposed and the biological agent will not be identified until a large segment of the population is ill or dying.

If physicians know the symptoms associated with the various biological agents or toxins, and are able to identify that a widespread pattern of disease is present among their patients, it may be possible to reduce the effects of a biological attack. Very few doctors currently practicing medicine have ever seen a case of smallpox, pneumonic plague, or the affects of other possible biological warfare agents. It is likely that the initial diagnosis will not accurately identify the true cause of the disease, and, as in the hypothetical scenario at the beginning of this chapter, infected patients will have considerable interaction with other people before appropriate infection-control measures can be initiated, thereby further enabling the disease to spread among an unprotected population (O'Toole & Inglesby, 2002).

In general, a biological agent may have either a rapid progression of illness, but little or no potential for spreading further, or it may have a high potential for spreading further, but with a longer incubation period or slower progression of illness. In either circumstance, it is important that the agent is quickly identified and corrective treatment initiated. Since it is important to respond quickly, it may not be practical to confirm a diagnosis by a laboratory test; instead, it may be necessary to initiate medical treatment based only on the recognition of high-risk syndromes or possible exposure.

Most biological agents have vague, nonspecific symptoms during the early stages of illness that are identical or similar to those of many naturally occurring diseases. Physicians, and key physician office staff, need to be familiar with the symptoms patients will exhibit when exposed to biological agents. They should be able to distinguish between an epidemic of natural origin and a biological attack (MMWR, 2001). Familiarity with symptoms will assist the office staff in preparing and using triaging questionnaires, and other methods of preparedness that will be discussed at greater length in this chapter.

Most biological agents have incubation periods, and symptoms may not be apparent until days or even weeks after initial exposure. Therefore, even in a mass biological attack, the first indication that an attack has occurred may be when a large number of patients simultaneously present with similar symptoms. The Centers for Disease Control and Prevention (CDC) defines three categories of biological agents with potential to be used as weapons, based on ease of dissemination or transmission, potential for major public health impact (e.g. high mortality), potential for public panic or social disruption, and requirements for public health preparedness (Kortepeter et al., 2001).

Category A agents are of the highest concern because these organisms are most easily disseminated or transmitted from person to person, result in high mortality rates, have the potential for major public health impact, might cause public panic and social disruption, and require special action for public health preparedness. Category B agents are those that are moderately easy to spread, result in moderate morbidity rates and low mortality rates, and require specific enhancements of the public health system's diagnostic capacity and enhanced disease surveillance. Category C agents include emerging pathogens that could be engineered for mass dissemination in the future because of availability, ease of production and dissemination, and potential for high morbidity and mortality rates and major health impact (Centers for Disease Control, 2003a, b).

Table 1 describes the Centers for Disease Control and Prevention Category A and Category B biological agents and toxins, with information on the incubation period, early and late symptoms, how the biological agent could be spread, the level of contagiousness with the level of clinical precautions needed to minimize spread of the disease, and the untreated mortality rate for the disease (Centers for Disease Control, 2003a, b).

Familiarity with these agents and toxins is an important first step in preparing a physicians office for bioterrorism. For example, during the week of January 6, 2003, agents in London arrested several suspected terrorists who had Ricin in their possession, a Category "B" toxin public health planners have considered to be among the more commonly expected substances for use by terrorists.

It is important for physicians and medical office staff to understand the information presented in this chart. If a bioterrorism alert occurs, the physician practice must respond to presenting symptoms of patients and provide appropriate facility precautions identified in Exhibit E (see Appendix, Bioterrorism Infection Control: Guidelines for Patient Management).

Table 1. CDC Category "A" and "B" Biological Agents and Toxins.

Agent	Incubation	Symptoms	Likely Dissemination	Contagiousness	Untreated Mortality
Category A					
Anthrax (Inhalational)	1–6 days	*Early* – nonspecific prodrome (i.e. fever, dyspnea, cough, and chest discomfort) *Late* – respiratory failure and hemodynamic collapse ensue, thoracic edema	Aerosol	Very low (Standard precautions)	>90%
Anthrax (Cutaneous)	1–12 days	*Early* – An area of local edema becomes a pruritic macule or papule, which enlarges and ulcerates *Late* – Small, 1–3 mm vesicles may surround the ulcer. A painless, depressed, black eschar usually with surrounding local edema subsequently develops. The syndrome also may include lymphangitis and painful lymphadenopathy	Direct contact	Very low (Standard precautions)	5%
Plague (Pneumonic)	1–6 days	*Early* – High fever, headache, malaise *Late* – Cough (often with hemoptysis), chest pain, dyspea, stridor, syanosis, death	Aerosol	High (Airborne and droplet precautions)	>90%
Plague (Bubonic)	2–10 day	*Early* – High fever, headache, malaise, myalgias, abdominal pain *Late* – Buboes (swollen, very painful, infected lymph nodes)	Infected insect	Low (Standard precautions)	60%
Botulinum toxin (Inhalational)	12–72 hours	*Early* – Ptosis, progressive muscular, and respiratory weakness, diplopia, blurred vision, slurred speech, difficulty swallowing, dry mouth, and muscle weakness *Late* – Respiratory failure, paralysis of arms and legs	Aerosol Ingestion	Very low (Standard precautions)	5%

Smallpox	7–17 days	*Early* – high fever, malaise, head and body aches, vomiting, possible delirium. Onset of rash on tongue and mouth spreading first to face and extremities and then to entire body *Late* – Lesions progress from macules to papules and eventually to pustular vesicles. From 8 –14 days after onset, pustules form scabs that leave depressed depigmented scars upon healing	Aerosol direct contact	Very high (Airborne, contact and droplet precautions)	70%
Tularemia	1–21 days	*Early* – fever, chills, headaches, muscle aches, joint pain, dry cough, progressive weakness, and pneumonia *Late* – Persons with pneumonia can develop chest pain and bloody spit and can have trouble breathing or can sometimes stop breathing. Other symptoms include ulcers on the skin or mouth, swollen and painful lymph glands, swollen and painful eyes, and sore throat	Aerosol	None (Standard precautions)	5–35%
Viral hemorrhagic fevers (VHF)	2–19 days	Include a diverse group of illness caused by RNA viruses such as Ebola hemorrhagic fever, Lassa fever, congo-Crimean hemorrhagic fever, Marburg hemorrhagic fever, etc. *Early* – Specific signs and symptoms vary by the type of VHF, include fever, fatigue, nausea and vomiting, dizziness, muscle aches, diarrhea, chest pain, cough, and pharyngitis, loss of strength, and exhaustion *Late* – Bleeding under the skin, in internal organs, or from body orifices like the mouth, eyes, or ears. Severely ill patient cases may also show shock, nervous system malfunction, coma, delirium, seizures, and renal failure	Aerosol direct contact	Very high (Airborne and contact precautions)	5–90%

Table 1. *(Continued)*

Agent	Incubation	Symptoms	Likely Dissemination	Contagiousness	Untreated Mortality
Category B					
Brucellosis	5–60 days	Fever, headache, malaise, myalgias, back pain, fatigue	Aerosol	Very low (Standard and contact precautions)	5%
Food safety threats (e.g. Salmonella, escherichia coli, shigella)	1–6 days	Diarrhea, fever, and abdominal cramps	Aerosol, food contamination	Very low (Standard and contact precautions)	<5%
Epsilon toxin of Clostridium perfringens	1–6 hours	*Early* – An aerosol delivery can produce a severe pulmonary capillary leak, resulting in adult respiratory distress syndrome (ARDS) and respiratory failure *Late* – Absorbed toxin can lead to intravascular hemolysis, thrombocytopenia and liver damage	Aerosol	Very low (Standard and contact precautions)	90%
Glanders	10–14 days	*Early* – fever, rigors, sweats, myalgias, pleuritic chest pain, granulomatous or necrotizing lesions, generalized erythroderma, jaundice, photophobia, lacrimation, and diarrhea *Late* – skin pustules, abscesses of internal organs, such as liver and spleen, and multiple pulmonary lesions	Aerosol	Low (Standard precautions)	90%
Melioidosis	10–14 days	*Early* – fever, rigors, sweats, myalgias, pleuritic chest pain, granulomatous or necrotizing lesions, generalized erythroderma, jaundice, photophobia, lacrimation, and diarrhea	Aerosol	Low (Standard precautions)	90%

Disease	Incubation period	Signs and symptoms	Mode of transmission	Infection control	Case fatality rate
Psittacosis	5–14 days.	*Late* – skin pustules, abscesses of internal organs, such as liver and spleen, and multiple pulmonary lesions *Early* – fever, chills, headache, malaise, myalgia *Late* – pneumonia	Aerosol	Low (Standard precautions)	15–20%
Q fever	10–40 days	*Early* – high fever, headache, myalgias, chills, sweats, sore throat, abdominal pain or chest pain *Late* – One-third to one-half of patients with Q fever will develop pneumonia	Aerosol	Very low (Standard precautions)	<5%
Ricin toxin from Ricinus communis	18–24 hours	*Early* – Acute onset of fever, chest tightness, cough, dyspnea, nausea, and arthralgias *Late* – Airway necrosis and pulmonary capillary leak resulting in pulmonary edema, followed by severe respiratory distress and death from hypoxemia in 36–72 hours	Aerosol	Very low (Standard precautions)	50–90%
Staphylococcal enterotoxin B	3–12 hours	*Early* – fever, chills, headache, myalgia, and nonproductive cough. Some patients may develop shortness of breath and retrosternal chest pain	Aerosol food or water contamination	Very low (Standard precautions)	5%
Typhus fever	1–7 days	Fever, headache, rash	Aerosol	Low (Standard precautions)	15–20%
Viral encephalitis (Venezuelan equine encephalitis, eastern equine encephalitis, western equine encephalitis)	1–21 days	*Early* – Acute systemic febrile illness, generalized malaise, spiking fevers, rigors, severe headache, photophobia, and myalgias	Aerosol	Low (Standard precautions)	<5%

Table 1. (Continued)

Agent	Incubation	Symptoms	Likely Dissemination	Contagiousness	Untreated Mortality
		Late – Nausea, vomiting, cough, sore throat, and diarrhea			
Water safety threats e.g. vibrio cholerae, cryptosporidium parvum)	1–3 days	Watery diarrhea, vomiting, leg cramps, dehydration	Aerosol food or water contamination	Very low (Standard precautions)	<5%

If a practice knows that a biological attack has occurred, even if it is in a distant location, the level of awareness on the part of both staff and physicians needs to increase. If the biological warfare agent is characterized as highly contagious, such as smallpox, the medical practice must assess its risk and employ appropriate safeguards to protect staff and patients from a contaminated patient or object. The columns in the chart, "Likely Dissemination," "Contagiousness," and "Untreated Mortality," provide information to assist physicians assess the risk of exposure and to evaluate, for each agent, how the practice should respond.

Even if the likelihood of exposure to a bioagent is small, it may be appropriate for the practice to vaccinate staff members and their immediate families for the diseases caused by an agent or to prescribe prophylaxis antibiotics for staff members. At a minimum, if a bioterrorist attack occurs anywhere in the country, doctors and staff need to be instructed about the agent, how it is spread, how to recognize the early and late symptoms, and how to protect themselves against the hazards presented by the agent. Through education and appropriate preventive therapies, the medical office physicians and employees will be best capable of identifying and treating bioterrorism victims, while minimizing the possibility that they will become a casualty.

Categorization of Patients Related to a Bioterrorist Attack

The initial fallout of a bioterrorist attack generates three patient categories:

Category 1: Infected patients treated in the practice that are not easily identified as victims of a bioterrorist event until hours or days after visiting the facility;

Category 2: Infected patients treated in the practice who can be identified with a high degree of probability as victims of a bioterrorist event; and

Category 3: Worried well patients, whose reaction to the news of a bioterrorist attack will create pseudo symptoms and a demand to be tested.

As a result of the random and un-planned nature of a bioterrorist event, each patient category creates unique challenges to the physician office for triaging, treating and tracking patients. Further, the practice is burdened with the requirement to track anyone who was visiting or working in the facility during this time period. It is also probable that the practice could be overwhelmed, particular with the worried well. The practice should not assume help would be available or arrive immediately from outside community-based sources. The Brentwood anthrax event demonstrated the lack of government coordination and support during a bioterrorist attack (Reaves, 2000). The ability of local, state and federal entities to provide resources may be preempted by other, more serious demands. In reality, the practice may be required to work through the crisis for some time without additional resources.

The major question facing the physician practice is how best to balance the need for patient and staff safety against the use of scarce resources. It will not be practical for most physician facilities to maintain a state of readiness or preparation to meet the needs of all categories of patients who will present to the clinic. If the practice knows in advance that a bioterrorist event has occurred, it can implement appropriate policies to isolate Category 2 infected patients and prepare for the large volumes of the Category 3 worried well patients. The major challenge to the practice will be developing and maintaining the appropriate level of oversight to reduce the risk associated with Category 1 infected patients who are not immediately and/or easily identified as victims of a bioterrorist event.

In practical terms, the following areas within the physician practice require review and modification to properly respond to the randomness of a bioterrorist event:

(1) Operation procedures;
(2) Organization policies; and
(3) Education programs.

Within these broad categories are more specific organizational elements that are found in most practices, all of which require change to prepare for the possibility of bio-events, or the outbreak of unusual illnesses that may mimic other known illnesses of lesser severity (e.g. SARS). These elements are described in Fig. 1.

Operational Procedures for the Physician Office

Major changes are necessary in operational procedures to enable physician practices to respond safely and appropriately to a bioterrorist event, including the appointment triage function, patient referral process, patient information tracking system, and facility capabilities.

Appointment/Triage Functions

Appointment triaging for patients will require changes both in phone protocol and clinic check-in procedures.

As described in the beginning of this chapter, receptionists who make appointments, and outside answering services hired by physicians, should follow a script that expands the traditional information captured from the patient, to include recent out-of-country travel, military service, mass transit utilization, recent attendance at large gatherings, as well as current symptoms and length of illness. The information captured is then processed through the patient information system that generates daily data for the assigned medical staff to red flag potential patients

Key Elements in
Bio Terrorism
Management

A Appointment
Scheduling and
Triage

B Patient
Registration

C Patient
Education

D Medical
Records and
Health
Information

E Patient
Checkout.

F Physician and
Staff Cleaning
Techniques

G Administra-
tive, Employees
Education and
Event Preparedness

H Review of
BioHazardous
Material
Outsourcing

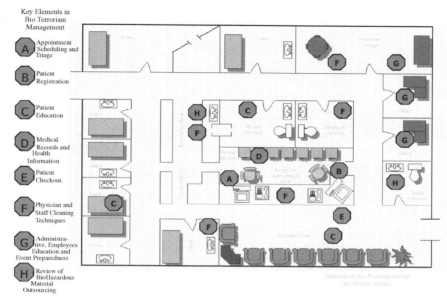

Fig. 1.

at high-risk for a biological agent exposure. The targeting of patients at potential high risk might be those that are listed as having been out of the country, in military service, or other key parameters that reflect the unique environment of the practice.

Exhibit A provides a sample script a receptionist can use to screen patients.

Exhibit A: Sample Script for a Receptionist to Use

To Assess the Risk of a Bioterror Exposure

Receptionist:

– May I ask you for some information?
– What symptoms do you have? (Score 1 point for each symptom)
   Fever?
   Fatigue?
   Sore throat?
   Difficulty swallowing?
   Nausea?
   Vomiting?
   Coughing?

Headache?
Back pain?
Chest pain?
Abdominal pain?
Chills?
Sweats?

Note: Common symptoms related to bioterrorist agents include fever, malaise (vague feeling of physical discomfort or uneasiness), fatigue, sore throat, difficulty swallowing, anorexia (lack of appetite), nausea, vomiting, backache, coughing, substernal (below the sternum) chest tightness, pleuritic chest paid (difficult, painful breathing), myalgia (muscle pain), hemorrhage (bleeding), chills, sweats, abdominal (stomach) pain, chest pain, weight loss and depression.

– How long have you had these symptoms? (Score: Less than week = 2 points, More than a week =1 point)

– Have you traveled outside the United States recently? If yes,
  When did you leave the U.S.?
  What countries were you in? (Score 1 point for each country)
  When did you return? (Score: Less than a week = 2 points, More than a week = 1 point)

– Are you currently, or have you been in the military? If yes,
  Were you stationed outside the U.S.? (Score 1)
  When were you discharged? (Score: Less than a month = 2 points, More than a month but less than year = 1 points)

– Have you recently utilized mass transportation? If yes, (Score: If used all the time or within the past weeks = 2 points, otherwise 1 point)
  Subways?
  Buses?
  Trains?
  Airplanes?

– Have you attended any large gatherings recently? If yes, (Score 1 point for each positive answer)
  Where?
  Purpose of gathering?

If the Score Equals 8 or Greater, Consider the Patient a Risk for a Biological Agent Induced Disease.

A score in the range of eight (8) or more might indicate that a patient has been the victim of a bioterrorism event and should be referred to a physician or triage nurse who can better assess the situation and either direct the patient to come immediately to the doctor's office, go to another health care facility, or activate the Emergency Medical Response (EMR) system by calling 9–1–1. If the patient is asked to come to the physician's office, upon arrival they should be isolated immediately. Walk-in patients also should be asked the appointment triage questions at the front desk and isolated from public areas based on the scoring.

Registration/Check-in
At the time of patient check-in, the receptionist should follow a consistent protocol to establish correct patient addresses, phone numbers and emergency contacts. This step is necessary to provide accurate information to track patients in the event further emergent notification is required, or for health agency monitoring. This protocol requires the receptionist to ask the patient for an answer to be compared against the current patient information and to provide information about those members of family or friends who have come to the clinic with the patient. The physician practice cannot assume the patient data in the information system is correct. Examples of the manner in which some questions should be asked, and some key questions related to suspected bioagent induced illness follow:
    Examples of correct and incorrect check-in questions:

> Correct question: "What is your phone number so I can check it against our records?"
> Incorrect question: "Has your phone number changed since your last visit?"
> Correct question: "What is your current address?"
> Incorrect question: "We have your address as 4014 W. Quail Ridge, is that correct?"

Additional Questions for the Check-in Protocol:

Have you visited any other healthcare providers in the past few days for this illness?
YES____NO____
If YES, who _____

Have you worked since feeling ill? YES____ NO ____
If YES, what are the addresses and phone numbers of your employer(s)? ____
_____

Is your work confined to one site? YES____ NO ____
If YES, how many sites? _____

Have you visited with family or friends since feeling ill? YES _____ NO_____

If YES, validate again the correct contact information for patient.
(Securing names and addresses of family and friends will occur after the illness is
    differentiated as bioterrorism agent induced.)

Certain key questions should be part of the check-in protocol. These are specifically designed to provide exposure and tracking information to health agencies.
As part of the check-in protocol, the response and scoring of the appointment
triage questions should be included. This requires coordination between the
appointment/triage function and patient check-in.

The goal of the check-in protocol is to compare the patient response with the
patient record the receptionist is viewing. Patient information should be checked
each time the patient comes to the office. Although patients may view this as
an inconvenience, the goal for the practice is to have each receptionist following
a consistent protocol for every visit, thus eliminating the need to decide if the
protocol should be used. The more a receptionist might diverge from a protocol,
the greater the probability for gaps and errors in the data, thus creating potential
risk for patients and staff. Therefore, the protocol should be followed, with no
flexibility allowed.

The practice should be sensitive to the issue of non-medical employees asking
specific and intrusive medical questions of patients. All patient health information
gathered by the practice should follow federal (HIPAA) and state laws that
govern the solicitation, use, and storage of sensitive medical and personal
information.

Patient Referral Process

All patients who present to the practice for treatment should be tracked carefully if
they are referred to other specialists or medical facilities. Tracking patients sent to
specialists is appropriate under normal circumstances, and is particularly critical
during the period when the patient is not suspected to have had exposure to a
bioterrorist agent, but ranks high on the biological agent risk scale. Therefore,
the patient referral protocol requires the capture of current information during
the visit, including the date, time, physician, and location to which the patient is
referred. This can be accomplished within the patient information system or by
using a database program that allows easy access by staff members to key patient
information. For information to have value during a bioterrorism crisis, it must be
quickly accessed and analyzed.

Patient Information Tracking Systems

Practice employees are usually not trained to analyze clinical or statistical data or recognize disease patterns. Most practice information systems summarize practice performance, usually days or weeks after the fact, e.g. patient visits, financial graphs, etc. Disease patterns that will emerge from a bioterrorist event require daily data generation and evaluation, not currently tracked by physician practice systems. Since most practices do not emphasize or provide rapid analysis of epidemiological patterns, this will require a major change in physician practice information systems. Further, significant education will be required to sensitize practice employees to extract from practice data, when it is available, potential patterns of infection that might signal a biological agent exposure.

A patient disease tracking system, either manual or computer based, should be available to monitor the symptoms of patients seen during the day, and correlate those symptoms to final clinical diagnoses. The patient disease profile data, generated daily, should be reviewed during the day by a medical staff member trained in the characteristics of biological agent induced disease to identify anomalous patterns. This includes identifying and reporting suspicious trends that may be attributed to bioterrorist agents. For example, a July rise in flu-like symptoms may indicate biological agent infection. On the other hand, a February increase in flu-like symptoms may not represent bioagent-induced disease, but may only represent increased volume of patients with flu. Therefore, pattern and volume comparisons of diagnoses in the practice from time period to like time period in prior years are important.

Further, all health care providers, whether nurse or physician, must be sensitized to reviewing patient information from the perspective of potential biological agent activity. Physician practices should not be seduced into thinking the probability is low or null for such an event to occur. Such thinking is shortsighted and dangerous.

Facility Capabilities

The physician practice should develop prospectively written emergency response procedures. Those procedures include a Disaster Plan Checklist, and an Emergency Preparedness Plan (Exhibits B and C-1), which will assist the practice in dealing with all types of emergencies, and disasters, including bioterrorist events. The Disaster Plan Checklist encompasses the significant operational, systems, patient safety, financial, mechanical and other considerations that should be initiated in a practice when responding to the outcomes of a bioterrorist action.

Exhibit B: Disaster Plan Checklist

Disaster Plan Checklist

Patient Safety
  Evacuation Plan
    With power
    Without power
      Elevators/escalators
      Phone system
      Public address system
      Automatic doors
    Who triggers plan (key employee[s])
      Partial evacuation
      Total evacuation
    Person identified who is responsible to assure practice is completely evacu-
    ated and premises secured
    Safe gathering areas identified
    Items to be removed, if any
      Person responsible identified
    Employee training
      Role-playing and practice exercises
  Power Outages
    Backup power source for procedures that require power for completion
  Contingency Plans for Patient Disposition to Other Facilities
    Hospitals
    Clinics
    Designated public facilities
  Contingency Plans for Medical Supplies
    Vendor capability determined (national, regional and/or local)
    Emergency inventories identified and stocked

Employee Security
  Chain of Command
    Loss of key employee[s] contingency
    Responsibility of managers/supervisors to track employees
    Employee training
      Identify and practice duties
      Identify who steps in to assist management

  Staff Safety
    Central location identified off-premises for physicians and staff to meet
    following an emergency

       Employee check-off list
       Preventative vaccinations
       Personnel protective equipment

    Employee/Family Safety
       Communication protocol
       Staff & families understand/trained
         (Note: Worry about family during emergencies will reduce employee effectiveness)

    Aftermath Counseling
       Counselors identified
         Group therapy
         Individual counseling

Practice Capability
    Communication
       Spokesperson[s] designated
         Practice scenarios to develop comfort responding to press and public
    Patient Charts
       Duplicate charts off premises for patients with
         Chronic illness
         Complicated treatment plans
         Potential or pending liability claims
    Patient Billing System
       Plan for duplication of practice system with vendor/other clinic with similar systems
       Off premises storage of data
         Passwords & access codes
         Tested data restoration
    Downtime
       Patient referral plan for short-term disruptions
       Contingency plan for temporary clinic
         Facility
         Patient charts
         Data systems
       Facility clean-up strategy
         Establish priority with disaster clean-up company
    Public Health Issues
       Define clinic/practice role within local community disaster plan
       Training for key personnel

Participation in community disaster drills

Financial Survivability
Cash Issues
Back-up financial institutions identified
Disruption of banking system contingency plan
Safe to hold cash
Financial Records
Duplicate records stored off-site
Financial
Personnel
Personnel identified who can access off-site records
Plan in place if key employee lost
Employee compensation
Plan for partial wages, if necessary
Insurance Coverage
Liability
Business interruption
Property
Buildings
Business personal property
Equipment breakdown

The twelve steps for developing an Emergency Response Plan are profiled in
Exhibit C-1

Exhibit C-1: Steps to Develop an Emergency Response Plan

1. Keep the plan simple.
2. Obtain copies of disaster plans from primary hospital providers.
3. Obtain copies of community disaster plans.
4. Use hospital and community plans as a model to build practice plan.
5. Let the plan reflect uniqueness of practice staff and resources available.
6. Distribute draft plan to physicians, nurses, and other employees for input.
7. Review plan with appropriate hospital/community entities, defining role of practice within framework of community disaster plans.
8. Finalize plan, including group governance approval.
9. Implement emergency response plan.

10. Develop training modules.
11. Conduct initial and on-going physician/staff training program, including testing and simulation.
12. Re-evaluate and update emergency response plan annually.

These steps should lead to the development of an emergency response plan that is customized to the specific needs of the practice (see Exhibit C-2, Appendix). The size of the physician practice is immaterial to developing an emergency response plan. The challenge for the physician office is to develop a plan that identifies the unique characteristics and resources available within the practice, and how those resources will be used in response to a bioterrorism event. In a small practice, fewer people will be available to partition duties, but all duties need to be assigned, with appropriate training and annual review.

Specific responses related to bioterrorism should be developed as part of the emergency response plan. These include isolation procedures, facility lock-down criteria, evacuation concerns, biological agent identification procedures, communication protocols, patient placement, referral management, patient transport and the cleaning and disinfecting of rooms and equipment. If a bioterrorism alert occurs, the physician practice must respond to presenting symptoms of patients and provide appropriate facility precautions identified in Exhibit E (see Appendix, Bioterrorism Infection Control: Guidelines for Patient Management). This requires employee understanding and implementation of safety policies identified in the emergency response plan developed in concert with these guidelines. If the bioterrorist agent is suspected, the practice quickly can implement the patient management guidelines (Exhibits D and E, see Appendix) to minimize the possibility that staff or other patients will be exposed to the disease.

Suspected but Unconfirmed Biological Agent Exposure

If a biological agent exposure is suspected but unconfirmed, the practice should take the action identified in the following rules. The rules assume a biological agent exposure is present but not identified. As the agent is identified, the practice can revert quickly to the patient management guidelines for that biological agent.

Rule 1: Assume the worst and respond accordingly
Until the disease/agent is identified, maximum precaution should be exercised within the facility. Once disease/agent is identified, procedures

may be modified using patient management guidelines that fit the disease profile.

Rule 2: Implement Isolation Precautions

Identify the medical team that will provide medical oversight.

Isolate the patient and exposed area using appropriate isolation protocol based on the suspected disease/agent.

Keep patient in private room. If more than one patient is involved, you may cohort patients together.

Keep door closed at all times.

If symptoms of smallpox are presented, use negative pressure room if available.

No family access to patient without approval of medical team.

No staff access to patient or exposed areas without approval of medical team leader.

If smallpox symptoms present, disable ventilation systems (heating and air conditioning) and notify the landlord if in a multi-office building.

Isolation Precautions – Suspected Biological Agent Exposure

If a biological agent exposure is suspected, the practice should utilize the appropriate level of isolation precaution to protect staff and other patients. The isolation precautions summarized below are explained in greater detail in Exhibits D and E (Appendix).

Standard Precautions

Standard precautions prevent direct contact with all body fluids (including blood), secretions, excretions, non-intact skin, rashes, and mucous membranes. Standard precautions routinely practiced by healthcare providers include: Hand washing, gloves, mask/eye, protection/face shield, while performing procedures that cause splash/spray, and gowns to protect skin and clothing during procedures that may involve splashing/spraying.

Contact Precautions

Contact precautions are designed to minimize the spread of disease by isolating the patient from other patients and staff, and the use of gowns and gloves to limit potential exposure. Any patient care items and equipment with which the patient may have had contact, such as instruments, examination room equipment, or furnishings, should be cleaned with a disinfectant recommended for neutralizing the identified disease.

Droplet Precautions
Droplet precautions involve the use of a surgical mask to prevent the provider or staff from inhaling particulate from the patient.

Airborne Precautions (if symptoms of Smallpox or Glanders suspected)
Airborne precautions are the most extensive, and are usually beyond the capability of a doctor's office. They require a negative pressure room, sophisticated air filtration systems, and require respiratory protection by all individuals in the room with a patient.

Rule 3: Activate Communication Protocol.
Inform Disaster and Public Health Agencies.
Centers for Disease Control and Prevention, state Health Department, County/Community Health Department, local hospitals, police, emergency medical response, Federal Bureau of Investigation, including the building landlord if it is necessary to disable ventilation systems to prevent spread of infectious diseases.

Be prepared to address the media.
Use the same spokesperson (Previously identified individual such as the office manager, administrator or lead physician) for all communications. Use a scripted announcement based on known facts for all communications. Do not speculate! Using a script will enable the spokesperson to reconstruct all communications at a later time. The spokesperson must be prepared and trained to talk with patient families, staff families, as well as the media. A practice, as part of its planning, may wish to consult a public relations firm to assist in script development and the development of guidelines for speaking with the public. Large practices also may wish to retain a public relations firm to represent them in any media contacts, but not necessarily with the families of patients or staff because these communications are of a more personal nature. Any suspected bioterrorism event will have extreme media interest and intense scrutiny.

Exhibit F, contains examples of scripts to use to communicate to patients or the public and to the medical practices employees, and a checklist for media contacts.

Exhibit F: Script Examples and Media Checklist
Script to Notify Patients or the Public

Example statement by spokesperson: "A possible case of _____ was identified in our office, and as a result we are applying maximum precautions within our facility. As the agent or disease is identified, more detailed information regarding possible exposure will be immediately provided. We are maintaining a safe environment for patients and staff. Please direct questions to me at (phone number, e-mail address)."

Script to Notify Staff

Example statement by spokesperson: "A possible case of _____ was identified in our office and we are applying maximum precautions within our facility. As the agent or disease is identified, more detailed information regarding personal exposure will be immediately provided to you. We are maintaining a safe environment for patients and staff. Do not approach exposure areas without approval or direction of senior management or the medical team. Dr. (or Nurse) _____ is the medical person responsible for handling all clinical questions. We are arranging an immediate group meeting for non-involved staff and follow-up meetings for all staff to keep you informed. If the media or anyone outside of the medical practice contacts you, you should direct them to contact me at (phone number, e-mail address) for information. You do not have to answer any questions posed by the media or by anyone outside of our medical practice."

Media Strategy Checklist

– Alert the spokesperson
– Gather who, what, where, when and why of the situation
– Confirm the facts
– Clarify and verify technical information, if provided
– Prepare a summary statement
– Prepare a fact sheet
– Notify key people in the organization
– Tell patients, staff and public about changes in services/operations
– Respond to the media
– Keep a media log of callers and questions
– Update media as situation develops
– Follow-up information to prevent inappropriate backlash
– Evaluate and adjust the system

Rule 4: Compile Patient/Staff Roster

Compile a list of all patients, employees, and visitors who have been in the facility during the suspected period of exposure, including their addresses and phone numbers. As identified earlier, patient information is not always current or accurate, but in compiling this list it is critical that the review process ensures accuracy at the highest level possible. Use this list to account for all patients, staff and visitors during the exposure period, and provide to appropriate health agencies for follow-up. Notify physicians to whom patients have been referred.

Organizational Policies

Each practice will need to develop or modify organizational policies to assure appropriate response to a bioterrorist event. As stated in an earlier section, many practices do not have detailed policies and procedures, and, as a result, the creation of organizational policies related to bioterrorism may represent a departure from typical administrative practices. This includes identifying the chain of command, which specifies from the top to the bottom of the organization the responsibilities for each physician, administrator, manager, supervisor and employee along the chain. Critical to this process is defining who steps in if a person is absent or incapacitated.

A major outcome of a bioterrorist event is the probable loss of key employees through exposure or isolation. The chain of command policy must assume the loss of key employees and provide contingency managers/employees to continue implementing the Emergency Response Plan. Particularly important is identifying the leadership roles of the medical providers and the non-clinicians, and providing practice simulations to familiarize each employee with their role. Further, employees throughout the organization should be identified and trained to step in, as appropriate, to assist management during a crisis.

Current job descriptions (often not extant in a practice) require review and modification to reflect the employee's role in neutralizing a bioterrorist attack. Key to implementing revised job descriptions is the need for constant review by management and the employee. The practice cannot adequately meet a bioterrorist event if the staff is not conversant with polices and their responsibility.

An organizational policy that is required relates to the manner in which a practice will collaborate with the community and community agencies. If structured carefully, collaborative efforts with other health care organizations can be an effective way to conserve resources, identify appropriate responses, access specific crisis resources not cost effective for the practice to maintain, and share risk. The steps for the physician practice in collaboration with the community are: (1) Identify organizations with whom the practice should, and can collaborate; (2) Confirm compatibility among the organizations; (3) Understand motivations for collaborating; (4) Conduct due diligence to assure the proposed collaboration is appropriate; (5) Clarify expectations of the parties; and (6) Put it in writing (Exhibit G has a Checklist for a Memorandum of Understanding to use for community collaboration).

Education

Educating employees and patients on bioterrorism will require a major emphasis for physician offices. An appropriate education protocol requires initial employee orientation, on-going training, and continual testing to assure employees understand and can respond correctly to a bioterrorist event. Talking about such issues once a year will not suffice. On-going training and testing should occur every six months, with event simulations planned to assess the readiness of the practice to respond. Being prepared in this area will require a major commitment of time and resources for education with the recognition that follow-up training and testing is not a current practice for physician offices.

Further, a concerted effort is required to educate staff on bio-agents. This accomplishes two objectives: (1) Employees attain an understanding of bio-agents and how they might be personally affected; and (2) Employees are properly trained in correct procedures. This creates a comfort level with the randomness of potential exposure by understanding how to reduce the risk through utilizing appropriate procedures in day-to-day clinical operations.

Patient education is an important aspect of practice preparation, both during medical intervention and prior to any bioterrorist attack. Identifying physicians and staff who are able to speak to the community on preparedness offer a real opportunity to reduce the impact of a bioterrorist event, particularly the worried well. Numerous resources are available to the practice from local hospitals, universities, community, state and Federal health agencies. A patient education program will provide positive public relations program for the community and reinforce Homeland Security's plan for citizen education.

Responding to a Bioterrorism Attack

Notwithstanding educating physicians and office staff and routine practice drills, involvement in a real or suspected bioterrorism attack will test the medical practice's leaders as it did during the Brentwood anthrax outbreaks in 2002 (Borio et al., 2001). Clinical and administrative staff members will look to the physician and administrative leaders of their practice for both instruction and for example. The manner in which the practice's leaders respond will determine how the staff will cope with the emergency and how patient care will be delivered during a crisis or emergency.

During a crisis, leaders must ensure that employees and health providers feel confident about their personal safety and the safety of the families. In the midst of crisis activity, rest, hydration, and nutrition are often overlooked; however, good leaders will ensure that their and their employee's bodily and security needs are met (*Field Manual*, 2002, pp. 2–5).

A medical practice should consider the following ten-step process in preparing for the possibility of bioterrorism (Kortepeter et al., 2002).

(1) *Maintain an index of suspicion.* Health-care providers should be alert to illness patterns and diagnostic clues that might indicate the infectious disease outbreak associated with a biologic agent. Without a high index of suspicion, it is unlikely that a provider, especially those without sophisticated laboratory and preventive medicine resources, will promptly arrive at a proper diagnosis and institute appropriate therapy.

(2) *Protect thyself.* Medical personnel must first take steps to protect themselves. These steps may involve a combination of physical, chemical, and immuno-logical forms of protection.

(3) *Assess the patient.* Patients with a suspected biological agent exposure should have an initial assessment to insure airway adequacy and to address breathing and circulation problems before attention is given to specific management. Patient historical information of potential interest includes: illnesses in family members and others, their presence during unusual events such as smoke clouds, exposure to food and water procurement sources, vector exposure, immunization history, travel history, and occupation. The preliminary assessment should concentrate on the pulmonary and neuromuscular systems, as well as unusual dermatologic and vascular findings.

(4) *Decontaminate as appropriate.* Decontamination plays a very important role in managing patients with exposure to chemical agents and biological toxins, but will be minimal for patients exposed to biological agents. Where decontamination is warranted, simple soap and water bathing will usually suffice. A dilute 0.1% bleach solution reliably kills anthrax spores, the hardiest of biological agents. Routine use of caustic substances, especially on human skin, however, is rarely warranted following a biological attack.

(5) *Establish a diagnosis*. With decontamination accomplished, a diagnosis should be established. This will involve a combination of clinical, epidemiologic, and laboratory examinations. In larger, more sophisticated practices, a full range of laboratory capabilities could enable definitive diagnosis. In other practices diagnostic specimens should be taken for further analysis. Diagnostic specimens that should be taken include: nasal swabs (important for culture and PCR, even if the clinician is unsure *which* organisms to assay for), blood cultures, serum, sputum cultures, blood and urine for toxin analysis, throat swabs, and environmental samples.

(6) *Render prompt treatment*. Unfortunately, it is precisely in the "Initial Symptom Stage" of many diseases that therapy is most likely to be effective.

(7) *Practice good infection control*. Standard precautions provide adequate protection against most biological agents. However, suspected smallpox victims should be managed using airborne precautions, pneumonic plague warrants the use of droplet precautions, and certain viral hemorrhagic fevers require contact precautions.

(8) *Alert the proper authorities*. Both the local health department and local police or other law enforcement agency should immediately be notified if there is suspected exposure to biological agents. The local public health officials will notify their counterparts in their state health department who will notify the Center for Disease Control and Prevention. Local and state health officials also should immediately notify the Federal Bureau of Investigation of all incidents of apparent or threatened bioterrorism (*Centers for Disease Control*, 2003a, b). Additionally, if specimens are sent to a clinical laboratory, the clinical laboratory also should be informed that the specimens could be from a patient with exposure to a biological agent. This notification will enable laboratory personnel to take proper precautions and will allow the optimal use of various diagnostic modalities.

(9) *Assist in the epidemiologic investigation*. A sound epidemiologic investigation is essential to confirm the pathogen and institute the appropriate medical response. The medical practice needs to document information about the patient and assist public health officials determine possible means and routes of exposure.

(10) *Maintain proficiency*. The threat of biological attack has remained a theoretical one for most medical personnel. Inability to practice casualty management, however, can lead to a rapid loss of skills and knowledge. It is imperative that physicians maintain proficiency in dealing with this low probability, but high consequence problem. It is only through ongoing training that you will be ready to deal with the threat posed by biological weapons.

CONCLUSION

The hypothetical office scenario at the beginning of the chapter demonstrates how simple it would be to face a situation in which an individual, or a family that has been exposed to a bioagent, can in a contagious stage risk the health and lives of office staff and other patients in a physician's office. Clearly, if a bioterrorist event is of great magnitude, causing exposed and sick patients to seek medical care, it is likely that hospital emergency rooms and emergency clinics will be the first line of response.

However, if the infiltration and use of bioagents by terrorists is more intermittent, "subtle," and geographically dispersed, the physicians' office is more likely to be the first to experience patients that have been exposed. Under this set of conditions, it may take public health officials a greater amount of time to recognize that bioterrorist activities have taken place.

As low as the likelihood may be that there will be bioterrorist activities in the United States, should they occur and be compounded by a lack of preparation in physician's offices, the results can be devastating. The hypothetical scenario also demonstrates that physician offices, by their nature, will generally not be prepared to recognize and organizationally accommodate patients with symptoms related to exposure to bioterrorist activities. The hypothetical scenario points to a range of common weaknesses inherent in many physician offices including patient triaging, collection of key patient information, need for detailed operating policies and procedures, infection control processes, and better understanding among clinical and office staff regarding symptoms of diseases arising from bioterrorist activities.

The purpose of this chapter was to identify a range of preparations in which the physician's office should be engaged so that it is better prepared to *consistently and uniformly*:

(1) Recognize the symptoms of bioterrorist related diseases, and to effectively triage for them over the phone and in the office.
(2) Maintain accurate patient information so that patients that may have been exposed to infected patients can be contacted using reliable information.
(3) Collect disease pattern information so that analysis can be conducted to determine if there are unusual patterns of what appear to be common diseases such as flu, which actually are more threatening diseases such as smallpox.
(4) Develop internal disaster plans, emergency preparedness plans, and isolation procedures.
(5) Utilize internal education and preparedness drills to continue to maintain a level of proficiency regarding bioterrorist-induced diseases.

While public health agencies and public emergency response teams will play a major role in these bioterrorist intrusions, the physician's office, particularly those of primary care doctors, may be the first to recognize that these events have taken place. Preparation is integral to a successful practice response when a bioterrorist attack occurs.

REFERENCES

Borio, L. et al. (2001). Death due to bioterrorism-related inhalational Anthrax. *JAMA*, *286*(20), 2554–2559. Available at: http://www.jama.ama-assn.org/cgi/content/short/286/20/2554.

Boyce, J. M., & Pittet, D. (2002). *MMWR (Morbidity and Mortality Weekly Report)*, "Guideline for Hand Hygiene in Health-Care Settings Recommendations of the Healthcare Infection Control Practices Advisory Committee and the HICPAC/SHEA/APIC/IDSA Hand Hygiene Task Force". Centers for Disease Control and Prevention, October 25, 2002/51(RR16), pp. 1–44.

Centers for Disease Control and Prevention (2003a). Biological diseases/agents list (August 29). Available at: http://www.bt.cdc.gov/Agent/agentlist.asp#categorydescriptions.

Centers for Disease Control and Prevention (2003b). Public health emergency preparedness and response (August 29). Available at: http://www.bt.cdc.gov/emcontact/index.asp.

(2002). *Field Manual (FM) 4–02.7 (FM 8–10–7), Health Service Support in a Nuclear, Biological, and Chemical Environment, Tactics. Techniques and Procedures*, Headquarters, Department of the Army, Washington, DC, pp. 2.5–2.6 (October 1).

Jenks, S., & LaPoll, A. (2001). Anthrax was likely planted, Sen. Graham says. *Florida Today* (October). Available at: http://www.flatoday.com/news/local/stories/2001/oct/loc100901a.htm.

Kortepeter, M. et al. (2001). *Medical management of biological casualities handbook* (4th ed.). U.S. Army Medical Research Institute of Infectious Diseases, Fort Detrick, Frederick, MD, pp. 13–78 (February). Available at: http://www.usamriid.army.mil/education/bluebook.html.

Landers, S. J. (2002). CDC Asks Physicians to Come Clean with Gels. *American Medical (Association) News* (November 18), *45*(43), 1–2.

MMWR (Morbidity and Mortality Weekly Report) (2001). Recognition of Illness Associated with the Intentional Release of a Biologic Agent. Centers for Disease Control and Prevention (October 19), *50*(41), 893–897.

O'Toole, M. M., & Inglesby, T. V. (2002). Shining light on dark winter. *Center for Civilian Biodefense Strategies. Johns Hopkins University*. Baltimore, MD (January 25). Electronically published 19 February 2002.

Pasko, T., & Smart, D. R. (2003). *Physician characteristics and distribution in the U.S., 2003–2004 Edition*. American Medical Association Press (p. 8).

Reaves, J. (2000). Can anyone solve the anthrax mystery? (October 26). Available at: http://www.time.com/time/nation/article/0,8599,180777,00.html.

Simon, S. W. (2001). Permanente practice guidelines in wake of anthrax cases available to health care organizations (Rockville, Maryland) (October 26). Available at: http://www.kaiserpermanente.org/locations/midatlantic/newsroom/mas102601.html.

APPENDIX

Exhibit C-2: Custom Emergency Response Plan

Statement of Objective

The Emergency Response Plan applies to the environment of care for _____ (clinic/practice). The objective of the plan is to provide service to patients and staff in a safe environment in the event of an emergency.

The plan is designed to improve compliance with emergency management standards; provide education to staff; monitor effectiveness of the program; and identify opportunities to improve performance.

Scope of the Plan

The plan is designed to identify potential situations where the resources of _____ (clinic/practice) will be affected. The plan is continually evaluated for effectiveness and appropriateness.

Safe working conditions and practices are established by using knowledge of emergency management principles to educate staff, design appropriate work environments, and purchase appropriate equipment, furnishings and supplies.

Authority

The _____ (board, executive committee) has established the Emergency Response Plan and delegated authority for the Plan to _____ (administrator/manager).

Organization and Responsibility

The Emergency Response Plan guides emergency and disaster protocols for all _____ (clinic/practice) departments. Operational oversight is under the jurisdiction of the _____ (administrator/manager), who is responsible for maintenance and monitoring of the Emergency Response Plan. Included is the responsibility to examine problems, operational issues, monitor system testing, to report and follow up on incidents, and to institute corrective action. Each individual in the organization is responsible to be familiar with emergency and disaster safety protocols.

The Emergency Response Plan will provide assistance to normal operations in progressive stages as appropriate for the incident. If the incident requires a response beyond normal operations, the _____ (administrator/manager) or designee _____ (clinical nurse, president, physician coordinator) will determine the appropriate response.

Notification

In the event of an incident, notification will be given as follows: _____ (define for clinic/practice).

Emergency Response Plan Processes

Types of emergency responses: (Clinic/practices should develop a specific policy or complete description to handle each category)

(a) Natural Disasters (Earthquakes, storms, volcanic eruptions, floods, tidal waves, landslides, wild fire, etc.)
(b) Acts of Violence/Terrorism (Biological, chemical, explosive, system sabotage, etc.)
(c) Practice Emergencies (Facility damage through fire or water, illness or death of key employees, etc.)
(d) Public Health Outbreaks (Flu, hepatitis, plague, etc.)
(e) Management Alert (Developing or potential emergency situation/incident, including bomb and bioterrorism threats, etc.)

Emergency responses and planning activities are coordinated with local, regional and state agencies. Emergency Communication is coordinated through State Communications, county dispatch center. Plans include:

(a) County Mass Casualty Plan
(b) County Terrorism Plan
(c) County Hazardous Materials Response Plan
(d) Hospital Communications Plan
(e) State Communication Plan
(f) Facility Evacuation Plan
(g) Facility Communication Plan
(h) Emergency Communications Process
(i) Staff responsibilities during specific emergency incidents are outlined in the appropriate policy.

Management of patients during an incident is outlined in the appropriate policy.

(a) Evacuation of patients and staff is outlined in the Evacuation Policy.
(b) Alternate sites for service delivery; including patient care is directed by the _____ (administrator/manager) or designee.

(c) Transportation arrangements are in place to transfer patients from _____ (clinic/practice) to other facilities to assure appropriate medical care is provided.

(d) A Communications Plan contains contingency plans for power failure, equipment failure and system overload. Designated phones, including an analog phone, are strategically placed in the facility. A list of location and number is included in the plan.

Response to weapons of mass destruction, including chemical, biological, radiological and explosive devices are included in the Emergency Response Plan.

Education and training for Emergency Response is done initially during employee orientation and annually, coordinated by the _____ (administrator/manager or designee).

Annual disaster exercises are conducted _____ (when)

The Emergency Response Plan is evaluated annually by the _____ (administrator/manager), updated as appropriate and reviewed as part of education and training.

Exhibit D: Patient Isolation Precautions

Standard Precautions

- Wash hands after patient contact.
- Wear gloves when touching blood, body fluids, secretions, excretions and contaminated items.
- Wear a mask and eye protection, or a face shield during procedures likely to generate splashes or sprays of blood, body fluids, secretions or excretions
- Handle used patient-care equipment and linen in a manner that prevents the transfer of microorganisms to people or equipment.

Use care when handling sharps and use a mouthpiece or other ventilation device as an alternative to mouth-to-mouth resuscitation when practical.

Standard precautions are employed in the care of ALL patients

Airborne Precautions

Standard precautions plus:

- Place the patient in a private room that has monitored negative air pressure, a minimum of six air changes/hour, and appropriate filtration of air before it is discharged from the room.
- Wear respiratory protection when entering the room.

- Limit movement and transport of the patient. Place a mask on the patient if they need to be moved.

Biothreat Diseases requiring Airborne Precautions: Smallpox.

Droplet Precautions

Standard Precaution plus:

- Place the patient in a private room or cohort them with someone with the same infection. If not feasible, maintain at least 3 feet between patients.
- Wear a mask when working within 3 feet of the patient.
- Limit movement and transport of the patient. Place a mask on the patient if they need to be moved.

Biothreat Diseases requiring Droplet precautions: Pneumonic Plague.

Contact Precautions

Standard precautions plus:

- Place the patient in a private room or cohort them with someone with the same infection if possible.
- Wear gloves when entering the room. Change gloves after contact with infective material.
- Wear a gown when entering the room if contact with patient is anticipated or if the patient has diarrhea, a colostomy or wound drainage not covered by a dressing.
- Limit the movement or transport of the patient from the room.
- Ensure that patient-care items, bedside equipment, and frequently touched surfaces receive daily cleaning.
- Dedicate use of noncritical patient-care equipment (such as stethoscopes) to a single patient, or cohort of patients with the same pathogen. If not feasible, adequate disinfection between patients is necessary.

Biothreat Diseases requiring Contact Precautions: Viral Hemorrhagic Fevers.

Exhibit E: Administration Isolation Precautions

	Bacterial Agents								Viruses				Biological Toxins			
Isolation precaution	Anthrax	Brucellosis	Cholera	Glanders (Rarely Seen)	Bubonic Plague	Pneumonic Plague	Tularemia	Q Fever	Smallpox	Venez. Equine Encephalitis	Viral Encephalitis	Viral Hemorrhagic Fever	Botulism	Ricin	T-2 Mycotoxins	Staph. Enterotoxin B
Standard precautions for all aspects of patient care	X	X	X	X	X	X	X	X	X	X	X	X	X	X	X	X
Airborne precautions				X					X							
Droplet precautions						X				X						
Use of N95 mask by all individuals entering the room									X							
Contact precautions		X							X			X				
Wash hands with antimicrobial soap		X	X						X			X				
Patient placement																
No restrictions	X															
Cohort "like" patients when private room unavailable		X	X		X	X	X	X			X		X	X	X	X
Private room		X	X	X	X				X	X		X				

	Bacterial Agents								Viruses				Biological Toxins			
	Anthrax	Brucellosis	Cholera	Glanders (Rarely Seen)	Bubonic Plague	Pneumonic Plague	Tularemia	Q Fever	Smallpox	Venez. Equine Encephalitis	Viral Encephalitis	Viral Hemorrhagic Fever	Botulism	Ricin	T-2 Mycotoxins	Staph. Enterotoxin B
Negative pressure									X							
Door closed at all times				X					X							
Patient transport																
No restrictions							X	X								
Limit movement to essential medical purposes only		X	X	X	X	X			X	X	X	X	X	X	X	X
Place mask on patient to minimize dispersal of droplets				X		X			X	X						
Cleaning, disinfection of equipment																
Routine terminal cleaning of room with hospital-approved disinfectant upon discharge			X	X			X	X	X	X	X		X	X	X	X
Disinfect surfaces with bleach/water sol. 1:9 (10% sol.)	X	X			X	X						X				

	1	2	3	4	5	6	7	8	9	10	11	12	13	14	15
Dedicated equipment disinfected prior to leaving room					X			X			X			X	
Linen management as with all other patients	X	X	X	X	X	X	X	X	X	X	X	X	X	X	X
Routine medical waste handled per internal policy	X	X	X	X	X	X	X	X	X	X	X	X	X	X	
Discharge management															
No special discharge instruction necessary	X	X	X	X	X	X	X	X	X	X	X	X	X		
Home care providers should be taught principles of standard precautions					X				X		X				
Patient not discharged from hospital until determined to be no longer infectious								X	X		X				

	Bacterial Agents								Viruses				Biological Toxins			
	Anthrax	Brucellosis	Cholera	Glanders (Rarely Seen)	Bubonic Plague	Pneumonic Plague	Tularemia	Q Fever	Smallpox	Venez. Equine Encephalitis	Viral Encephalitis	Viral Hemorrhagic Fever	Botulism	Ricin	T-2 Mycotoxins	Staph. Enterotoxin B
Patient generally not discharged until 72 hours of antibiotics completed						X										
Post-mortem care																
Follow principles of standard precautions	X	X	X	X	X	X	X	X	X	X	X	X	X	X	X	X
Droplet precautions						X										
Airborne precautions									X							
Use of N95 mask by all individuals entering the room									X							
Negative pressure room									X							
Contact precautions									X			X				
Routine terminal cleaning of room with hospital-approved disinfectant upon autopsy	X	X	X	X			X	X	X	X	X		X	X	X	X

	Bacterial Agents								Viruses				Biological Toxins			
	Anthrax	Brucellosis	Cholera	Glanders (Rarely Seen)	Bubonic Plague	Pneumonic Plague	Tularemia	Q Fever	Smallpox	Venez. Equine Encephalitis	Viral Encephalitis	Viral Hemorrhagic Fever	Botulism	Ricin	T-2 Mycotoxins	Staph. Enterotoxin B
Disinfect surfaces with bleach/water sol. 1:9 (10% sol.)	X				X	X						X				

Standard precautions: Prevent direct contact with all body fluids (including blood), secretions, excretions, non-intact skin (including rashes) and mucous membranes. Standard precautions routinely practiced by healthcare providers include: Handwashing, gloves when contact with above, mask/eye protection/face shield while performing procedures likely to generate splashes or sprays of blood, body fluids, secretions or excretions. Handle used patient-care equipment and linen in a manner that prevents the transfer of microorganisms to people or equipment. Use care when handling sharps and use a mouthpiece or other ventilation device as an alternative to mouth-to-mouth resuscitation when practical.

Airborne precautions: Place the patient in a private room that has monitored negative air pressure, a minimum of six air changes/hour, and appropriate filtration of air before it is discharged from the room. Wear respiratory protection when entering the room. Limit movement and transport of the patient. Place a mask on the patient if they need to be moved.

Droplet precautions: Place the patient in a private room or cohort them with someone with the same infection. If not feasible, maintain at least 3 feet between patients. Wear a mask when working within 3 feet of the patient. Limit movement and transport of the patient. Place a mask on the patient if they need to be moved.

Contact precautions: Place the patient in a private room or cohort them with someone with the same infections if possible. Wear gloves when entering the room. Change gloves after contact with infective material. Wear a gown when entering the room if contact with patient is anticipated or if the patient has diarrhea, a colostomy or wound drainage not covered by a dressing. Limit the movement or transport of the patient from the room. Ensure that patient-care items, bedside equipment, and frequently touched surfaces receive daily cleaning. Dedicate use of noncritical patient-care equipment (such as stethoscopes) to a single patient, or cohort of patients with the same pathogen. If not feasible, adequate disinfection between patients is necessary.

Exhibit G: Checklist for a Memorandum of Understanding

- Define Overall intent – reflects what the parties are intending to do.
- Describe the parties – name, type of organization, city and state of the headquarters for the organization.
- Define the time period – a start and end date of the partnership.
- Describe assignments and responsibilities – describe each organization's responsibilities separately, beginning with those that are the sole responsibility followed by any shared responsibilities.
- Define any disclaimers – employee relationship's to each partner, and what the partnership is not intended to do, guarantee or create.
- Describe financial agreements – spell out in detail, including which entity will pay for each item and when payment is due.
- Specify risk sharing – describe who will bear risk of a mishap. Never assume responsibility for something over which you do not have control. Ideally indemnification provisions should be mutual: each party is responsible for its own acts or omissions. (Make certain each partner is not only willing but is able to pay.)
- Obtain Signatures and date of execution – by each partner's representative who is authorized to bind the organization contractually.

RESPONDING TO BIOTERRORISM: A LESSON IN HUMILITY FOR MANAGEMENT SCHOLARS

Donna Malvey, Myron D. Fottler, George W. Buck Jr. and Robert S. Fry

ABSTRACT

Although we have yet to experience a major bioterrorism event in the U.S., we are nevertheless preparing for such an event. In this paper, we consider the nature of bioterrorism and the threats and challenges it brings to managing health care organizations. Because existing managerial theory may be inadequate in responding to bioterrorism events, management scholars are advised to approach their research with a great deal of humility and openness. Inasmuch as emerging theoretical frameworks based on complexity science and chaos theory are not fully developed, we propose that stakeholder management theory may be the best approach at this juncture.

In the biological arena, we have not yet experienced a major bioterrorism event in the U.S. Consider the following noteworthy specific occurrences of bioterrorism: (1) the blending of ricin by a Minnesota antigovernment, tax-protest group whose members were convicted for violating the Biological Weapons Anti-terrorism Act in 1996; and (2) the sending of anthrax-laden letters through the mail to Florida, Washington, DC, New York, and New Jersey in October 2001. In no manner

Bioterrorism, Preparedness, Attack and Response
Advances in Health Care Management, Volume 4, 259–286
ISSN: 1474-8231/doi:10.1016/S1474-8231(04)04010-8

could the severity of either of these occurrences classify them as either a public health disaster or major catastrophic event. In the case of the anthrax mailings of 2001, the widespread national and media attention belied the fact that only five deaths resulted and twenty-two persons were actually contaminated. We have automobile accidents that result in higher mortality and morbidity rates. So, why did we react as though the sky were falling? Was it because we were dealing with a phenomenon with which we have had little or no experience and consequently were unsure how to respond?

According to experts such as Dr. Michael Richardson, the Senior Deputy Director for Medical Affairs at the Washington, DC Health Department, because the anthrax events unfolded chronologically, it was actually easier to coordinate federal and local resources and mitigate the bioterror attack. However, if the anthrax events had occurred simultaneously in numerous cities across the nation, the response most likely would not have been as swift or as effective (Interview with Dr. Richardson, 2002). So, even though we had no past experience on which to base our response to this bioterrorism event, the timing of events reduced the possibility of more disastrous outcomes. Clearly, we were fortunate in this particular instance. However, our lack of experience also means that we consequently have limited knowledge of how a major bioterrorism event has been or could be managed.

Until very recently, the management literature has focused little attention in this area. For example, even though the management literature on environmental jolts has been useful over the years in studying the impact of such things as employee strikes and natural disasters, it has not been applied to major bioterrorism events prior to the publication of this research volume. In this research volume, Friedman and colleagues augment our theoretical and practical knowledge in this area by examining the impact of organizational responses to environmental jolts related to bioterrorism events. Similarly, until this research volume, there had been little consideration of complexity theory in responding to bioterrorism events. The paper by McDaniel and colleagues, also in this volume, gives attention to these issues. However, as much promise as complexity theory offers, the theory is in its infancy and has not yet fully developed (Begun, 1994). The result is that we are still in search of management theories that can aid managers in responding to bioterrorism.

Our paper proposes that stakeholder management theory actually may be more useful for managers in responding to bioterrorism than existing managerial frameworks or emerging constructs. Our paper suggests that stakeholder management theory affords managers tools and techniques for managing relationships that exist outside the organization's hierarchy and are subsequently unpredictable, unstable, and uncontrollable. As such, stakeholder management theory is expected to be consistent with the bioterrorism challenge in which the nature, timing, and severity of events produces unpredictable and rapidly changing relationships. Stakeholder management theory may also serve as a theoretical bridge between traditional management theories that focus on control and predictability and emerging theories such as complexity theory, which considers how order can arise from complex and dynamical interactions and relationships (McDaniel & Driebe, 2001).

In this paper, we will:

- Consider the nature of bioterrorism and the threats and challenges it represents for managing in health care organizations, including the sequence of events associated with a bioterrorism event;
- Look at traditional planning responses to a bioterrorism event and consider their adequacy;
- Consider the need for integrative response planning; that is, planning which requires a network of relationships among a wide variety of organizations that are at the forefront of a community response to bioterrorism;
- Examine the role of the hospital as a bioterrorism respondent;
- Address the relevance of management theory for responding to bioterrorism. We compare and contrast the management of day-to-day operations with that of a bioterrorist event. We consider the traditional management approaches along with emerging theories;
- Select stakeholder management theory as a proposed management prototype for responding to bioterrorism, and we develop an example of a managerial response to bioterrorism using the stakeholder framework. We include stakeholder identification as well as stakeholder management prescriptions for selected key stakeholders;
- Suggest recommendations for future research in this important health care management area.

BIOTERRORISM AND THE HEALTH CARE CONTEXT

In this section we consider the nature of bioterrorism and the threats and challenges it raises for managing health care organizations. We examine delivery systems and develop a flow chart that depicts the response process to a bioterrorism event. We also look at traditional planning responses, which are based on the four phases of emergency management.

"How does a country prepare for a man made disaster such as a release of a biological agent into its population?" The answer is complicated, expensive, and not yet fully known. After the hijacked planes crashed into the World Trade Center and the Pentagon on September 11, 2001, it was apparent, that the U.S. was not as prepared as it could have been for a terrorist disaster of that magnitude. The important word here is *magnitude*. Overall the emergency and public health system did a great job on September 11th and the days following, especially when confronted shortly thereafter with the release of a biological agent known as anthrax.

However, as the days following September 11 passed, it also became apparent that the U.S. was not prepared for a biological event of a large scale *either*. This became evident when the first victim of inhalation anthrax died on October 5, 2001, and over-reaction resulted in thousands of people waiting in line to receive antibiotics. Moreover, delivery of mail was all but suspended on the East Coast, and many government offices were officially closed for several weeks (Frist, 2002). We were not adequately prepared to respond to a bioterror attack.

Bioterrorism preparedness subsequently has emerged as one of the critical challenges to protect and care for our communities. A recent survey of hospital chief executive officers (CEOs) revealed that most hospitals were preparing to respond to bioterrorism threats in a variety of ways, including discussions with other hospitals to learn about resources for response, development of disaster plans, and equipment acquisition. However, the survey also revealed that not all hospitals, staffs, and communities were aligned in terms of what they saw as necessary for responding to bioterrorism, especially with regard to inoculations against biologic agents (Research Notes, 2003).

Bioterrorism Delivery Mechanisms

In order to respond to bioterrorism, it is important to understand more about biologic agents and how they are dispersed into the population. Biological emergencies are an actual or imminent set of conditions in which biological agents are intentionally introduced. Biological agents can be delivered by a variety of means including dispersal via explosive devices such as bio-bombs or dirty bombs, mechanical devices such as crop-dusting aircraft, mosquito-control trucks, garden spray devices, dispersal through a building's water or ventilation system, or in the food system. Additionally, conventional and stand-off attack on bio-engineering facilities, or goods in transit (i.e. an intentional HAZMAT incident) must be considered. Such routine shipments offer a prime target, either for hijacking to appropriate the materials or a more conventional attack aimed at extortion or the release of the agents. A bio-attack would most likely involve aerosol dispersal and would afford no easily discernible signature (i.e. the on-set of symptoms will typically occur within days to weeks after the initial exposure making detection difficult).

Before developing a response plan for biological terrorism, the agents that may be used for an intentional release must be examined because they present different management challenges. Many planning agencies reference the following five biological agents: Smallpox (Variola virus), Plague (Yersinia pestis), Anthrax (Bacillus anthracis), Tularemia (Francisella tularensis), and Brucellosis (Brucella sp.). For the purpose of this paper we will only consider the three agents, anthrax, small pox, and botulism, which are described in Table 1.

The medical and emergency management concerns are complex, since little experience exists in coping with the impact of biotoxins on a large scale. Sports arenas, concert halls, department stores and malls, transportation terminals, and office buildings – in short, any location capable of housing a large or small number of people in an enclosed or nearly enclosed space – are amenable to aerosol dispersal and thus are at greatest risk in a bioterrorism incident. Specialized response and management capabilities are required to effectively mitigate the impact of a biological terrorism situation.

Specifically, if there were an early warning of a bioterror attack, we might be able to avert mass casualties by way of syndromic surveillance. Syndromic surveillance involves monitoring patients based on symptoms that they report. Some of the deadliest biologic agents such as smallpox, anthrax, and plague can be treated successfully if diagnosed early, but they progress rapidly to more serious illness and eventually death if left untreated (Crenson, 2002).

Local Response Flowchart

However, response to a biological release may be difficult until it has been identified as such. One of the primary challenges associated with biological weapons is that their release may be covert: there may be no "K-boom" involved. The first sign that a biological agent has been released may be the illness it causes in the exposed population, as seen in a passive and active surveillance. As a result, the most important component of an effective response to the release of a biological agent is recognition. Early recognition of an event creates a larger "window of opportunity" during which effective treatments and public health measures, such as quarantine or emergency public education program, can be implemented (Buck, 2001).

The flowchart of a local response to a bioterrorism event (Fig. 1), indicates the sequence of events that can be expected to follow the release of a biological agent. From a management perspective, this flowchart underscores the need for early recognition and diagnosis systems. Historically, health departments have relied on knowledgeable physicians to identify bioterror attacks by diagnosis.

Table 1. Summary of Biological Agents.

Biologic Agent	Definition	Symptoms	Prevention/Treatment
Anthrax	Acute bacterial disease that usually affects the skin, but which may very rarely involve the oropharynx, mediastinum or intestinal tract (Chin, 2000)	Symptoms vary according to type. • Skin: itching followed by a lesion • Inhalation: fever, mild cough, chest pain, and malaise • Gastrointestinal: abdominal distress followed by fever, signs of septicemia and death (Chin, 2000)	Antibiotic treatment may prevent infection in those who are exposed to anthrax. Anthrax is usually susceptible to penicillin, doxycycline and fluoroquionolones (Chin, 2000). Anthrax vaccine can also prevent infection, but is not recommended for general public to prevent disease and is not readily available
Small pox	Caused by variola virus, the incubation period is about 12 days following exposure. In 1980, the World Health Assembly certified that the world was free of naturally occurring small pox	Initial symptoms include high fever, fatigue, and back aches. Characteristic rash, most prominent on face and arms follows within 2-3 days and then lesions and scabbing. Majority of patients recover, but death occurs in up to 30% of cases	There is no proven treatment for small pox. Antiviral agents are currently under investigation. Patients may benefit from supportive therapy and antibiotics for any secondary bacterial infections that occur.

| | | | Vaccines are available and have been a highly effective immunizing agent enabling the global eradication of small pox. Quarantine regulations have also reduced risk of spread of small pox. Vaccination generally has not been recommended to prevent the disease in the general public. Interactions with immune system functions in the elderly and in immune-impaired patients such as those suffering from Hepatitis and HIV could be fatal |
| Botulinum toxin | Substance released from the bacteria clostridium. Incubation period ranges from 2 hours to 8 days. | Difficulty speaking, seeing, and swallowing. Can lead to muscle paralysis and in severe cases a need for mechanical respiration (Johns Hopkins, 2000) | None |

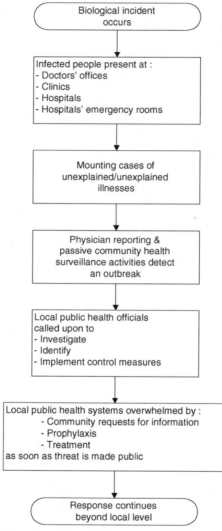

Fig. 1. Flowchart of Local Response to a Bioterrorist Incident.

In fact, the initial anthrax case in Fall 2001 was surfaced by a physician at JFK Medical Center in Atlantis, Florida.

Yet, physicians may not always be equipped to detect other potential bioterror agents and rare diseases such as Q fever or bubonic plague. Furthermore, one diagnosis would not inform public health officials about the scope, location or timing of a bioterror attack. As a result, *everything* must be monitored. The monitoring system would extend to such things as emergency room visits, physician office visits, school absenteeism, and surges in purchases of over-the-counter medicines that might alleviate symptoms associated with exposure to biologic agents (Crenson, 2002; Frist, 2002). In the next section, we identify the four phases of management of medical emergencies as developed by FEMA. Obviously, different management tactics are associated with different phases.

Traditional Planning Responses: Four Phases of Emergency Management

In the past, traditional planning responses for an *intentional* release of a biological agent have included the four phases of comprehensive emergency management. These phases are: mitigation, preparedness, response and recovery (Drabek & Hoetmer, 1989). The Federal Emergency Management Agency (FEMA) has described the four phases of emergency management as having a "circular relationship to each other. Each phase results from the previous one and establishes the requirements of the next one" (Federal Emergency Management Agency Emergency Management Institute, n.d.). The first phase of emergency response is mitigation, which refers to activities undertaken that eliminate or reduce the effects of a disaster. The second phase is preparedness, referring to the plan of how to respond in case an emergency occurs and working to increase resources available to respond. The third phase, response, refers to activities that occur during and immediately after an emergency has occurred. These activities will provide emergency assistance to victims and reduce the likelihood of additional damage. The final phase is recovery. In the short term, this means returning vital life support systems to minimum operating standards. In the long term, recovery includes returning an area back to normal. Recovery should also include a review of ways to avoid future emergencies. This illustrates the overlapping of phases where mitigation for future disasters may be part of the response effort.

Mitigation for the threat of a biological event can in many cases parallel those needed for any natural and man made disaster. Special needs shelters will be required for the ill and anyone who lives in the area of contamination. The local HAZMAT teams will have protective clothing and equipment for chemical spills. More specifically for biological terrorism is the maintenance of the National Pharmaceutical Stock Pile which can help to reduce the number of casualties during a biological event, which is managed by the Center for Disease Control and Prevention (CDC) (2001, October 19, November 9, September, and June 22).

Preparedness is undertaken before an event will happen, to build emergency management's capacity to respond to disasters. In the case of biological warfare, preparedness means training of medical personnel and first responders, public awareness, vaccine and pharmaceutical stockpiles, and planning for the care of, sheltering of, isolation and decontamination of the affected population. Because the key to dealing with a biological incident is to identify the event, it is critically important for medical professionals to be trained to recognize the epidemiological aspects of an event.

The following are a number of indicators that health care professionals should be trained to be alert to:

- Large numbers of simultaneous cases with similar symptoms or disease.
- Large number of cases of a rare or unusual disease.
- A single confirmed case of smallpox.
- The unexplained diagnosis of two or more diseases in a single patient.
- The presence of a disease with unusual geographic or seasonal distribution (i.e. flu-like illnesses in the summer or Venezuelan Equine Encephalitis in Washington State).
- Unusual pattern of illness or death among animal populations that precedes or accompanies human illness outbreaks (Crenson, 2002).

Once an event has been identified it is important to take steps to limit the spread of the disease by limiting the movements of those who may have been exposed (Buck, 1998). For example, it may be necessary to quarantine exposed individuals until it is determined whether the disease is communicable or not. It also may be necessary to dispense medications and vaccines to a large number of people in a short period of time. This will require a number of nurses, physicians, pharmacists, and other providers who are familiar with the medications. It will require documentation for all of the victims and those who are exposed to the agent. Response will also include dealing with mass casualties and deaths. Storage and disposal of contaminated bodies must be incorporated into response planning because of the potential to spread disease.

In the next section, we examine integrative response planning. This type of planning considers the need to "integrate" and coordinate planning efforts such that participating agencies and health care organizations are functioning as a team. For example, integration involves coordination of a wide variety of agencies to disseminate information to the public. All participants must work together to assure that important messages and news are transmitted. This is in contrast to response planning that would rely on individual organizations and agencies to function independently without consideration of exactly what it is others are engaged in.

INTEGRATIVE RESPONSE PLANNING

From a management perspective, integrative response planning should include the following:

- Coordination of public safety, law enforcement and health agencies. Public information specialists.
- An assessment of likely targets in the community. (Water treatment plants, public transportation, public buildings, etc.)
- An incident command system, including communication of all agencies involved.

Decontamination procedures in a biological incident may not need to be done if the incident was a covert release, unless the area of contamination is discovered. Isolation of victims may be necessary if it is determined that the disease is contagious, as in the case of small pox. Anthrax is not spread from person to person so isolation would not be indicated. However, in 2002 less than half of our nation's hospitals had adequate decontamination capabilities or the ability to manage noncontaminated casualties (Frist, 2002). Consequently, part of the plan should include the sheltering of individuals outside of the hospital environment.

The plan should also include assistance for the population in the community who were not affected but who may be worried and afraid that they may be infected. This may be accomplished by establishing a pre-triage center or Casualty Collections Points (CCP) outside of the hospital to keep the emergency room from being overwhelmed by people who are not ill. Part of the plan should also include management of the medications received from the National Pharmaceutical Stock pile. Local health authorities must do the dispensing of medications and immunizing people as needed. The plan should also include the storage and disposal of large numbers of casualties.

Integrating resources from many levels requires the commitment of agencies to mitigate against, prepare for, respond to, and recover from an incident of biological terrorism. A comprehensive approach to terrorism requires coordination with a variety of federal agencies. For example, prior to the establishment of the Department of Homeland Security (DHS), the Federal Bureau of Investigations (FBI) has been the lead federal agency for crisis management and terrorism investigations. Similarly, the President had designated FEMA as the lead agency for consequence-management efforts. The Terrorism Incident Annex of the Federal Response Plan implements Presidential Decision Directive 39 (PDD-39) and defines federal roles in terrorist incidents.

Despite these designated roles, local agencies logically will be in the forefront of emergency response to terrorist incidents. Local jurisdictions have the responsibility to manage the consequences of terrorist incidents occurring within their borders. Consequently, it is incumbent on agencies at the local level to establish a committee for the planning and management of incident response. At the local level, this committee can address themselves as they see fit. For example, groups around the country have named themselves Terrorism Stakeholders Group (TSG) or Terrorism Task Force (TTF). This committee should comprise but is not limited to representation of the following:

- Local and regional federal agencies and departments such as those that we have previously mentioned, including FBI and FEMA, plus the Department of Health and Human Services (DHHS), which oversees the U.S. Public Health Service, and the Department of Defense (DoD);
- State agencies and departments of emergency management, public health and fire service;
- Local agencies and departments of health, fire service, and law enforcement; churches; airports; schools; port services; the coroner, and Geographic Information Systems personnel.

The committee's primary role should be planning, training, and development and coordination of inside and outside agencies for technical assistance in developing measures to address biological terrorism. During a biologic incident, federal agencies and departments will augment these efforts, particularly in regard to crisis management. In the areas of intelligence sharing and situation monitoring, the committee should coordinate with the lead local law enforcement agency. Local law enforcement in turn should monitor open source information for trends and potential criminal intelligence operations to assess the capabilities and intentions of known terrorist groups. Local and regional law enforcement should then coordinate their activities with the FBI.

In the area of response and consequence management, local law enforcement should be the lead point of contact for marshalling during consequence management efforts at the area of responsibility (AOR). The fire service will have direct consequence-management roles. The Emergency Operations Center (EOC), which is responsible for strategic resource management, will coordinate support to local consequence-management activities with FEMA through the state's Division of Emergency Management (DEM), Office of Emergency Management (OEM).

In the realm of investigations, the FBI is the lead investigative agency. Local law enforcement investigations will be the primary supporting investigative entity, coordinating investigative efforts with the FBI. The representative of local law enforcement assigned to the EOC should coordinate law enforcement support to the incident site or AOR with state law enforcement.

The strategic activities of all of the response efforts will be coordinated through the local EOC. The state's Division of Emergency Management will support incident site or area of responsibility efforts by helping to obtain the resources necessary to manage an incident that exceeds local capabilities. In addition, the Division will generally act as the point of contact for obtaining routine mutual aid or resource support from within the state and federal government. Due to the FBI's direct federal crisis management responsibilities, national security considerations, and the need for rapid access to federal or military resources to manage a weapons of mass destruction incident, the local division of the FBI may directly request federal resources for biological incident(s) independently or at the request of the emergency operations center.

In the next section, we shift from a more general or community overview of bioterrorism to consideration of the role of the individual community hospital. In other words, we move from the general to the particular and identify practices that contribute to "state of the art."

ROLE OF HOSPITALS

According to Bill Frist (2002), the Senate's Majority Leader and also a heart transplant surgeon, hospitals must play a key role in bioterror attacks, but most are currently unprepared to manage a bioterrorism event. They must be prepared to assume a key role. Prior to September 11, 2001, only one out of five hospitals actually had a response plan that was specifically designed for events of a biochemical nature. Less than half of all hospitals had decontamination units that were useful, and less than one-third of them had an antidote for the common chemical agents such as nerve gas. Furthermore, although 10% of hospitals have performed chemical disaster drills, only 3% were found to have conducted biological disaster drills (Frist, 2002).

Since November 1, 2002, states have been required to report their progress in preparing for bioterrorism. Among the key questions asked were: how would they distribute medicines, where could they provide 500 hospital beds in case of mass casualties, and how their hospitals would plan to isolate highly contagious patients. Only Florida was deemed ready for a bioterrorist attack, but that achievement was marginalized by the fact that the state still must conduct drills to assure implementation. Most states were not prepared to dedicate 500 beds in an emergency; much less the 1500 that they were expected to have in place in 2003. Furthermore, most hospitals have no isolation rooms for infectious victims. Also, some states' plans relied on the National Guard. Because the guard might be unavailable during times of emergencies, states need to think of alternatives or other options (Associated Press, 2002).

The American Hospital Association (AHA) reports a great increase in awareness on the part of U.S. hospitals and suggests that the terrorism risk has been factored into hospital disaster plans. In addition, the AHA believes that hospitals are working with community providers, public health agencies, and emergency services to coordinate planning efforts. The AHA conducted a survey (AHA 2002 Survey) of hospitals in February 2002, five months after September 11 attacks. Although only a third of hospitals responded, the majority of respondents (69%) reported incorporating a bioterrorism response into their disaster plans with another 28% expecting to achieve the same goal within the coming year (Goedert, 2002).

The question of who will pay for bioterrorism readiness is unclear, and the additional expense could threaten the survival of hospitals, many of which exist on slim operating margins. State health department budgets are especially overwhelmed by increased surveillance and monitoring activities and also by the challenges of training, coordination, and education. The poor performance of the economy nationally has affected state budgets, and states are looking for federal dollars to fund the bioterrorism activities. However, because of budget reductions, many states have hiring freezes that will not allow them to spend money even if they receive federal monies.

The AHA contends that U.S. hospitals require at least $11.3 billion to address all areas of disease management in cases of terrorism, especially since hospitals are expected to deal with mass casualties on their own for up to 48 hours prior to the arrival of government resources. In addition, 78% of respondents of the AHA 2002 survey reported that they simply do not have the resources to perform the necessary disease surveillance functions or even to train laboratory personnel in the identification, handling, and reporting of chemical and biological agents. They also cannot afford to stockpile adequate protective garments and equipment for hospital staff nor can they establish community-wide communication systems among cohort hospitals and other emergency services (Associated Press, 2002; Frist, 2002; Goedert, 2002). Furthermore, hospitals and major health facilities are going to have to be built differently. Ventilation systems and quarantine systems, and systems that provide for decontamination of individuals must be programmed into the designs of these structures (Interview with Dr. Benjamin, 2002).

Figure 2 summarizes bioterrorism preparedness in 2003, according to a recent survey of U.S. hospitals by the American College of Health Care Executives (ACHE) in 2003. Of the survey respondents, which included 295 U.S. hospitals, 85% of hospitals reported that they currently have in place discussions with other hospitals to learn about resources available for responding to bioterrorism. The area of preparedness that was shown to be least developed is hospital wide surveillance equipment. Only 51% of hospitals reported having such equipment in place. Another area of weakness was hospital decontamination units with 40% of hospitals surveyed reporting that they do not have these units in place.

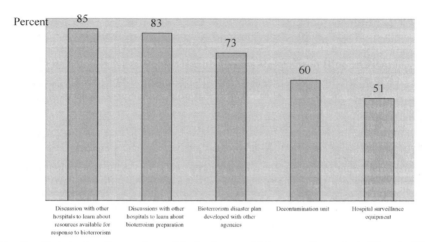

Percent

| 85 | 83 | 73 | 60 | 51 |

| Discussion with other hospitals to learn about resources available for response to bioterrorism | Discussions with other hospitals to learn about bioterroism preparation | Bioterrorism disaster plan developed with other agencies | Decontamination unit | Hospital surveillance equipment |

Fig. 2. Bioterrorism Preparedness: Mechanisms and Measures Currently in Place in 295 U.S. Hospitals, 2003. *Source:* Adapted from "Bioterrorism Preparedness: Mechanisms and Measures." *Healthcare Executive* (May/June 2003), 64.

Hospital Response Efforts Using a Severity System

A sample system is proposed for hospitals to respond to bioterrorism events. Because bioterrorism events will vary according to the numbers of patients involved, this system is based on three stages of severity of a bioterrorism event. These severity stages are defined as below with corresponding patient volumes:

Stage I: Developing Public Health Crisis – 0 to 100 patients
Stage II: Public Health Disaster – 101 to 1,000 patients
Stage III: Catastrophic Public Health Event – 1,001 to 10,000 or more patients

The patient numbers associated with each stage are to be considered guidelines since the release of a highly contagious agent may require Stage III procedures even if the number of patients is far less than 1,001.

Each stage of severity requires a different strategic response that is implemented through a suitable general action plan. Decision makers must anticipate when or if a bioterrorism event will progress from one stage of severity to the next. By anticipating this progression, resources in place can be redirected, or additional resources can be requested in a timely, progressive manner. Analysis of decision factors can help decision makers to anticipate increasing response requirements. When any decision factor is reached, leaders must look at initiating actions

Table 2. Summary of Decision Factors Associated with Stages of Severity and Possible Strategies and Actions for Hospitals.

Stage and Patient Cohorts	Strategy	General Action	Decision Factors
I. Developing public health crisis (0–100 patients)	Assess needs	Utilize existing response structure	>100 patients probable Hospitals approaching maximum capacity Hospital triage approaching maximum capacity
II. Public health disaster (101–1000 patients)	Establish alternate response structure	Augment resources within region	>1000 patients probable Cohort hospital system approaching maximum capacity Casualty collection points (CCPs) approaching capacity Logistics using all regional staff, security, or transportation resources
III. Catastrophic public health event (1001–10,000 or more patients)	Utilize all available federal and state resources	Integrate resources into response structure	Alternate care facilities (ACFs) at maximum capacity Federal resources delayed for more than 24-hours

suitable for the next higher stage of severity. In addition to these specific decision factors, leaders must be aware that implementing any functional action normally associated with a stage of severity requires analyzing all other functional areas to determine if the next level of action is necessary in those areas.

Table 2 summarizes sample decision factors associated with each severity stage and illustrates possible strategies and actions for each stage. Decision makers must be alert to the possibility that information provided through critical information channels may present additional decision factors at any time and thus affect implementation.

RELEVANCE OF MANAGERIAL THEORY TO BIOTERRORISM EVENTS

Is existing managerial theory relevant for bioterrorism events? If managers rely on traditional management theories to guide them in responding to bioterrorism events, the answer to this question is probably, *no*. From the perspective of traditional management processes, it is highly unlikely that these processes, which are aimed at predictability and control, will serve managers well in attempting to respond to bioterrorism events.

Table 3 provides a comparison of the outcomes of traditional management processes used in day-to-day operations management and in responding to bioterrorism events. As indicated, the processes typically employed by managers in day-to-day operations reflect order, stability and control. Traditional management theories assume that managers can dispel ambiguity and uncertainty by taking control, and control occurs through implementation of these management processes. However, the very opposite is true of bioterrorism in which the nature and time and severity of events is unknown and unpredictable. In coping with bioterrorism, managers must ultimately learn to adapt to ambiguity and surprise rather than viewing these things as obstacles to be overcome in order to set things

Table 3. A Comparison of the Outcomes of Traditional Management Processes Used in Day-to-Day Operations and in Response to Bioterrorism Events.

Traditional Management Processes	Outcomes: Management of Day-to-Day Operations	Outcomes: Management of a Bioterrorism Event
Planning	Understandable planning objectives Rational relationships among planning variables Organizational survival goals typically related to financial objectives	Nature, timing, and severity of event lead to unpredictability and ambiguity in establishing planning objectives Opportunistic relationships developed Non-financial survival goals for the organization
Organizing	Coordination of necessary resources, both internal and external	Limited ability to coordinate resources necessary to respond to unknowable and unpredictable demands
Directing/Leading	Organization has autonomy to lead and to make decisions internally and also decisions that involve key external stakeholders	Leadership and decision making authority are subordinated to higher authorities
Controlling/Evaluating	Measuring and monitoring organizational progress is possible	Measurement and monitoring become subordinate to other goals such as responding to immediate threats and survival

right. Thus, traditional management analysis that instructs managers to anticipate, organize, and plan for events before they occur may be limiting in circumstances such as bioterrorism. During bioterrorism events, managers will be required to do what is unknown and to take action as events unfold. Past experience will not be informative (McDaniel, 1997; McDaniel & Driebe, 2001).

Beyond the basic management processes, are there other management theories that can guide managers in responding to bioterrorism? A review of some existing theoretical frameworks also suggests that most management theories are not focused on handling unforeseen situations. For example, population ecology theory attempts to predict which organizations will survive and why – in the long run. However, this theory does not provide managers much assistance in making decisions in situations in which the threats are immediate and unknown, such as bioterrorism. Resource dependency theory has guided many managers in the past because it examined how and why organizations shift their dependence over time to ensure their survival. Responding to an immediate bioterrorism event, however, involves different dynamics than how an organization will generate and sustain support in the long-run. Institutional theory predicts that once certain managerial fads and practices are adopted by major health care organizations, other health care organizations will follow. Again, this theoretical assumption does not provide much guidance in determining how and why organizations respond in particular ways to catastrophic events such as bioterrorism. The theory underpinning environmental jolts may provide some insight for managers, and as mentioned earlier, this topic is explored elsewhere in this research volume.

Even though these managerial theories are inadequate, there are emerging theories that may offer some promise and understanding for responding to bioterrorism. For example, complexity theory may be useful because it assumes the unknown and the unpredictable and focuses on the interaction of non-linear events and interactions. Furthermore the theory assumes a macroscopic viewpoint and examines how order can emerge from a volatile and complex system (McDaniel, 1997; McDaniel & Driebe, 2001). But such theoretical assistance may be limited at this time because the theory is not fully developed. Indeed, Begun (1994) and others have indicated that while theories of chaos and complexity offer a new perspective that is deserving of attention, these theories remain works in progress.

We believe that stakeholder management theory may actually offer the most potential at this moment because it serves as a bridge or conduit between traditional managerial theories that aim to control and predict and emerging theories such as chaos and complexity theories. Stakeholder management theory provides this important linkage because it assists managers in managing relationships that are outside of the manager's direct control and are therefore unpredictable. It allows external relationships to be made "knowable" or "understandable" through a variety of diagnostic typologies. Furthermore, the stakeholder framework ultimately suggests strategies for managing these relationships in an equilibrium that recognizes stakeholder wants and needs as well as those of the organization. Therefore, during a bioterrorism event, managers will have the ability to manage a variety of external relationships, including many at the state and federal level, without jeopardizing organizational functions and needs.

Stakeholder Applications

Within the stakeholder framework, managers have a variety of tools and techniques that they may call upon to organize and plan. Preeminent among them is stakeholder analysis. Stakeholder analysis offers managers unique diagnostic abilities, which provide for a comprehensive assessment of the environment while simultaneously assessing the abilities of organizational stakeholders to fulfill the needs of external stakeholders. Furthermore, stakeholder analysis provides mechanisms for developing managerial prescriptions for responding to stakeholder needs and achieving balance or equilibrium among the stakeholders and the organization (Blair & Fottler, 1990, 1998). We offer the following example of how stakeholder analysis can assist hospitals in identifying organizational stakeholders related to a bioterrorism event.

Regardless of the level of severity of a bioterrorism event, the hospital must respond to a large number and a wide variety of external stakeholders. External stakeholders fall into three categories in their relationship to the healthcare organization. Some provide inputs into the organization, some compete with it, and some have a particular special interest in how the organization functions. Unless both the organization and the stakeholder believe such an agreement will be mutually beneficial and of fair value (relative to alternatives), agreement will not be reached and/or sustained.

Obviously, some agreement among the various managers concerning *who* (singular or plural) will manage which *stakeholders* on *which issues* is necessary to assure all key stakeholders are managed. This process typically involves internal negotiations and the development of organizational policies and procedures. The advantage of the stakeholder map is to *clarify* relationships and responsibilities.

As Table 4 indicates, the list of key stakeholders that would be included in a bioterrorism event encompasses a wide variety of stakeholders that represent local, state, and federal agencies. The list of stakeholders included in this table is not meant to be exhaustive, but instead illustrates the range of stakeholders that would be essential in responding to the many challenges associated with bioterrorism events.

Table 4. Hospital Stakeholders in the Preparedness and Response Planning for
Bioterrorism.

Local
  Neighboring local public health agencies
  Local Hazardous Material (HAZMAT) teams
  Local environmental health agencies
  Local emergency management officials
  Local Emergency Planning Commissions
  Fire, police, and EMS
  Local 9-1-1 Centers
  Local Metropolitan Medical Response System teams
  Other local hospitals
  Health plans
  Other healthcare providers (i.e., physician practices)
  Local laboratories
  Poison Control Centers
  School officials
  Elected officials
  Local news media
  Area businesses
  Terrorism Stakeholders Group (TSG)

State
  State Health Department
  State Division of Emergency Management (DEM)
  Office of Emergency Management (OEM)
  State emergency management officials
  National Guard
  State laboratories
  Other volunteer groups

Federal
  Department of Homeland Security (DHS)
  Surgeon General
  Office of Emergency Preparedness (OEP)
  Department of Health and Human Services (DHHS)
  Centers for Disease Control and Prevention (CDC)
  Food and Drug Administration (FDA)
  Department of Justice (DOJ), including the Federal Bureau of Investigation (FBI)
  Federal Emergency Management Agency (FEMA)
  Environmental Protection Agency (EPA)
  Department of Defense (DoD)

Table 5. Stakeholder Management Prescriptions for Selected Key Stakeholder Respondents Based on Sample Stages of Severity of Bioterrorism Events in U.S.

Selected Key Stakeholder Respondents by Level of Response	Management Prescriptions According to Severity Stage		
	Stage I: Developing Public Health Crisis (0–100 Patients)	Stage II: Public Health Disaster (101–1000 Patients)	Stage III: Catastrophic Public Health Disaster (1001–10,000 or More Patients)
	Example: Anthrax in Florida, Washington, DC, and New York – October 2001	Example: Not yet occurred in U.S.	Example: Not yet occurred in U.S.
Local-level	• Assess needs	• Establish alternate response structure and augment resources within region • Establish triage stations	• Utilize all resources available at all levels, including federal and state
Physicians practices	• Anticipate response requirements • Utilize existing response structure		• Integrate both federal and state resources into response structure • Widespread prophylaxis campaign
Hospitals		• Cohort hospitals will internally augment by transferral of existing patients to area hospitals • Begin prophylaxis campaign	
Clinics	• Provide or receive training & education to recognize and report symptoms • Conduct public information/education campaign		
Local public health offices			
State-level	• Provide technical assistance in developing measures to address specific manpower and resource needs	• Augment response efforts by ensuring rapid access to a wide-variety of state resources to manage event	• Directly request federal resources to enhance response efforts and capabilities

Table 5. (Continued)

Selected Key Stakeholder Respondents by Level of Response	Management Prescriptions According to Severity Stage		
	Stage I: Developing Public Health Crisis (0–100 Patients)	Stage II: Public Health Disaster (101–1000 Patients)	Stage III: Catastrophic Public Health Disaster (1001–10,000 or More Patients)
Health Departments	• Implement monitoring system that includes local level respondents	• Augment epidemiological investigation	• Significantly augment epidemiological investigation
Dept. of Emergency Management		• Provide additional training and other manpower needs	• Enhance communication with both local and federal level respondents
		• Enhance security with use of National Guard	
Federal-level	• Monitor event	• Support local and state-level stakeholder efforts	• Assume overall management responsibility
DHS[a]	• Provide recommendations, resources, and other assistance as required	• Coordinate strategic management activities to ensure effective deployment of resources	• Direct resource coordination efforts
FEMA	• Expand education campaign to nationwide	• Integrate monitoring activities with state and local levels	• Assess allocation of resources and re-direct as necessary
DHHS		• Augment existing manpower, including security with National Guard	• Expedite access to federal sources of communication and consultation
CDC			• Mobilize additional manpower

[a] At the time of publication, the DHS had just been established and its integrative relationships with other federal agencies were unclear.

We have developed a set of management prescriptions that will aid in managing these key stakeholders. Table 5 presents sets of generic managerial prescriptions for selected key stakeholders. We developed these generic prescriptions using the sample system of severity stages developed earlier in this paper and presented in Table 2. We also organized the prescriptions according to the variable requirements of each level of selected stakeholder respondent, including local, federal and state.

Although we have not yet experienced Stage 2 or Stage 3 bioterrorism events, it is appropriate to speculate on possible management actions, especially because such speculation encompasses proactive planning that managers must undertake if they are to be adequately prepared to respond to future disasters and catastrophic levels of bioterrorism activities. Furthermore, Table 5 prescribes management actions that integrate response efforts at each level. This effort is particularly useful because historically much of the planning and preparedness activities have not been well-integrated and have subsequently led to conflicts and uncertainty in terms of the roles, especially those of local level responders.

Table 5, however, directs management attention to specific levels of severity, from crisis to catastrophe and the involvement of stakeholders at particular environmental levels. Thus, for a Stage 1 event (developing public health crisis), managers will focus on providing education and training efforts for their local level respondent key stakeholders such as physicians' practices, clinics, or hospitals. Such activities will be important for surveillance, detection, and monitoring. However, at the state level of respondent, management's role shifts to one of assisting stakeholders, such as the state health department, in their measurement or monitoring activities. Finally, Table 5 shows that as the stages of severity increase, management's involvement with state and federal level respondents also increases, especially with regard to manpower needs and funding.

CONCLUSIONS AND RECOMMENDATIONS

The study of bioterrorism should humble management scholars because most of their theories were developed in more stable organizational circumstances and are not readily applicable to catastrophic events and disasters such as bioterrorism. Clearly, management's response to bioterrorism will require new ways of thinking and writing about management that go beyond immediate organizational concerns of predictability and control and getting things right. Existing managerial theories may be inadequate in guiding management's responses to bioterrorism. New and emerging theories of chaos and complexity offer promise for coping with the unknown, but at this time such theories are still evolving. Stakeholder management theory may be the most useful at this juncture because it enables administrators to manage external relationships while sustaining equilibrium among stakeholders and the organization.

Are we better prepared today than we were in 2001 to manage a bioterrorist event? The answer from a variety of sources is also *yes*. A recent survey of hospital CEOs' opinions about bioterrorism indicates improvement (Research Notes, 2003). Figure 3 represents that 84% of CEOs surveyed by ACHE in 2003 believed that their hospitals have worked more closely with public agencies on this issue and 70% of CEOs now consider their hospitals to be a *safer* place. However, only a third of these CEOs or less feel that the media has inaccurately reported their bioterrorism preparedness. Fewer still, 19%, reported that they felt bioterrorism preparedness funding is being distributed fairly. In addition, the CEO respondents overwhelmingly believe that their staff and people in their communities are unwilling to be inoculated against biologic agents.

The AHA believes that hospitals are proactively preparing to meet the challenges of bioterror attacks (Goedert, 2002). Key politicians such as Bill Frist, U.S. Senate Majority Leader, believe that funding specifically targeted to improve hospitals' readiness, including hospital infrastructure improvements, training exercises, and comprehensive regional planning will not only assist in preparing to meet future attacks but will also decrease the likelihood of those attacks (Frist, 2002). In 2002, $135 million was targeted for hospitals specifically and President Bush's 2003 budget boasted an investment of an additional $591 million to assist in hospital readiness (Frist, 2002). However, these figures fall far short of AHA estimates that call for $11.3 billion for hospital preparedness (Goedert, 2002). Reconciling payment for bioterrorism preparedness clearly will be a major challenge in the coming years.

Dr. Georges Benjamin, secretary of Maryland's Department of Health and Mental Hygiene, believes that the basic infrastructure such as communications and tracking systems, clearly are better than they were a year ago. The medical community will be the first responder to the event because the first cohort of people who get sick will go to a local hospital or clinic or physicians office. A bioterrorism event will quickly expand outside of the medical community and create the need for an organized incident command structure. This structure must necessarily integrate a wide variety of key stakeholders who have the ability to communicate across local, state, regional, and federal boundaries (Interview with Dr. Benjamin, 2002). The present paper has outlined how this might be achieved in some detail.

In this paper, we have used stakeholder analysis to both identify key stake-holders and also to develop managerial prescriptions according to stage of

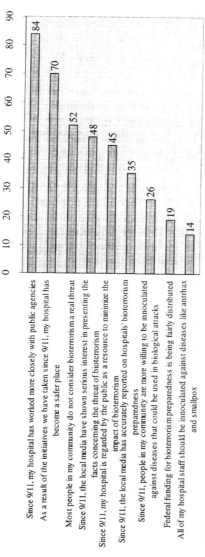

Fig. 3. CEO Opinions About Bioterrorism ($n = 295$ U.S. hospital CEOs). *Source:* Adapted from "Bioterrorism Preparedness: Mechanisms and Measures." *Healthcare Executive* (May/June 2003), 64.

severity of the bioterror event. In so doing, we have identified many of the areas where communication, coordination, and training will be needed. Our stakeholder framework has also functioned as an umbrella of sorts for the myriad of stakeholders with whom managers must attempt to integrate response efforts while maintaining organizational equilibrium. For example, under this umbrella will be the cohort of hospitals in the community, local law enforcement and emergency planners, local public health organizations, state health departments, and a variety of federal agencies who are charged with planning and preparedness activities.

Communication among key stakeholders will be challenging, especially because of the need to talk with one another in real time. Although much communication can be enhanced through the use of Internet technology, protocol issues still remain to be resolved. Web sites have given organizations the ability to communicate accurate information and to alert and advise health providers quickly. The aftermath of the most recent anthrax incidents has been the establishment of effective communication linkages among a variety of community, state and or regional health providers, as well as federal agencies (Interviews with Drs. Richardson and Benjamin, 2002).

Yet clearly much is to be done if managers are to be prepared to meet the challenges of managing future terrorist events. Management research should build upon the work of investigators such Meyer (1982) and Meyer, Brooks and Goes (1990) who have examined organizational responses to discontinuous change and what has been identified as "environmental jolts." Their findings suggest that discontinuous change presents opportunities to redefine structural niches and environmental space for new strategies and structures. In this research volume, Friedman and Marghella (2004 this volume) contribute to our knowledge and understanding in this area and subsequently advance theoretical applications of the work done on environmental jolts. In addition, the important work of McDaniel (2004 this volume), also included in this research volume, continues to enhance our knowledge and understanding of complexity and chaos theories while suggesting methods for application for responding to bioterrorism.

Ultimately, however, the academic community has had little experience applying management theory to bioterrorism events. We therefore encourage researchers to further investigate to determine whether it will be necessary to develop new and unique theoretical constructs to augment our understanding of how organizations and their members respond to major bioterrorism events. These new constructs could serve as the foundation for future empirical studies in this area. Ultimately, we must know all that we possibly can if we are to successfully manage and respond to bioterror attacks. Possible research topics relating management theory to bioterrorism events might include the following:

- How do multidisciplinary teams function under the stress of bioterrorism events or other significant environmental jolts?
- How effective are emergency plans in dealing with unexpected catastrophic events?
- What skills do health care executives need to manage information flows across multiple organizational boundaries?
- How do managers coordinate the allocation and deployment of resources across organizational boundaries? What problems arise in allocating and deploying these resources?
- Who has authority to make and execute decisions when organizational boundaries are fluid?
- What are the skills, knowledge, and abilities required of effective leaders in responding to bioterrorism events?

REFERENCES

Associated Press (2002, November 2). Only Florida ready for bioterrorist attack. *St. Petersburg Times*, A1 and A13.

Begun, J. W. (1994). Chaos and complexity: Frontiers of organization science. *Journal of Management Inquiry*, *3*, 329–335.

Blair, J. D., & Fottler, M. D. (1990). *Challenges in health care management: Strategic perspectives for managing key stakeholders*. San Francisco: Jossey-Bass.

Blair, J. D., & Fottler, M. D. (1998). Effective stakeholder management. In: W. J. Duncan, P. M. Ginter & L. E. Swayne (Eds), *Handbook of Health Care Management* (pp. 19–54). Malden, MA: Blackwell.

Buck, G. (1998). *Preparing for terrorism: An emergency services guide*. Albany, NY: Delmar.

Buck G. (2001). *Preparing for biological terrorism: An emergency services guide*. Albany, NY: Delmar.

Chin, J. (Ed.) (2000). *Control of communicable diseases manual* (17th ed.). Washington, DC: American Public Health Association.

Centers for Disease Control and Prevention (2001, October 19). Update: Investigation of Anthrax Associated with Intentional Exposure and Interim Public Health Guidelines. *MMWR, Morbidity and Mortality Weekly Report*, *50*(41).

Centers for Disease Control and Prevention (2001, November 9). Update: Investigation of Bioterrorism-Related Anthrax and Adverse Events from Antimicrobial Prophylaxis. *MMWR, Morbidity and Mortality Weekly Report*, *50*(44).

Centers for Disease Control (2001, September). Facts about Anthrax, Botulism, Pneumonic Plague and Small Pox [Web Page] http://www.bt.cdc.gov/DocumentsApp/FactsAbout/FactsAbout.asp.

Centers for Disease Control (2001, June 22). Recommendations of the advisory committee on immunization practices (ACIP), *50*(RR10), 1–25. Vaccinia (Smallpox) Vaccine.

Crenson, M. (2002, November 4). Sneezes could trigger alert: Routine symptoms bioterror warning? *The Tampa Tribune*, A3.

Drabek, T. E., & Hoetmer, G. J. (Eds) (1989). *Emergency management: Principles and practice for local government*. Washington, DC: International City Management Association.

Federal Emergency Management Agency Emergency Management Institute (No Date of Publication). Federal Emergency Management Program Manager, Independent Study Course. Washington, DC: Author.

Friedman, L., & Marghella, P. (2004). Environmental jolt of likely bioterrorism. In: J. Blair, M. Fottler & A. Zapanta (Eds), *Bioterrorism Preparedness, Attack and Response, Volume 4, Advances in Health Care Management*. London: JAI/Elsevier.

Frist, B. (2002, December 23–30). A time for preparedness: With funding on its way, hospitals need to get ready for bioterror attacks. *Modern Healthcare*, 19.

Goedert, J. (2002). Hospitals make progress with disaster preparedness. *Health Data Management*, *10*(10), 14 and 24.

Interview with Dr. Georges Benjamin on December 5, 2002, retrieved on line from http://www.pbs.org/newshour/bb/health-july-dec02/benjamin_bioterror.html.

Interview with Dr. Michael Richardson on December 5, 2002, retrieved on line from http://www.pbs.org/newshour/bb/health-july-dec02/richardson_bioterror.html.

Johns Hopkins University, Center for Civilian Biodefense Studies. Botulinum Toxin (2000). [Web Page]: http://www.hopkins-biodefense.org/pages/agents/agentbotox.html.

McDaniel, R. (2004). Chaos and complexity in a bioterrorism future. In: J. Blair, M. Fottler & A. Zapanta (Eds), *Bioterrorism Preparedness, Attack and Response, Volume 4, Advances in Health Care Management*. London: JAI/Elsevier.

McDaniel, R. R., Jr. (1997). Strategic leadership: A view from quantum and chaos theories. *Health Care Management Review*, *22*(1), 21–37.

McDaniel, R. R., Jr., & Driebe, D. J. (2001). Complexity science and health care management. In: M. D. Fottler, G. T. Savage & J. D. Blair (Eds), *Advances in Health Care Management* (Vol. 2, pp. 11–36). Amsterdam, Holland: JAI/Elsevier.

Meyer, A. D. (1982). Adapting to environmental jolts. *Administrative Science Quarterly*, *27*, 515–537.

Meyer, A. D., Brooks, G. R., & Goes, J. B. (1990). Environmental jolts and industry revolutions: Organizational responses to discontinuous change. *Strategic Management Journal*, *11*, 93–111.

Research Notes (2003, May/June). Executive news: Bioterrorism preparedness. *Healthcare Executive*, 64.

BIOTERRORISM PREPAREDNESS AND RESPONSE: A RESOURCE GUIDE FOR HEALTH CARE MANAGERS

Cynthia A. Holubik and Steven R. Tomlinson

ABSTRACT

The threat of future biological attacks within the United States forces the additional responsibility of preparedness and response onto health care managers. An endless amount of information on this topic can be readily obtained from a variety of sources. The purpose of this resource guide is to provide health care managers with a well-organized, up-to-date listing of credible sources that can be accessed electronically. This resource guide is designed to facilitate the retrieval of relevant information that is crucial in the process of health care managers designing a bioterrorism preparedness and response plan.

The implications of a bioterrorist attack within the United States were well observed in the 2001 anthrax attacks. The effects of these attacks were not only felt by the primary victims but also reached into communities and health care organizations that treated patients. The concerns for future attacks have spurred an enormous amount of interest and forced health care administrators to implement safety measures and develop plans encompassing preparedness and response. With respect to this topic, a major problem facing health care managers is the ability to obtain relevant, up-to-date, reliable information.

Many health care managers are aware of the endless amount of resources that are available pertaining to preparedness and response planning. It

Bioterrorism, Preparedness, Attack and Response
Advances in Health Care Management, Volume 4, 287–308
ISSN: 1474-8231/doi:10.1016/S1474-8231(04)04011-X

should also be noted that the information concerning bioterrorism rapidly changes and the available information can be overwhelming. To that end, the authors have compiled an array of selected websites that provide excellent, up-to-date, and unique resources in attempts to facilitate the gathering of knowledge and make that experience more efficient.

The objective of the following resource guide is to provide a comprehensive list of topics in a "user-friendly" format that can be obtained electronically. To facilitate information retrieval, electronic resources have been categorized according to topic. For example, up-to-date information concerning the multitude of biological agents can be obtained by using the links contained within the biological agents category. Under this topic heading you will find a listing of important resources. Each listing contains a heading that identifies the sponsoring organization, a short explanation of information and materials that can be expected, and a web address identifying its location.

The authors' compilation was not intended to be exhaustive nor do the authors necessarily endorse or warrant the reliability of the information/ products. Websites may change regularly; please visit www.bioterrorism@ba. ttu.edu to obtain an electronic version of this resource guide and for updates to web addresses.

BIOTERRORISM PREPAREDNESS AND RESPONSE CATEGORIES

General Preparedness
Biological Agents
Checklists/Templates
Communication (network & interoperability)
Community Resources
Containment (isolation & quarantine)
Credentialing
Crime Scene Management
Decontamination & Personal Protective Equipment (PPE)
Detection/Surveillance
Incident Command System (ICS)
Infrastructure (security: cyber, physical/environmental)
Mass Casualty Incident (MCI)
Mass Inoculation
Medical Materiel (Supplies & Equipment)
Psychological Issues
Public Policy & Law

Terrorism Threat/Vulnerability/Scenarios
Training

GENERAL PREPAREDNESS

9–1–1 Reference Center
Information and products for responding to public health, natural disasters, and
   household emergencies. Free downloads for Emergency Response with chapter
   summaries from the New York Times Best Seller, *The Survival Guide.*
www.9–1–1center.com/

Agency for Healthcare Research and Quality (AHRQ)
"Assessing Hospital and Health System Preparedness and Response" >
   teleconferencing and PowerPoint resource
www.ahcpr.gov/news/ulp/bioteleconf/

Agency for Health Care Policy and Research
Hospital assessment questionnaire
www.ahcpr.gov/about/cpct/bioterrtxt.htm

American Hospital Association
Chemical and Bioterrorism Preparedness Checklist Tool to describe/assess
   present state of preparedness.
www.aha.org

Hospital Connect
Resources – Hospital Readiness, Response, and Recovery Resources. Excellent
   array of preparedness and other operational resource web-links with content
   descriptions.
www.hospitalconnect.com

American Medical Association
JAMA and Other Journals. Extensive list of journal articles pertaining to
   biological agents, and hospital preparedness. www.ama-assn.org

AORN
Bioterrorism Resources.
Bongiovanni & Hatfill. Answering the Chemical and Biological Warfare Threat,
   Surgical Services Management, 5:9, 31–36. Full text. Presents The Life
   Support for Trauma and Transport system for hospitals and other medical
   facilities. Other full text articles and web-links to other resources.
www.ssmonline.org/Links/bioterrorism.asp

Center for Disease Control
Updates on bioterrorism and emergency preparedness
www.bt.cdc.gov/

Department of Health and Human Services
Updates on bioterrorism and the nation's emergency preparedness plans and
    web-links to other worthy bioterrorism and emergency preparedness sites.
    www.hhs.gov/hottopics/healing/biological.html

Department of Homeland Security
www.homelandsecurity.org/

Emergency Response and Research Institute
Disaster Operations Archive. EMS focus.
www.emergency.com/disaster.htm

Infectious Diseases Society of America (IDSA)
Bioterrorism Information and Resources.
www.idsociety.org/bt/toc.htm

Johns Hopkins University Center for Biodefense Strategies
The Center draws upon the expertise of a multidisciplinary staff to provide
    resources for preparedness and response. An excellent website.
    www.hopskins-biodefense.org

Medscape, Inc.
A variety of resources including news, clinical articles, consensus statements,
    and conference summaries.
    www.medscape.com

National Academies
Responding First to Bioterrorism. Searchable web-shelf collection of
    expert-selected resources.
    www.nap.edu/shelves/first/sites.html

National Association of City and County Health Officials (NACCHO)
Elements of Effective Bioterrorism Preparedness: A Planning Primer for Local
    Public Health Agencies, January 2001. Though aimed primarily at PH officers,
    hospital administrators and/or first responders may find it useful.
    www.naccho.org/files/documents/Final_Effective_Bioterrorism.pdf
      List of web-links for articles, organizations, mental health, and other resources.
      www.naccho.org/UsefulLinks.cfm

National Environmental Health Association
www.newscientist.com

National Governors' Association (NGA)
www.nga.org

National Homeland Security Knowledgebase
This virtual library links to 100 websites and documents.
www.twotigersonline.com/resources.html#biological-warfare-2

Managed Healthcare Executive
www.ManagedHealthcareExec.com

MGMA
Response plan template for administrators
http://bioprn.advancepcsmdnet.com/admin.html

Office of Emergency Preparedness
Metropolitan Medical Response System (MMRS) Field Operations Guide. Book
    contains position descriptions and duties for each of the major sectors in a
    WMD response: field medical operations, HAZMAT, hospital operations,
    medical information, law enforcement, logistics and equipment.
http://ndms.dhhs.gov/CT_Program/Response_Planning/response_planning.html

Pam Pohly's Net Guide
Toolbox for Health Managers and Administrators. Provides thirteen pages of
    articles, info, and hyperlinks to other websites.
www.pohly.com/admin.shtml

Risk Management Internet Services Library
Disasters – Health Services. Numerous web-links and quick access current
information.
www.rmis.com

*St. Louis University School of Public Health – Center for the Study of
    Bioterrorism*
www.slu.edu/college/sph/csbei/bioterrorism/quick.htm

St. Petersburg College, National Terrorism Preparedness Institute
Web-cast program series for first responders and administrators. Topics include
decontamination protocols and incident recovery, with archives and new
    programs regularly.
http://terrorism.spjc.edu/

BIOLOGICAL AGENTS

*American College of Physicians/American Society of Internal Medicine
(ACP-ASIM)*
Bioterrorism Resources. Provides web-links for general information on
bioagents,
Including biotoxins, nerve agents, and toxic gases.
www.acponline.org/bioterro/index.html?hp

Association for Professionals in Infection Control and Epidemiology, Inc (APIC)
Exhaustive resource – click "Bioterrorism Resources." Updated regularly.
www.apic.org/

Center for Disease Control
www.bt.cdc.gov

Chemical and Biological Arms Control Institute (CBACI)
www.cbaci.org

Center for Infectious Disease Research and Policy (CIDRAP)
Medical Fact Sheets on Diseases of Bioterrorism. Team mines and disseminates
comprehensive, accurate, and up-to-the-minute biomedical knowledge.
Assesses current/anticipated problems to develop state-of-the-art policy
recommendations, program strategies, and/or clinical practice guidelines.
www1.umn.edu/cidrap/umn.edu

Jane's
Internationally renown British website for defense industry. Click "Products" to
order Bio-Chem Defense Handbook
www.janes.com

Infectious Diseases Society of America (IDSA)
Resource for wide array of bioagents.
www.idsociety.org

Johns Hopkins University – Center for Civilian Biodefense Studies
Biological Agent Information.
www.hopkins-biodefense.org/pages/agents/agent.html

Journal of American Medical Association (JAMA)
Anthrax as a Biological Weapon: Medical and Public Health Management
http://jama.ama-assn.org/issues/v281n18/ffull/jst80027.htm

Smallpox as a Biological Weapon: Medical and Public Health Management
http://jama.ama-assn.org/issues/v281n22/ffull/jst90000.html

Plague as a Biological Weapon: Medical and Public Health Management
http://jama.ama- assn.org/issues/v283nl7/ffull/jst90013.html

Botulinum Toxin as a Biological Weapon: Medical and Public Health Management
http://jama.ama-assn.org/issues/v285n8/ffull/jst00017.html

Tularemia as a Biological Weapon: Medical and Public Health Management
http://jama.ama-assn.org/issues/v285n21/ffull/jst10001.html

McGraw-Hill
Well-known publisher provides complimentary access as a public service to
  relevant
bioagent chapters in printable format.
www.accessmedicine.com/amed/public/amed_news/news_article/281.html

Medline
Chemical and biological weapons section.
www.nlm.nih.gov

Rapid Response Info System (RRIS)
Comprised of several databases. Searchable by NBC agent name: characteristics,
  signs and symptoms of exposure, PPE & decontamination (decon), first aid,
  detection equipment.
www.rris.fema.gov/

St. Louis University School of Public Health – Center for the Study of Bioterrorism
Quick References for disease/agent specific info, Personal Digital Assistants
  (PDA)
Applications, algorithms, and tables.
www.slu.edu/colleges/sph/csbei/bioterrorism/quick.htm

U.S. Army Medical Research Institute of Infectious Diseases (USMIIRD)
www.usamriid.army.mil/eudcation/bluebook.html

U.S. Army Medical Research Institute of Chemical Defense
Downloadable documents for PDAs include Medical NBC Battle Book, Medical
  Management
of Chemical Casualties, Field Management of Chemical Casualties, and
  Textbook of Military

Medicine: Medical Aspects of Chemical and Biological Warfare
http://ccc.apgea.army.mil/products/pda_docs/pda_downloads.htm

U.S. Army Surgeon General's site on NBC Defense
Field Manual (FM) 4.02.7 Health Service Support in a NBC Environment –
  Tactics,
Techniques, and Procedures, October 2002, Headquarters, Department of the
  Army
www.nbc-med.org

CHECKLISTS AND TEMPLATES

Association for Professionals in Infection Control and Epidemiology, Inc. (APIC)
Bioterrorism Readiness Plan – A Template for Healthcare Facilities
www.apic.org/edu/readinow.cfm

Kaiser Permanente
Bioterrorism Preparedness and Response Plan. Contact CME coordinator.
www.kp.org

Risk Management Internet Services Library
Click for quick access to samples, forms, and manuals.
www.rmis.com

COMMUNICATION:

Interoperability

Agency for Healthcare Research and Quality (AHRQ)
User Liaison Program (ULP). January 2002 audio teleconference shares findings
  to assess and strengthen capacity of healthcare systems.
www.ahrq.gov

National Institute of Justice AGILE Program– Task Force on Interoperability
Why Can't We Talk? A Guide for Public Officials, Feb 2003. (104 pages).
Raises interoperability issues and problems. Presents short/long-term strategies
for wireless telecommunications and IT applications.
www.agileprogram.org/ntfi/ntfi_guide.pdf

Network

Center for Disease Control (CDC)
Health Alert Network (HAN)
www.phppo.cdc.gov/han/

Kaiser Permanente of California.
Contingency Plans for Communicating with Local, State, and National
  Authorities.
www.kp.org

COMMUNITY RESOURCES (VOLUNTEERS)

Community Emergency Response Teams (CERT)
www.citizencorps.gov/cert.html

FEMA
Community Emergency Response training materials.
www.fema.gov/emi/cert/mtrls.htm

Public Entity Risk Institute
Community Response to the Threat of Terrorism: Issues and Ideas Symposium
  (65 pages). Presented November 2001.
www.riskinstitute.org/ptrdocs/CommunityResponse-Terrorism.pdf

U.S. Department of Defense (DOD)
Homeland Defense: Neighborhood Emergency Help Center Pamphlet: A Mass
  Casualty Care Strategy for Biological Terrorism Incidents, May 2001. Lays out
  a strategy localities might use. Published by the U.S. Army Soldier Biological
  Chemical Command (SBCCOM).
www2.sbccom.army.mil/hld/bwirp/bwirp_nehc_green_book_download.htm

Local Emergency Planning Committees (LEPCs)

Environmental Protection Agency
LEPCs and Deliberate Releases: Addressing Terrorist Activities in Chemical
  Emergency Preparedness and Prevention Office.
www.epa.gov/ceppo/factsheets/lepcct.pdf

LEPC Database
www.epa.gov/ceppo/lepclist.htm

CONTAINMENT

Joseph Barbera, et al.
Large-*Scale Quarantine Following Biological Terrorism in the United States:*
  Scientific Examination, Logistic and Legal Limits, and Possible Consequences,
  December 5, 2001 (abstract)
http://jama.ama-assn.org

The Brookings Institute
Toward a Containment Strategy for Smallpox Bioterror: An Individual-Based
  Computational Approach. Epidemic movie with model of smallpox epidemics
  in a two-town county helps viewers to develop strategies.
www.brookings.edu/dybdocroot/dynamics/models/bioterrorism.htm

CREDENTIALING

Center for Disease Control (CDC)
www.phppo.cdc.gov/Docs/BTNatTrainPlanExecSum0202.pdf

Healthcare Marketplace
Online store for healthcare management professionals. Free e-zines.
www.hcmarketplace.com

Joint Commission on Accreditation of Healthcare Organizations (JCAHO)
New Emergency Privileging Hospital Standard during emergencies and
  emergency management planning in healthcare.
www.jcaho.org

CRIME SCENE MANAGEMENT

Community Research Associates
Provides training courses.
www.community-research.com/services.html

Department of Defense (DOD)
Outbreak and Investigation Toolbox.
http://141.2236.12.246/outbinv.asp

DECONTAMINATION AND EQUIPMENT

3M
Detailed info regarding their products and safe practices for PPE. Links to training
resources.
www.3M.com/occsafety/html/respirator_notice.html

International Safety Equipment Association
Lists PPE manufacturers and related websites. Includes new CDC
recommendations.
www.safetyequipment.org

National Institute of Justice
Guide for the selection of chemical and biological decontamination equipment
for emergency first responders, October 2001
www.ojp.usdoj.gov/nij/pubs-sum/189724.htm

Rapid Response Info System (RRIS)
Comprised of several databases. Searchable by NBC agent name.
www.rris.fema.gov/

U.S. Army Medical Research Institute of Chemical Defense
Appendices have lists of agents, equipment to detect them, information on
decontamination, patient management, PPE equipment list, and recipes for
patient decon solutions.
http://ccc.apgea.army.mil/

U.S. Department of Justice
Equipment guides for chemical and biological contaminant situations.
www.ojp.usdoj.gov/terrorism/whats_new.htm

*Guide for the Selection of Chemical and Biological Decontamination Equipment
for Emergency First Responders*
Part of a series of guides distributed by NIST, the Office of Law Enforcement
Standards (OLES), and the National Institute of Justice.
www.ncjr.org/pdffiles1/nij/189724.pdf

Introduction to Biological Agent Detection Equipment for Emergency First
  Responders
NIJ working draft.
www.ojp.usdoj.gov/nij/DRAFTGuide101.pdf

DETECTION

Center for Disease Control
Health Alert Network
www.phppo.cdc.gov/han/

National Institute of Justice
Domestic Preparedness Equipment Technical Assistance Program (DPETAP).
  Fact sheet on a teaching program to select, use, maintain chemical and
  biological attack equipment such as detectors and suits.
www.ojp.usdoj.gov/odp/library/../docs/dpetap.htm

Guide to Selection of Detection Equipment
Provides info about detecting chemical agents and toxic industrial materials and
  selecting equipment for different applications.
www.ojp.usdoj.gov/nij/pubs-sum/184449.htm

U.S. Army Medical Research Institute of Chemical Defense
Appendices have lists of agents, equipment to detect them.
http://ccc.apgea/army.mil/reference_documents/reference.asp

INCIDENT COMMAND SYSTEM (ICS)

Community Research Associates
Provides comprehensive assessment and training for emergency operations plans
  (EOP): strategic planning, coordination/integration of response resources,
  decision-making process, event logistics, and solutions.
www.community-research.us.services.html

FEMA – IS-195 Basic Incident Command System – Independent Study
A FEMA independent-study course in the basics of the ICS. Includes course materials and exam.
www.fema.gov/emi/is195.htm

Greater New York Hospital Association
Hospital Emergency Incident Command System (HEICS) modules for
community and providers. Comprehensive guides, briefings, Power Point
presentations, templates/policies, articles, and guidelines include topics:
lessons learned, communicating with public/media, security, emergency
contact information activation.
www.gnyha.org/eprc/general/ics

Hospital Incident Command System (HEICS III)
Latest downloadable version, originally funded by LA EMS Authority.
www.heics.com

National Wildlife Coordinating Fire-Fighters Group
National Training Curriculum: basic, advanced, command & general staff, execs,
and interagency modules developed by those who pioneered this general
concept. Power Points included. Revision in 2003.
www.nwcg.gov/pms/forms/ics_cours/ics_courses.htm

INFRASTRUCTURE

The Terrorism Research Center, Inc. (TRI)
Click "services." Analysis of vulnerabilities for critical protection of cyber and
physical systems. www.terrorism.com

Cyber Security

Center for Strategic and International Studies (CSIS)
Assessing the Risks of Cyber Terrorism, Cyber War, and Other Cyber Threats,
December 2002.
www.csis.org/tech/0211_lewis.pdf

EmergencyNet (ERRI)
"Hot" website with 24 hr. coverage of news, late-breaking info, analysis. Click
on v-watch section, computer virus section, and hacker reports.
www.emergency.com

Physical Security

Department of Energy – Lawrence Berkeley National Labs
Building Protection Guidance and Resources (WMD events). Pre-, current, and
post-event advice for biological/chemical attack.
http://securebuildings/bl.gov/

National Institute for Occupational Safety and Health (NIOSH)
Guidance for Protecting Building Environments from Airborne Chemical,
   Biological, or Radiological Attacks, May, 2002. Specific recommendations for
   physical security, ventilation and filtration, maintenance, administration, and
   training, plus things not to do.
www.cdc.gov/niosh/bldvent/2002–139.html

OSHA
How to Prepare for Workplace Emergencies
www.osha-sic.gov/Publications/Osha3088.pdf

Protecting the workplace against anthrax.
www.osha-slc.gov/SLTC/journals/bioterrorism_biblio.html
Anthrax Risk Reduction Matrix > basic advice/suggestions for protective
   measures
www.osha.gov/bioterrorism/anthrax/matrix/index.html

U.S. Army Soldier and Biological Chemical Command (SBCCOM)
Basic Information on Building Protection. Basic strategies for airborne hazards,
   positive pressure, air filtration, pressurization and levels of protection, filter
   unit sizing, detection-based approaches, internal filtration (recirculation units),
   operational measures, physical security measures, and airlocks.
http://buildingprotection.sbccom.army.mil/basic/index.htm

MASS CASUALTY INCIDENT (MCI)

9–1–1 Reference Center
Mass Transportation Preparedness
www.9–1-center.com/

American Hospital Association (AHA)
Hospital Preparedness for Mass Casualties. Final report of the invitational forum
   held by AHA with support from OEP. Includes recommendations on
   community-wide preparedness, staffing, training, support, internal/external
   communications, and public policy.
www.aha.org

Association for Professionals in Infection Control and Epidemiology (APIC)
Mass Casualty Disaster Plan Checklist: A template for healthcare facilities
   designed to provide facilities with questions that stimulate assessment/dialogue
   with key stakeholders.
www.apic.org/edu/readinow.cfm

U.S. Army Soldier Biological and Chemical Command (SBCCOM)
Neighborhood Emergency Help Center Pamphlet: A Mass Casualty Care Strategy
for Biological Terrorism Incidents (22 Pages), May 2001.
www2.sbccom.army.mil/hld/downloads/bwirp/nehc_green_book.pdf

Triage

American College of Emergency Physicians
Comparative Analysis of a Multiple-Casualty Incident Triage (algorithms)
www.acep.org/1,4858,0.html

EMedProfessional
Comprehensive resources for emergency medical professionals menu covers
hundreds of resource categories, A-Z.
www.emedprofessional.com/index.cfm?task=overview

Jane's
Renown British resource in defense industry. Click "Products" to order Mass
Casualty Handbook, Hospital-Emergency Preparedness and Response, and/or
Facility/Workplace Security
www.janes.com

MASS INOCULATION

Anthrax Vaccine Immunization Program (AVIP)
Forms, resources, educational toolkit for commanders, clinicians, and individuals.
www.anthrax.mil

Associated Press
Article: Plan Helps Guide States Should Mass Smallpox Vaccinations Be Needed
Describes federal plan for states to administer mass smallpox vaccinations.
www.nlm.nih.gov/medlineplus/news/fullstory_9527.html

Center for Disease Control (CDC)
Smallpox Vaccination Training for Health Professionals Adverse Events Training
www.bt.cdc.gov/agent/smallpox/index.asp

McGraw Hill
Complimentary access as a public service to relevant chapter contents about
vaccines, updates, response plans, and guidelines. Printable format. Related
resources and news stories.
www.accessmedicine.com/amed/public/amed_news/news_article/281.html

MEDICAL MATERIEL

Mutual Aid Agreements

Joint Commission of Accreditation of Healthcare Organizations (JCAHO)
www.jcaho.org

Strategic Pharmaceutical Stockpile

Center for Disease Control (CDC)
www.bt.cdc.gov/

Inventory Lists

Medical-Surgical Supply Formulary by Disaster Scenario (March 2003)
http://66.77.118.213/ahrmm/files/disaster_formularies.pdf
can be accessed using www.hospitalconnect.com

U.S. Army Medical Research Institute of Chemical Defense
Medical supply list and calculation tool.
http://ccc.apgea/army.mil/products/medsupplies/Med_Sup_Data.htm
http://ccc.apgea.army.mil/products/medsupplies/select.asp

PERSONAL PROTECTIVE EQUIPMENT (PPE)

AORN
Sawicki, Jack. Protection from Chemical and Biological Warfare, *Surgical
    Services Management*, 5:9, Sept 1999, 11–17. Full text, outlines choices
    available and offers suggestions to ensure that healthcare facilities have
    adequate protection.
www.ssmonline.org/SSMOnlineMedia/Documents/Sawicki.pdf

*DOJ Fact Sheet: Office of Domestic Preparedness Defense Supply Center
    Philadelphia.*
First Responder Equipment Purchase Program.
Program description and contact info helps buy equipment
www.ojp.usdoj.gov/odp/docs/fs-padef.htm

International Safety Equipment Association
Lists PPE manufacturers on related websites. Includes new CDC
  recommendations for protecting Workers.
www.safetyequipment.org

MSA & other vendors
National Institute of Justice Working draft of a PPE guide.
www.ojp.usddoj.gov/nij/DRAFTGuide102VI.pdf

Rapid Response Info System (RRIS)
Several databases, searchable by NBC name.
www.rris.fema.gov/

SBCCOM – Army Biological and Chemical Warfare Command
Extensive list of reports on equipment for WMD.
www2.sbccom.army.mil/

PSYCHOLOGICAL ISSUES

*American College of Physicians/American Society of Internal Medicine
  (ACP-ASIM)*
Bioterrorism Resources. Website links for psychological issues and interventions.
www.acponline.org/bioterro/index.html?hp

American Psychiatric Association
Disaster Psychiatry. Web-site specific to psychological components of
  bioterrorism.
www.psych.org/pract_of_psych/disaster_psych.cfm

When Disaster Strikes. A 22-page report with sections on coping with tragedy,
  national disaster, disaster psychiatry, history, barriers, and stress management.
www.psych.org

Kaiser Permanente
Dealing With Stress and Uncertainty. Special articles.
www.kp.org/newsroom/articles.html

National Institute of Mental Health
Information About Coping With Traumatic Events.
www.himh.nih.gov/outline/responseterrorism.cfm

National Partnership for Workplace Mental Health
Disaster, Terror and Trauma in the Workplace: What Did we Know Before 9/11
    and What We Learned Since Then, April 25, 2002. Conference proceedings.
    Click "Lessons Learned" hyperlink.
www.workplacementalhealth.org

U.S. Department of Health and Human Services
www.mentalhealth.org/publications/allpubs/ADM90–537/default.asp

Post-Traumatic Stress Syndrome

American Medical Association
Index of Bioterrorism Resources. AMA link for Post-Traumatic Stress Syndrome.
www.ama-assn.org

Department of Veterans Affairs
Disaster Mental Health Services: A Guidebook for Clinicians and
    Administrators. (The National Center for Post-Traumatic Stress Disorder
    Education and Clinical Laboratory)
www.ncptsd.org/disaster.html

National Center for Post-Traumatic Stress Disorder
Managing Aftereffects of Terrorism.
www.ncptsd.org/terrorism/index.html

U.S. Department of Health and Human Services
Mental Health and Traumatic Events. Includes PTSD, war and terrorism, and
    anxiety.
www.hhs.gov/mentalhealth

PUBLIC POLICY & LAW

*Center for Law and the Public's Health at Georgetown & Johns Hopkins
    Universities*
CDC Collaborating Center Promoting Health Through Law. A resource on
    public health law/ethics/policy for PH practitioners/lawyers/policy-makers,
    and others.
www.publichealthlaw.net

Kaiser Permanente
Healthcast: The Public's Health and the Law in the 21st Century, June 2002.
www.kaisernetwork.org/healthcase/phl/jun02

National Governor's Association Center for Best Practices
Emergency Management and Terrorism Page contains many links to bioterrorism
  information, especially regarding policy formulation and program support.
http://www.ng.org/center/topics/1,1188,C_CENTER_ISSUE^D_854,00.html

National Security Institute
Security Resource Net – Counter-Terrorism.
Terrorism Legislation and Executive orders, Commentaries, Precautions, and
  related sites.
http://nsi.org/terrorism.html

Risk Management Internet Services (RMIS)
Quick access to case law, laws, sample policies, and procedures.
www.rmis.com

*Center for Biodefense, Law and Public Policy, Texas Tech Universtiy School of
  Law*
http://www.ttu.edu/biodefense/

SCENARIOS

Association for Professionals in Infection Control and Epidemiology, Inc.(APIC)
Presentation of real and potential cases – click "Bioterrorism Resources"
www.apic.org

National Defense Magazine
Article/editorial from 1999: Anti-Bio Terrorism Training Needs Realistic
  Simulations. Broad overview of training simulations.
www.potomacinstitute.com/press/NationalDefense.htm

TERRORISM THREAT/VULNERABILITY

Association for Professional in Infection Control and Epidemiology, Inc. (APIC)
Topical discussions of emerging and asymmetrical threats, domestic vulnerability,
  and historical trends. Click "Bioterrorism Resources."
www.apic.org

Center for Infectious Disease Research and Policy (CIDRAP)
Background paper on bioterrorism: history, likely agents, perpetrators, and
   dissemination.
www1.umn.edu/cidrap/umn.edu

Community Research Associates
Provides vulnerability surveys, strategic planning, and scenario development.
www.community-research.com/services.html

Emergency Response and Research Institute (ERRI)
EmergencyNet – Counter-Terrorism Operations Page. Foreign and domestic
   terrorism reports and analysis.
www.emergency.com

Health Sciences and Human Services Library
Terrorism Resources for the Healthcare Community.
www.hshs.umaryland.edu/resources/terrorism.html

National Governor's Association
Homeland Security and Emergency Management
www.nga.org

RAND
Overview – Terrorism. Research group presents news, hot topics, publications,
   and projects/programs.
www.rand.org/terrorism_area/

The Terrorism Research Center, Inc. (TRI)
Threat Management and Analysis Tool. Check "services." Independent institute
   dedicated to research, information warfare and security, critical infrastructure
   protection, homeland security, and other issues.
www.terrorism.com

WMD Consulting
Terrorism Vulnerability Self-Assessment. Worksheet can be customized to
   organizations.
www.wmdconsulting.us/pagetwo.htm

TRAINING

Agency for Healthcare Research and Quality
"Training of Clinicians for Public Health Events Relevant to Bioterrorism
   Preparedness" – a Johns Hopkins' evident-based report summary.
www.ahcpr.gov/clinic/epcsums.bioterror.pdf

Center for Disease Control (CDC)
Public Health Training Network (PHTN). List of distance learning websites.
www.phppo.ced.gov/phtn/sites.asp

Catalog for wide array of courses and distance learning resources.
www.phppo.ced/gov/PHTN/catalog.asp

*Center for Infectious Disease Research and Policy (CIDRAP) at the University of
   Minnesota*
Provides wealth of information and educational opportunities. Additional
   web-links.
www.cidrap.umn.edu

Community Research Associates
Provides training and technical assistance services for domestic preparedness,
   particularly WMD – for over 50,000 first responders across the U.S.
   Comprehensive selection of workshops, training courses, and exercise design,
   development, delivery, and evaluation.
www.community-research.com/services.html

Federal Emergency Management Agency (FEMA)
Emergency Response to Terrorism Self-Study Manual (1999).
www.usfa.fema.gov/downloads/pdf/publications/ertss.pdf

Medical Group Management Association
Bioterrorism Practical Readiness Network (Bio-PRN)
Free web-based tool for administrators, physicians, and medical office staff.
   Emergency
www.mgma.com

Response Planning template guides administrators through critical steps in
   creating an effective plan.
www.mgma.com

National Association of City and County Health Officials (NACCHO)
Bt Create (tabletop CD). Utilize as is or customize to create scenarios to meet
  your organization's needs. One of many training tools, resources available.
www.naccho.org

*St. Louis University School of Public Health – Center for the Study of Bioterrorism
  Education and Training.*
Website list for resources.
www.slu.edu/colleges/sph/csbei/bioterrorism/education.htm

*Texas Engineering Extension Service, National Emergency Response and Rescue
  Training Center*
Offers many courses for first responders: training and search catalog. Extensive
  programs.
http://teexweb.tamu.edu/nerrtc/

Training Finder
Provides information on over 40 distance learning courses for public health
  professionals.
www.TrainingFinder.org

University of Alabama at Birmingham
Bioterrorism and Emerging Infections Site. Collaborative site provides new CME
  modules.
www.bioterrorism.uab.edu

*U.S. Army Medical Command and the Department of Veterans Affairs'
  Emergency Management Healthcare Group*
Interactive satellite broadcast from November 2002 will educate health
  professionals about proper medical response. Experts from USAMRIID,
  USAMRICD, and other organizations.
www.biomedtraining.org/proginfo.htm

U.S. Soldier Biological Chemical Command (SBCCOM)
Federal training sites.
www2.sbccom.army.mil/hld/about us.htm

WMDFirstResponders
First responders information for training and research.
www.wmdfirstresponders.com/index.htm

PART IV:
DEFENDING THE HOMELAND:
CHANGES AND CHALLENGES

THE ROLE OF THE RESERVE FORCES IN DEFENDING THE HOMELAND

Albert C. Zapanta, Richard O. Wightman Jr. and Mari K. Eder

ABSTRACT

The purpose of this chapter is to describe the role of the Reserve Components of the Armed Forces of the United States during the September 2001 terror attacks and the role they play in the war on terror, both in the United States and in foreign nations. Reserve Component forces will operate with the Department of Homeland Security to provide security within the United States. Reserve Component units possess many unique skills and knowledge of local conditions that will contribute greatly to maintaining homeland security. Reserve Component forces also participate in major military operations overseas.

The 21st century didn't begin with the official countdown clock, the champagne toasts, and streamers on December 31st 1999. Even as one looked bleary-eyed out the windows that next day, January 1st our world remained essentially the same. The United States as a nation focused on Europe and the policy of engagement that had driven our relations with Russia and the former Eastern Bloc since the Berlin Wall crumbled nearly ten years earlier. Our world had changed dramatically then. Foreign policies and national military strategies followed.

The United States recognized the world was even more dangerous without the focus on a Russian superpower. Yet our methods and our military changed little.

Bioterrorism, Preparedness, Attack and Response
Advances in Health Care Management, Volume 4, 311–317
© 2004 Published by Elsevier Ltd.
ISSN: 1474-8231/doi:10.1016/S1474-8231(04)04012-1

We still trained in Western Europe and faced the North in Korea. Our enemies were established, our protocols routine. America's domestic focus was positive. The economy was good; the nation seemed strong, impervious.

Then on a clear September day in 2001, 19 suicidal hijackers changed the paradigm. The U.S. was attacked, at home, and the 21st century finally began in earnest. Everything changed in those first few days. The unthinkable was reality; television images were of our neighbors and not fictional characters; there were tanks and troops in the streets and fears multiplied. For the professionals, from medical doctors to police officers and the military, each recognized quickly how it could have been so much worse. They felt the chill that they may not have had sufficient resources to cope with a worse scenario and, as the feelings of anger set in, began to plan for better information, intelligence, communications, coordination, and support. Then another thought: could biological agents and weapons of mass destruction be next? Next, anthrax arrived in the mail and the chaos of unplanned and uncoordinated response began anew.

Reserve component contributions to the War on Terrorism since Sept 11th have been both immediate and lasting. The mobilizations, deployments to Operations Enduring Freedom, Noble Eagle, and Iraqi Freedom drove issues covered in the Reserve Forces Policy Board's Annual Report to the President and Congress for 2002. The lessons learned will help ensure we revise and create policies that support our Reserve component members, enhance their support to the steady state in the war on terrorism, and provide for continuity of operations in the years ahead.

Below are just a few examples from our seven Reserve components of the commitment and dedication displayed by our men and women in uniform. These are the stories of those soldiers, sailors, airmen and marines who serve part-time. In many cases their service following Sept 11th, began in a volunteer status. They were there then and many of them remain on duty for America today (Reserve Forces Policy Board Annual Report, 2002).

Nearly 10,000 National Guard men and women were already on duty across the country when President Bush approved the call up order September 14, 2002. Among them were members of the Manhattan-based 1st Battalion, 69th Infantry, supporting the massive recovery operation at "ground zero." The attacks hit close to home for many unit members. "We're taking this personal," said SGT Dave Perez while issuing respirators, gloves, goggles, and other supplies to those working in the rubble. It was personal for many members of the National Guard with Guardsmen in every state called upon to boost aviation security the past two years, serving in airports across the country through May 31, 2002.

Major Billy Hutchinson and his wingman were among the first ones there at the Pentagon on Sept 11th, flying above the building's burning west wing. Hutchinson is a member of the DC Air National Guard, recalled that morning from a training mission in North Carolina. When he checked in by radio, he was told to take off immediately. He had no weapons on board, nothing. He and his fellow guard pilots in the air were quickly jointed by F-16s from Langley Air Force Base, all looking for a big airliner, what they would later learn was Flight 93. All three Guard pilots had limited options for bringing down a high jacked aircraft, but were prepared to disable it as quickly as possible, even if that meant ramming the plane to bring it down.

Port Security Unit 305 is based at Ft. Eustis, VA, a unit comprised of about 140 personnel, including 135 reservists. PSU 305 established waterside security in New York City within 48 hours of the attacks on the World Trade Center and the Pentagon. Various missions included escort for the hospital ship USNS Comfort, the conduct of anti-terrorist patrols along Newark Bay and rivers, and providing security for shore-side fuel storage depots and refineries, power plants, and ship loading terminals.

By September 15, the 311th Mortuary Affairs Company, U.S. Army Reserve, was on the ground at the Pentagon, having mobilized, conducted an overseas deployment from home station in Puerto Rico. Many of these soldiers were in their teens or early twenties; some were just joining the unit following Basic and Advanced Individual Training. They performed this most difficult and painful of missions with absolute professionalism and with respect and concern for the dignity of those individuals who perished in those attacks.

After 10 months in support of operation enduring freedom, the 40th Expeditionary Bomb Squadron completed its 1,000th B-52 sortie on July 23rd. The unit was deployed shortly after Sept 11th with both its Active and Reserve members from just about every job skill, from pilots and navigators to ammo and weapons loaders. Flying over 100 combat sorties in its first three months, the unit made a major contribution to the air war in Afghanistan.

Navy Command Center 106 is a unit designated as a surge responder during crises, and supports daily coverage of the Navy duty officer watch, keeping the Chief of Naval Operations and senior Navy staff informed of daily incidents and activities of the Navy worldwide. Two members of the unit were on duty on September 11, 2001, and were among the casualties that left the active duty Navy command center with 40% of its staff either dead or injured. The unit immediately self-mobilized to replace the casualties and manned all of the command center's billets throughout Sept and Oct.

In November, 2001, Company B, 1st Battalion, 23rd Marine Regiment from Bossier City, Louisiana was the first Marine Reserve unit called to active duty in Operation Enduring Freedom. Their call was to Guantanamo Bay, Cuba to provide security at the installation. "I can't wait to go and make a difference," said

Lance Corporal Brandon Allbritton. "But, I have a wife and kid to think about, so I'm sure once I get there I will be looking forward to getting back."

Lance Corporal Allbritton's response was typical of that of many members of the Reserve components. It was personal for us all and we have all wanted to make a difference. But we have also learned there is no going back. A new future and new focus on the homeland is ahead now and new roles, missions and focus for the Reserve components. This book examines how far we have come in planning to meet an uncertain future. As Americans we have made great strides in a very short period of time. Obviously, there is still much to do.

On the national level we have established the first new department in the Federal Government since World War II. The Department of Homeland Security is settling in to its new role, in integrating intelligence, securing our borders, and in preventing future attacks. The Department is also responsible for consequence management, coordinating the response and recovery from any such attack against our national security. This role is complex and multifaceted with inputs, actions, and coordination with multiple players, including the National Guard and Reserve, first responders at all levels, state and local governments, and the private sector.

In the summer of 2001, prior to September 11th, there was a great debate within the U.S. military, one that took place in classrooms at the Army War College and in senior-level Pentagon offices. What was the definition of homeland defense and how does it differ from homeland security? That debate still rages in a number of circles.

DoD's definition today is: homeland security is a concerted national effort to prevent terrorist attacks within the United States, reduce America's vulnerability to terrorism, and minimize the damage and recover from attacks should they occur. Homeland defense is in every part of that definition, the part that included the roles and missions of the U.S. military. But within the Department of Defense the issue of what that role means along with its numerous implications, is being addressed with resolve and deliberate process. In 2003 the position of the Assistant Secretary of Defense for Homeland Defense was established. This is the Department's main entry point for the development of policy and Defense integration and support to Homeland Security missions. These will be carried out through the new unified command, U.S. Northern Command, which will be fully operational 1 October 2003. The U.S. Northern Command has as its mission the defense of America; this also includes responsibilities for our borders with Canada and Mexico. The commitment of U.S. forces within the U.S. is the responsibility of this Command and originate with the direction of the Department of Defense, not the Department of Homeland Security.

Many relationships are still in the process of being defined. The coordination between DoD and DHS is likewise being refined and formalized at every level. Certainly one of the most important issues is that of the sharing of information and intelligence. In the arena of Chemical/Biological/Radiological threats our biggest challenge is one of early and distant detection of that threat and dissemination of that information throughout the network of responders. The authors of this article think we will soon see a network for information sharing, from first responders to a number of federal agencies involved (*2003 Reserve Forces Policy Board Symposium Proceedings Report Executive Summary*).

Our National Guard and Reserve are uniquely suited to this type of mission and support. America's "Home Team" the National Guard is reorganizing itself to support such missions, developing joint state headquarters and streamlining procedures to ensure rapid response capability. To provide assistance to states in the event of an incident involving weapons of mass destruction (WMD), Congress authorized Civil Support Teams for each state. These teams were established in 1999 and are still in the process of being set up. When fully operational the 55 teams will provide fully trained and equipped units to help local responders and local authorities deal with WMD incidents.

While in a state status (activated under Title 32 and not Title 10) state forces are not burdened by the restrictions of the Posse Comitatus Act and may be used in law enforcement. This would apply to a response to of an act of terrorism or any other such criminal activity occurring within the U.S. Whether it is a natural disaster such as a forest fire or flood, a criminal act such as the attack on the federal building in Oklahoma City, or a terrorist incident such as the September 11 attacks, our National Guard forces will be ready to respond. And that response will be coordinated. The Army Reserve provides Emergency Preparedness Liaison Officers (EPLO) to the National Guard in each state. These officers also coordinate with FEMA and other Federal agencies in consequence management. We have yet to take advantage of another of the Army Reserve's core competencies: training. The Army Reserve has a number of units dedicated to teaching the basics of Nuclear/Biological/Chemical (NBC) defense and response, an asset that could be of great utility to first responders and local fire and police departments.

Because the Reserve components live in local communities, they have a definite advantage in knowing local areas and local working relationships. That is one of our greatest assets and one not often formally acknowledged. The Coast Guard and Coast Guard Reserve for example, are renowned for their ability to develop private-public partnerships in support of port security efforts. During the Reserve Forces Policy Board field trip to the U.S. Pacific Command in January 2003 the Chairman of the Reserve Forces Board spoke to a number of Coast Guard Reserve personnel about their mobilized status. Several noted that Homeland

Defense missions constituted about 2% of their mission prior to Sept 11th. They now constitute over 60%. The USCG District in Honolulu expects that in several years it will settle at approximately 30%, still a major increase but one they feel should be manageable.

Our Marine, Navy, Air Guard, and Air Force Reserve units and personnel are likewise engaged in the HLD mission, whether through performance of their regular mission in fulfilling their statutory responsibilities in protecting the homeland, its people and critical assets or in other roles, every member of DoD is involved.

And they will not be alone. The above are but a few examples of some of the changes being made by our Reserve components in recognition of new missions. We know that in an emergency, response will naturally be overwhelming, and occur at all levels. Many Reserve components initiatives fall under the Military Assistance to Civil Authorities Program (MACA) including the state liaisons, and efforts to protect critical infrastructure. The establishment of a coordinated, tiered, well-communicated response remains everyone's goal.

Several members of the Reserve Forces Policy Board recently participated in the National Defense University strategic scenario or wargame, "Silent Prairie" at NDU's National Strategic Gaming Center. Also participating were representatives from the USDA's Animal and Plant Health Inspection Service, Maryland's Department of Agriculture, NDU's FEMA Chair, and the North Carolina Department of Agriculture and Consumer Services Director of Emergency Programs. In addition, Dr. John Blair, Director and Chief of Bioterrorism Studies at the Center for Healthcare Leadership and Strategy, Texas Tech University, and Board Senior Advisor, also participated. This exercise illustrated how rapidly spreading cases of foot and mouth disease could migrate from cattle herd to herd, quickly infecting and affecting food chains, transportation networks, tourism, state governments and ultimately whole economies, first regionally and ultimately nationally. The exercise is a superb illustration of the need for continuous communications and cooperation between agencies and organizations at all levels. NDU is in the process of exporting this exercise to state governments and it is a valuable learning tool (*2003 Reserve Forces Policy Board Symposium Proceedings Report Executive Summary*).

An exercise like Silent Prairie serves to illustrate the veiled fragility of our national food supply and our vulnerability to attack using biological weapons. We need to be able to defend against biological warfare, have the capability within DoD to create and sustain medical surge capabilities and to support prevention measures.

One of the outcomes of our 2003 Symposium, "*Strategic Challenges: Transforming the Total Force Vision for the 21st Century*" has been the establishment of four task groups within the Board's membership. The Board is composed of 24 flag and general officers, with ex-officio members now from the Deparment of Homeland Security, the Surgeon General and other agencies all concerned with the need for a coordinated effort in the war against terrorism. One of our task groups deals with defense education and we will be championing NDU's effort to develop and export strategic scenarios like "Silent Prairie" (*2003 Reserve Forces Policy Board Symposium Proceedings Report Executive Summary*).

The Chairman's good friend, VADM Richard Carmona, U.S. Surgeon General and a former Army Special Forces soldier, talked to us recently about the special challenges he faces as Surgeon General of the U.S., including the office's new role of preparedness and teaching resilience to the American people. He discussed some specific health issues including SARS and smallpox, emphasizing the importance of preventive medicine. Military members including the Guard and Reserve are already protected or will be, by a number of vaccines. The authors believe we need to expand this program nationwide, increasing emphasis on development of pre-symptomatic therapeutics and vaccines and broad-spectrum antibiotics (*Defense Science Board Study: Protecting the Homeland, 2003, pp. 5, 54*).

As the Chairman, Military Executive, and Chief of Staff of the Reserve Forces Policy Board, we strongly believe in the patriotism and dedication of the American Citizen Patriot. Since the beginning of this country it has been our citizens who have taken up arms in its defense. Our military grew from a militia tradition and many citizens today belong to unofficial associations, organizations like the Civil Air Patrol (CAP), Coast Guard Auxiliary and the Naval militia. Even neighborhood watch groups, such as the Guardian Angels continue to make a significant difference in increasing our security. Their role is an important one and each and every agency concerned with the vital defense of this nation against the terrorist threat should recognize that public/private partnerships are critical to our future. Our Citizen Patriots are both our strength and our best defense.

REFERENCES

Reserve Forces Policy Board (2002). The annual report of the reserve forces policy board. Available at: http://www.defenselink.mil/ra/rfpb/.

Reserve Forces Policy Board (2003). Strategic challenges transforming the total force vision for the 21st century (May 19–20). A Symposium at National Defense University.

CIVIL-MILITARY RELATIONS IN AN ERA OF BIOTERRORISM: CRIME AND WAR IN THE MAKING OF MODERN CIVIL-MILITARY RELATIONS

E.L. Hunter, Ryan Kelty, Meyer Kestnbaum and David R. Segal

ABSTRACT

The United States of America is on the verge of a possible revolution in civil-military relations in an era marked by increased defensive alertness stemming from the attacks of 11 September 2001. As we anticipate the normalization of terror as a way of life, we are witnessing a paradigmatic shift from the use of violence towards some political end to the use of violence as an end in itself (Jenkins, 2001).[1] And where, for most nations, homeland defense is the primary mission of the armed forces, the United States had to establish a new cabinet-level Department of Homeland Security due to the primarily expeditionary nature of American armed forces for the past half-century. The military has been a unique institution in modern societies. It has acted as the agent for the state's possession of a monopoly on the means of large-scale organized violence and war-making. The establishment of a second executive agency responsible for homeland security makes the equation more complex. As a result, ever greater attention must be given to the balance of civil-military relations in American society.

Bioterrorism, Preparedness, Attack and Response
Advances in Health Care Management, Volume 4, 319–344
ISSN: 1474-8231/doi:10.1016/S1474-8231(04)04013-3

CIVIL-MILITARY RELATIONS IN NORMAL TIMES

The formation of modern nation-states grew out of an historical process driven by the dynamic interaction of civilian rulers and men at arms. The current form of western democracies and their relationship with national militaries has been depicted as the result of a co-evolutionary process of military structures, technology, and societies (Andreski, 1968; Aron, 1968; Burk, 2002; Clausewitz, 1976; Hanson, 2001; Kestnbaum, 2002; Manicas, 1989; Segal, 1995; Tilly, 1996, 1992). The evolving structures of societies have facilitated changes in the form and function of militaries from noble classes waging war on each other, to mercenaries hired by princes and monarchs, to the professional military organizations of the modern nation-state (Andreski, 1968; Aron, 1968; Hanson, 2001; Howard, 1976; Huntington, 1957; Janowitz, 1960; McNeil, 1989; Paret, 1992; Tilly, 1992).

The relationship of a state's civil government to its military institution is constantly being negotiated, challenged, and redefined. The British Empire used its military, rather than a professional police force, to enforce laws and maintain civic order in its colonial possessions. As a result, at the founding of the United States, citizens were wary of empowering the military to act on domestic matters. These suspicions caused the framers of the U.S. Constitution to divide the powers of forming and leading the military; the Congress is responsible for raising and funding the federal military while the President acts as the Commander-in-Chief. Further, the Constitution specified the right of states to form militias, a direct attempt to ensure the subordination of the federal forces to civilian control (Weigley, 2001). Even so, there have been instances of conflict between American civilian political leaders and military leaders. Many definitions of "good" civil-military relations focus on the extent to which, when conflicts in civilian and military leadership views exist, military leaders follow the commands of their civilian leaders (Burk, 2002; Clausewitz, 1976; Desch, 1999). The history of the U.S. has shown, sometimes with severe consequences (e.g. Vietnam), that military subordination to civilian control is overwhelmingly the rule (Feaver, 1995).

Two of the most influential thinkers in civil-military relations are Samuel Huntington and Morris Janowitz. While they make different assumptions and each has different priorities, the essential question for each is how to sustain both military effectiveness and the primacy of democratic civilian norms. For Huntington (1989), the highest priority of the democratic state is to protect the rights and liberties of its citizens. How does a democratic *government* ensure that the coercive power of the state is not turned against its own people? Huntington's answer is a theory of "objective civilian control" of the military wherein democrat-ically elected civilian authorities would design national security policy but would leave military managers free to decide how best to meet its objectives. Civilian control is maintained by allowing the professional officer corps a large measure of autonomy in exchange for loyalty and obedience (Burk, 2002; Huntington, 1989). Huntington assumes separate and distinct military and political spheres. Burk (2002) notes that this model of civil-military relations requires a military staffed by professional soldiers willing to trade loyalty for autonomy. In this way, the roles and normative expectations of both the civilian and military elite become mutually reinforcing and interdependent.

Duty and honor, longstanding hallmarks of military service, are supported and enhanced through the professional autonomy granted the armed forces. Thus, the military is insulated from the waxing and waning of popular sentiment inherent in democracies. The military is viewed as a professional organization prepared to serve the government regardless of the person or political party occupying the White House or Congress. Because of its historically apolitical stance and desire to refrain from domestic intervention, the American military has enjoyed a long tenure of popular trust and support (Snider & Carlton-Carew, 1995).

The civilian government profits from this kind of relationship. In most western democracies military subordination to civil authorities is generally accepted as the normative standard. This has not always been the case. In the present era, there continue to be many countries whose militaries are not subordinated to the civil authorities. When the civil authorities are undermined or directly challenged by the military, highly unstable social and political situations arise.

If Huntington (1989) assumes that to *defend* democratic values the military needs to be subordinate to civilian government, but not enact those values, for Janowitz (1971) the military must identify with and embody the democratic values of the society it defends in order to *sustain* them (Burk, 2002). He argues that the most effective way to preserve such values is to ensure that the military is itself a reflection of the public body. For Janowitz, it was essential for the professional military to continue to think of itself as part of the broader society, creating a kind of subjective civilian control of the military (Burk, 2002; Janowitz, 1971). In this way, the interests of the military and of society overlap for the military is a part *of* society not, as Huntington argued, apart *from* society. To the extent that the military becomes homogenous and insulated from civilian society, civil-military relations and social stability will be threatened.

For much of American history, when it needed to raise an army for national defense, the U.S. instituted broad based conscription. Men from across the country of varying ages and demographic backgrounds were called up for national service. Once hostilities and threats to national security or national interests abated, the soldiers were discharged and sent home. It was not until the Cold War that followed the Second World War that a massive standing army was maintained by the U.S, sustained by a system of conscription supplemented by volunteers.

During the Vietnam War public dissatisfaction over the ways men were chosen for service by local draft boards or granted various types of deferments fueled social unrest. This tension contributed to the termination of conscription and the advent of the all-volunteer force in 1973. An undercurrent of the social concerns of the Vietnam era was perceived inequitable utilization of minorities. The nation was dealing with the realization that the military needs to be representative of the population from which it is drawn and which it serves.

One of the major problems with Janowitz, Huntington, and their intellectual progeny is that they assume that the relations to be explained are contained within sovereign states. However, in recent years the sphere of civil-military relations has expanded to include transnational concerns and relations with non-governmental organizations and private enterprise (Burk, 2002). Since the end of the Cold War in Europe, choices about how and when to risk the lives of American servicemen and women have become more complex for both domestic and diplomatic reasons (Nixon, 1994; Segal, 1994).[2] In an era of transnational civil-military relations, the problem continues to be who regulates national militaries and decides when and where to use force. The more important question no longer is how to use force to protect and sustain democratic values within states, but rather how to do so within a variety of transnational security communities and coalitions (Burk, 2002).

The 1990s witnessed what some perceived as increased tensions in civil-military relations. This increase has been attributed to manifestations of civilian and military separation. The military is viewed by some as becoming more socially and culturally isolated, existing within, yet not as part of, American society (Feaver & Kohn, 2001; Holsti, 2001). This may be attributed to factors such as reductions in force size, changes in force structure, and the aging of veterans serving prior to the advent of the all-volunteer force. The cumulative effect of these changes has several important consequences for U.S civil-military relations.

The past few decades have been marked by a decreased proportion of national political officeholders, including the President, having military experience, as the World War II and Korean War generations aged. This declining presence of veterans is thought by some to lead to increasingly fractious relations between the military and both the legislative and executive branches of the federal government (Bianco & Markham, 2001; McIsaac & Verdugo, 1995; Weigley, 2001). One potential consequence the military may face is elevated competition for funding by other departments such as the Department of Homeland Security. Another potential issue is that the military may be subject to increased requirements to perform lower-intensity missions such as humanitarian and peacekeeping missions (Bianco & Markham, 2001; Desch, 1999; McIsaac & Verdugo, 1995; Snider & Carew, 1995).

Some have also argued that increased isolation of the military is an unintended consequence of force reduction and closure of military installations nationwide.

Fewer men and women in uniform means less direct and indirect (via family and friends) citizen exposure to the military. This is exacerbated by the fact that many of the nation's domestic military bases are in relatively rural areas. Isolation of the military may perpetuate, or even increase, civil-military tension (Janowitz, 1971; McIsaac & Verdugo, 1995; Snider & Carew, 1995).

The end of the Cold War in Europe in the 1980s lead to an expectation of a "peace dividend" and, after the Gulf War, a continual downsizing of the American military reduced its visibility. The terrorist attacks of 11 September 2001 and subsequent military action in Afghanistan and Iraq have served to bring the military back into the public's eye. The military's resurgence as a publicly honored profession worthy of respect and support for the duties it performs on behalf of the nation is typical of public reaction in times of threats to national security. This has the effect of smoothing out relations between the two spheres. The terrorist attacks set into motion events that led both spheres to act according to normative expectations. Doing so simultaneously caused a convergence of the spheres and reinforced their respective roles in the normative model of civil-military relations.

The military holds the Huntingtonian model as its ideal. However, it is clear that the political and military have never been separate spheres (Burk, 2002; Cohen, 2001; Segal, 1994, 1995). In recent decades the demarcations between civil and military have become even more blurred. For example, increased deployments for humanitarian and peacekeeping missions, as well as military elites publicly voicing personal political views, are noted as highly visible instances of the blurring of these boundaries (Desch, 1999; Holsti, 2001; Roman & Tarr, 2001; Segal, 1994, 1995; Snider & Carew, 1995). As a result, military elites have become more insistent that they be given more autonomy in decisions on the use of military force rather than simply be expected to act in an advisory capacity (Desch, 1999; Holsti, 2001). Burk (2002) argues that this is not inherently problematic; the lack of total integration of the military into society is not necessarily a sign of poor civil-military relations. No institution or profession is perfectly integrated into American society. Institutions, such as the military and medical professions, are founded on a unique set of imperatives, beliefs, and goals. Good institutions successfully defend and advance their agendas within civil boundaries, often legally defined.

Democracy is a process, not a static state, falling somewhere on a continuum from more to less democratic (Segal, 1994). In a democracy individual rights and civilian political authority must be constantly negotiated relative to military necessity. Given the relatively rapid evolution of military technology and structure, international relations, and globalization of ideas and commodities the U.S. faces increasing difficulty in establishing contemporary understandings of constitutional powers for both the civil government and the military.[3] The question of constitutionality extends beyond the relation of the military and civilian authorities. The question inherently involves individual citizens' civil liberties as well. The result of civil-military negotiations of power is a relative balance between the elected leaders of the civil government and the military. By virtue of joining the ranks of "the professions" the military has become, de facto, institutionally politicized (Segal, 1995). The boundaries between the civilian and military spheres are blurred as the military becomes a recognized, distinct body invested with political power within broader society.

Professions view reality through a lens shaped by their collective ideals, ethics, motivations, and missions. The military is no exception. In their view, the job of the American military is to prepare for, fight, and win America's wars. The military also explicitly views ensuring the security of the United States against initial aggression as one of its primary missions. The oath taken by all soldiers on induction specifies their duty to protect America from all enemies, foreign and domestic. But the military elite are aware that their role in protecting American democracy and civil liberties is constrained. Military action within the borders of the U.S. has a strong history of suspicion. Military leaders are loath to subject the autonomy, legitimacy, and credibility of their institution to the scrutiny that would surely arise if they encroach on the jurisdiction of civilian authorities in matters of domestic security. Precisely because the U.S. military has such a strong professional ethic that accepts subordination to civilian authority, it has found itself, historically and in contemporary society, cast into numerous situations involving civilian unrest, security, and law enforcement.

CIVIL-MILITARY RELATIONS IN THE FACE OF A NATIONAL SECURITY CRISIS

The history of national security in the United States, and other western democratic states, has been marked by a continuing struggle to balance democratic values and ideals against the exigencies of national security. This conflict between the rule of law and the needs of national defense policy has, at times, spilled over to intrude on the individual freedoms of citizens (Segal, 1994). In times of war several U.S. presidents have surpassed their legal authority. Apologies, if offered at all, occurred only after threats to national security have been vanquished (Segal, 1995). In short, democracies often find themselves in the unenviable position of choosing to "either prepare for and fight its wars with one hand tied behind its back, or become a bit less democratic in the interest of effectively defending what democracy remains after the compromise" (Segal, 1994, p. 375).

Paret (1992) and Kestnbaum (2002) argue that a "revolution in war" occurred at the end of the 18th century, identifying the central role ordinary people came to play in war between states. Since then, technological innovations in both the means and modes of warfare, the "revolution in military affairs," that have shortened the time and expanded the space in which wars are waged have greatly exacerbated the conflict between expediency and constitutional limits on governmental power.[4] These strategic realities often confront American society with situations that the framers of the U.S. Constitution could not have anticipated, and which frequently pit perceptions of military necessity and individual rights against each other (Segal, 1994). Today, with the rapid deployment of troops, airpower and intercontinental missiles, hostilities demand quick and decisive action. Americans expect their President to act immediately to repel hostile attacks and to worry about congressional approval later (Epstein & Walker, 1995). Wars may evolve independently of legislative action, often doing so more rapidly than legislatures can respond (Segal, 1994). Indeed, in the face of a national security crisis, such as the war against terror, the Executive may find it necessary to take action to preserve the life of the nation that would be unlawful at other times, stretching the limits of the constitution to respond to the crisis (Epstein & Walker, 1995). In this formulation, legislative action is reduced to an increasingly reactive role as illustrated by the War Powers Act,[5] a Congressional act compelling the Executive to report on hostilities *after* they have been initiated (Segal, 1994).

The legal and judicial systems of the U.S. exist because social norms generally enjoy neither total consensus nor total conformity (Segal, 1994). The Civil War (a crisis of national scope) served as the primary catalyst for American judicial pronouncements on the tension between liberty, on the one hand, and governmental need to respond to crisis on the other. Civil War era Supreme Court cases serve as guide to how civil-military relations in the U.S. may work when burdened by an overwhelming national security crisis on the American homeland. It is generally recognized constitutional doctrine that the government may exercise more power during war and national emergencies than during times of peace (Epstein & Walker, 1995). What is not clear is at what point the extraordinary powers of the state can be activated. The answer, at least in part, was issued by the Supreme Court in the majority opinion, written by Justice Grier, of the *Prize Cases* of 1863.

> If war be made by invasion of a foreign nation, the President is not only authorized but bound to resist force by force. He does not initiate the war, but is bound to accept the challenge without waiting for any special legislative authority. And whether the hostile party be a foreign invader, or States organized in rebellion, it is none the less a war, although the declaration of it be "*unilateral.*" Lord Stowell observes, "It is not the less a war on *that account*, for war may exist without declaration on either side. It is so laid down by the best writers on the law of nations. A declaration of war by one country only, is not a mere challenge to be accepted or refused at pleasure by another" (Epstein & Walker, 1995, p. 249).

When national survival is at stake, societies are often found in an emotionally heightened state of alert and have historically been willing to tolerate encroachment on constitutional limits to governmental power, the observation of which they would otherwise insist upon. In America, the problem of when and where to commit military force in the face of a national security crisis is complicated by the fact that both the executive and legislative branches have powers that are interpreted as controlling the commitment of armed forces (Epstein & Walker, 1995). The basis for presidential control is found in Article II, Section 2 of the U.S. Constitution, which states, "The President shall be Commander in Chief of the Army and Navy of the United States, and of the Militia of the several States, when called into actual Service of the United States." The case for congressional control is derived from Article I, Section 8 (Epstein & Walker, 1995). "Congress shall have Power ... To provide for calling forth the Militia to execute the Laws of the Union, suppress Insurrections, and repel Invasions."

The American government has a tradition of exercising military might with little regard for civil liberties when faced with a national security crisis. Because national exigencies cannot be wholly foreseen, the authority to use force was given intentionally broad parameters to allow for flexibility in its application over time (Rakove, 1996). This tradition, reflected in such wartime statutes as the Espionage Act of 1917 and the Sedition Act of 1918,[6] has been the American norm. But governmental power is not absolute, even in the face of a national security crisis.

The arrests and trial of citizens by the military, even in the state of national emergency, is strictly prohibited by the doctrine of *habeas corpus*, a foundational American legal doctrine inherited from the English, which permits an arrested person to seek a judicial determination of whether the detention is legal or not (Epstein & Walker, 1995; Segal, 1994). Designed to protect citizens from the excesses of the Executive, the doctrine of *habeas corpus*, or rather its suspension, has been a contentious issue throughout our history. Article I, Section 9 of the U.S. Constitution states, "The privilege of the Writ of *Habeas corpus* shall not be suspended, unless when in Case of Rebellion or Invasion the public Safety may require it." The first problem evident with this formulation is that it is found in Article I, the legislative powers section, of the U.S. Constitution (Epstein & Walker, 1995). Secondly, what limits do executive power face in "providing for the common defense" in cases that involve neither rebellion nor an invasion, per se? President Lincoln chose the preservation of the Union over the legal and political rights of individuals, giving his military commanders sweeping power to arrest, detain, and try civilians suspected of treason (Epstein & Walker, 1995; McPherson, 1997; Neely, 1997). When a federal court found Lincoln's suspension of *habeas corpus* to be unconstitutional, the President defied the ruling in a message to Congress. The Supreme Court has yet to resolve the issue of who may suspend *habeas corpus* (Segal, 1994).

The legal doctrine of *Posse Comitatus* establishes the basis for the traditional separation of civilian and military authority, prohibiting the use of the military for the execution of civilian law while allowing it to assist civilian authorities by providing logistic support, etc. Exceptions to the law that have been formulated over the years, either formally or informally, such as military investigative authority in the wake of the Oklahoma City bombings and the empowering of the military in 1981 and 1989 to engage in continuous drug interdiction activities, have lead to a blurring of the lines between civilian and military authority (Hammond, 1997). The combined effect of these exceptions, plus the fact that *Posse Comitatus* does not govern the non-federalized National Guard, leads one to envision a scenario wherein both the National Guard (more easily) and the active-duty U.S. military (perhaps more problematically) can and will be called-up to assume civilian responsibilities in the face of a national security crisis.

MODELS OF CIVIL-MILITARY RELATIONS DURING NATIONAL SECURITY CRISES

Harold Lasswell's (1941) pivotal construct, the "garrison state," was meant to serve as a guide for clarifying expectations about the future state of civil-military relations given certain assumptions about a high level of strategic threat faced by a nation. Lasswell observed the emergence of the garrison state in nations characterized by the socialization of danger: the broad sharing of danger in a population. As we have seen, modern weaponry has become increasingly indiscriminate in its destructive power, placing civilians and uniformed personnel alike in harms way. The garrison state is theorized to be dominated by an elite core of specialists in the means of violence as it attempts to structure and fortify its society in the face of an imminent security crisis. To that end, propaganda and morale become important tools of the garrison state.

Stanley and Segal (1997) examine the social structure of nations in order to determine their susceptibility to garrison state emergence. Nations with volunteer armed forces tend to be less militarized[7] and therefore limit the ascendancy of specialists in violence. However, volunteer forces in technologically advanced military establishments require career military personnel who may be less likely to exhibit the civilianized sentiments of the citizen-soldier. In short, a modern force-in-being can potentially diverge from the society it is charged to defend, thus opening the door for the ascent of a garrison state. Given the recent parallels that have been drawn between the U.S. and Israel as America faces its current strategic threat, Stanley and Segal's (1997) depiction of Israel as a possible candidate for the garrison state is striking. Two elements of Israeli society are common denominators with Lasswell's (1941, 1997) garrison state construct: (1) socialization of civilians to accept the military model of society; and (2) the inability of civilian institutions to civilianize the military (Stanley & Segal, 1997).

As American society continues to adjust to the realities of the post-September 11 strategic world we ask whether the continuation of a threat-laden national security environment (the socialization of danger) holds the prospects of a garrison state for American society. America did not develop the worst elements of the garrison state during the Cold War, one of the longest periods of persistent strategic threat in American history. However, it would be pragmatically foolish and theoretically irresponsible to deny the implications for the current climate of U.S. civil-military relations. We must ask why America has remained true to its principles of democratic governance and civilian control of the military. For these answers, a look at America's judicial pedigree is in order.

Ex parte Milligan (1866) provides important guidance in clarifying the legitimate constitutional wartime powers of the executive. Lambdin P. Milligan, arrested, tried, and convicted by a military tribunal in 1864, sought a writ of *habeas corpus* from the federal circuit court at the conclusion of the Civil War. This court referred the case to the U.S. Supreme Court (Segal, 1994). The (slim) majority opinion held that "neither the president nor Congress, acting separately or in agreement, could suspend the writ of *habeas corpus* as long as the civilian courts were in full operation and the area was not a combat zone" (Epstein & Walker, 1995, p. 256).

On 2 February 1942 President Franklin Roosevelt issued Executive Order 9066 authorizing the Secretary of War to establish military zones, without Congressional approval, from which individuals might be restricted in order to defend against sabotage and espionage. Under Order 9066, all persons of Japanese descent, including 70,000 American citizens, were removed from the Pacific coast to various relocation centers (Segal, 1994). The constitutionality of the evacuation and detention orders were taken up by the Supreme Court in *Korematsu v. United States*, but not until the end of 1944. The Court affirmed the evacuation and detainment orders. However, in *Ex parte Endo* the Court modified its ruling by holding that, "an American citizen of Japanese ancestry whose loyalty had been established could not be held in a war relocation center, but must be released" (Segal, 1994, p. 378).

Huntington (1993, 1996) warned of a potential "clash of civilizations," pitting the west against the rest. Depending on the degree to which the religious/cultural nature of the current conflict is projected to and internalized by the American people, domestic approaches to the war on terrorism could take on ethno-religious dimensions. Huntington's cultural analysis can best be understood as an overly simple approach to geopolitics after the Cold War. Even so, its implications for how Muslims and Arabs, including those who are American citizens, might be perceived and treated by America's military, civil society, and courts as the nation "circles its wagons" in defense against all things non-western are disturbing. As is illustrated by *Korematsu*, America has historically been willing to mistreat its putatively identifiable minorities in times of national security crisis. To the extent that the American people perceive the War on Terror to be a war against Muslims, we could witness a crisis in civil-military relations as Arab and Muslim Americans become alienated from the military and broader American society. The importance of framing the War on Terror as a war against religiously radical terrorists rather than a war against all things (culturally) Islamic cannot be overstated. If the U.S. does not tread carefully, it may repeat the same mistakes made during World War II; Arab Americans may be separated out and subjected to "special" treatment simply because they look like the "enemy."

Yet the power of the government is not absolute. *Dames & Moore v. Reagan* (1981) recognized that the power of the president can be expanded during times of national crisis, though such an expansion is not without its own limits (Epstein & Walker, 1995). Ultimately, the question of who has the authority to suspend *habeas corpus* and when remains a debatable issue to which the courts have yet to provide definitive answers. However, existing constitutional precedent allows for stronger use of executive power when acting with assent of the Congress (Epstein & Walker, 1995). In sum, the Supreme Court has consistently ruled that military necessity justifies the loss of individual liberty (Segal, 1994).

If it is clear that military expediency and the needs of national security take precedence in times of national security crises in America, the post-September 11 era makes it unclear *who* will be the primary domestic guardians of national security. If broad-sweeping powers are assumed to be vested in the Executive in times of a national security crisis, should we expect to see the Department of Justice, the Department of Defense, or the Department of Homeland Security granted final authority to engage our terrorist enemies?

The major mission of the active duty U.S. military is external defense of the nation. This mission is not likely to change as a result of the recent terrorist attacks and the formation of the Department of Homeland Security (Shanahan, 2002). What is likely is that the National Guard will be called on to assist in homeland security functions. This is consistent with the historical utilization of the National Guard to protect against threats to domestic security. The National Guard has also been used for other domestic support roles including quelling civil unrest and natural disaster assistance. To be effective as support to civil agencies in homeland security the National Guard will have to make organizational and training modifications (Shanahan, 2002).

The role of the Coast Guard is also likely to change. The Coast Guard, during peacetime, has been a law enforcement agency under the Department of Transportation that has used paramilitary forces, tactics, and techniques to preserve coastal and waterway integrity. However, in times of war, it has contributed services such as harbor, coastal and shipping defense, and maritime interdiction operations under the operational control of the Navy. It has been relocated to the new Department of Homeland Security. Whether the Coast Guard will continue to serve as an element of the naval forces available to Geographic Combatant Commanders in times of national security crises is not yet known.

CRIME OR WAR AND BIOTERRORIST THREATS

The central question in the post-September 11 strategic environment is whether or not wars are "international events." The traditional aim of interstate war has been to impose one state's political will on another (Clausewitz, 1989; Luban, 2002). Can non-state actors, such as al Qaeda, wage war on nation-states? Correspondingly, was Timothy McVeigh's attack on the federal building in Oklahoma City a criminal act[8] or an act of war? While lacking the material resources of modern armies, terrorist organizations have proven their ability to mount large-scale attacks against major urban centers (Jacoby, 2002). In classic conceptualizations of war, killing the enemy was merely the means to a political end. In terrorist-initiated conflict, where killing the enemy appears to be the political end in itself, no capitulation is possible (Luban, 2002). The enemy becomes ambiguous. The conflict is no longer between militaries or states. If belligerency is not state-centered what are the implications for the calling of truces, the drafting of treaties, etc.? How do military organizations deal with non-state-affiliated entities?

As Americans review the consequences of the attacks of 11 September, the answers to these questions seem anything but clear. How a conflict is framed – as an orchestrated set of criminal activities or as a war – will have a significant impact on how the U.S. treats parties to the conflict, what roles are accorded the military and civil law enforcement agencies, and how each is received by the American public. The framing of these issues as either criminal or acts of war will be a critical determining factor for U.S. civil-military relations in the future.

The criminal model is most appropriate when civil *order* is threatened, the war model when civil *society* is threatened (Pfaff, 2002). While the military has traditionally protected society from *external* threat, police have assumed the responsibility of protecting society from *internal* threats. These traditional jurisdictions of the military and the police have been obscured by new functions associated with, and new objectives demanded by, the current counter-terrorism conflict (Dunlap, 2001; Jacoby, 2002). Framing the current threat in terms of war removes many of the problems associated with *posse comitatus* as related to the use of active-duty forces and, to the extent that they are called up, reserve forces.[9]

The civil-liberties question (*Habeas Corpus*) will lose much of its importance as well. The judicial litmus test appears to be whether or not civilian courts are still functioning. Presumably, in the most extreme case, the declaration of martial law would replace civilian with martial government in the area of conflict. Overall, framing the issue as a war greatly simplifies legal questions regarding civil-liberties *during* the time of conflict (Luban, 2002). In addition, the burden of proof on the government is lowered to merely "plausible intelligence." The use of lethal force is not only authorized, but is considered legitimate under the war model, as a mode of self-defense. The war model singles out not only those who have caused harm, but those that *might* cause harm as potential targets; in the war model, "conspirator" is equivalent to "combatant" (Luban, 2002).

Conversely, conceptualizing the current conflict in terms of war poses limitation on government forces. Under the war model, the enemy is entitled to special consideration once he is no longer an active combatant. Under international law, it is impermissible to punish captured combatants for their role in fighting a conflict; or to subject them to torture, forced interrogation, or place them on trial for criminal activity (Luban, 2002). These conditions apply only when the conflict is recognized as a legitimate military engagement and the actors are identified as legitimate combatants.

It seems likely that the legality of applying a war model to a conflict that does not meet the conventions of interstate conflict would come under fire. But if history is any guide, such scrutiny would generally surface only *after* the conflict has been resolved. Given the nature of terrorism, it seems questionable that the strategic threat can ever be totally contained or neutralized. Exit strategy thus becomes a very important issue when conceptualizing the framing of the conflict as a war on domestic territory. If the threat remains persistent we might see the institutionalization of martial law, i.e. trends denoting a more permanent condition than the language of emergency would lead us to believe. In short, the garrison state could become a reality. To the extent that public support depends on the short-term nature of domestic military engagement, a prolonged presence might lead to a straining of civil-military relations. If the engagement is short-lived, we might see a much higher level of civil support for the military.

During counter-terrorism campaigns, as in other forms of warfare, democratic states must, at times, impinge on the rights of their citizens (Jacoby, 2002; Segal, 1994). To the extent that counter-terror campaigns are waged on the home-front, the degree of sacrifice people are asked to make may be significant. To enable federal agencies, including the military, to be proactive in gathering information vital to national security, democratic societies will be asked to tolerate the monitoring of evolving threats, intrusive investigations, and increased scrutiny of a country's social, political, and religious institutions, all with limited public scrutiny of these activities (Jacoby, 2002). The risk that we run is that suspensions or limitations of civil liberties on the grounds of national security and military expediency over the long-term may result in the erosion of those liberties at the conclusion of the conflict, a delegitimizing of the armed forces, or the straining of civil-military relations.

Framing the current conflict as a criminal matter poses major problems for the extensive (and therefore effective) use of the active-duty armed forces and nationalized Reserves. Use of Reserves for essentially policing functions appears to meet the standards of legality. State governors can deploy National Guard units within their states for the enforcement of civilian law without materially violating *posse comitatus* (Mabry, 1988). However, if the line between active duty military support operations and the enforcement of civilian law gets blurred during times of conflict negotiation on the home-front, and if policing functions (Reserves) are popularly confused with martial law, *posse comitatus* concerns might become major issues.

In the criminal model of the conflict, civil-liberties questions might be much more contentious as well. The framing of the conflict in this way will pose much larger, and more intense, legal challenges if the armed forces are utilized. The issue of spheres, boundaries, or proper division of labor will be much more important. How people perceive these issues will dictate, to a large degree, how they perceive the armed forces. It is possible that in the absence of a declared state of war, the use of the armed forces on the home-front in the prosecution of terrorist threats will lead to major civil-military conflict as civilians perceive the armed forces as suppressing their civil liberties and the military views civilian society as ungrateful. The criminal model leaves the legal issues much more contentious and open to debate. To the extent that the military is actively engaged in such a contentious and ambiguous environment, it seems unlikely that they will enjoy good relations with the civilian population.

Strategically, the advantage of the law model is that it criminalizes acts of violence against American citizens and the destruction of American property. The law model conceives of captured terrorists as criminals who, contrary to the war model, can be tried and punished for their "crimes" (Luban, 2002).

One need only to turn on or read the news media, or walk through a college campus, to see that many seem genuinely concerned about how the United States has chosen to respond to the events of 11 September 2001. Critics claim, among other things, that the U.S. is using excessive force in its war against terror, disregarding international law and political and moral norms (Pfaff, 2002). In essence, critics are arguing about the conceptualization of the "war against terror" as a literal *war*, preferring to see terrorists pursued as criminals (Jacoby, 2002). The confusion over the kind of threat the terrorists represent and how best to deal with them makes it difficult for our uniformed personnel to resolve the tension between the imperative to accomplish missions and the prohibitions against breaching Americans' civil liberties (Pfaff, 2002). This confusion also complicates how civilians view the military. In the past, terrorists had primarily been regarded as *criminals* and were pursued by *law-enforcement agencies.* It is striking to note that following the attack on the USS Cole the FBI, not special forces teams and light infantry divisions, was dispatched to Yemen to pursue the terrorists (Pfaff, 2002).

In addition to concerns for the consequences for individual liberties in how we frame the current environment of conflict, and questions about the appropriate level of restraint on governmental power, there is the essential difference in the modes of force employed by civilian law-enforcement vs. military organizations. Civilian police organizations are committed to the minimum use of force necessary to achieve a given objective. Soldiers are committed to using the maximum amount of force allowable to meet a given objective (Dunlap, 2001). The significant distinction is the objective. Police maintain order; soldiers create order by degrading an enemy's ability to destroy it (Pfaff, 2002). These differences must be taken into account in formulating the division of labor between the military and law-enforcement agencies.

Currently, Washington's approach appears to be geared towards utilizing the most expedient portions of both the war and the crime models, regarding international terrorism "not only as a military adversary, but also as a criminal activity and criminal conspiracy" (Luban, 2002, pp. 9–10). Such a formulation allows Washington to maximize its ability to use lethal force while minimizing the traditional rights of captured combatants. Captured Al Qaeda "suspects" at Guantanamo Bay do not enjoy the benefit of a presumption of innocence or the right to a writ of *habeas corpus*. Additionally, these "suspects" are subject to indefinite confinement without the benefit of legal counsel (Luban, 2002).

One of the most interesting contemporary legal challenges in the unfolding civil-military dynamic centers around Jose Padilla, an American citizen accused of spending a substantial amount of time in Afghanistan and Pakistan with senior members of al Qaida, and implicated in an attempted "dirty bomb" plot as a continuation of the 11 September attacks. In *Padillia v. Bush* (2002), the government argued against Padilla's petition for a writ of *habeas corpus* on the grounds that he is an enemy combatant.

The Sixth Amendment to the U.S. Constitution, which guarantees the right to counsel, only applies in cases of criminal prosecutions. It is not extended to prisoners of war (Comey et al., 2002). Even under the Third Geneva Convention (1949), prisoners of war have no right to counsel in order to challenge their detention. The U.S. government argues that the fact that Padillia was not captured on the battlefield by the military is irrelevant. Citing *Ex parte Quirin* (1942), the government draws parallels between the defendants in that case who were taken into custody in New Orleans by the FBI and subsequently determined to be "enemy belligerents."[10] Indeed, the Court's ruling in *Quirin* highlights the defendants' attempt to pass themselves off as citizens in order to carry out their attacks on the United States as the very reason to categorize them as "unlawful combatants." But perhaps the most interesting aspect of the government's argument is a tacit deference to the military in determining whether an individual is a criminal or a combatant (Comey et al., 2002).

Since 11 September 2001 some have argued that the treatment of terrorists as criminals belies a fundamental misunderstanding of the nature of the *enemy* and the threat posed to U.S. national security concerns. Criminal acts are not aimed at the state but against individuals or groups within the state. Conversely, any act that violates the political sovereignty or territorial integrity of a nation-state constitutes an act of war (Jacoby, 2002; Pfaff, 2002). Though many nations have had to deal with the threat of terrorist activities for decades, the attacks of 11 September represented a fundamental shift in the nature of terrorist hostilities and, by extension, a fundamental shift in modern warfare. Previously, terrorist activities, such as the bombing of Pan Am Flight 103 over Lockerbie, Scotland, in which more than 300 people died and commercial property was destroyed, were aimed at instilling fear into a target population in an attempt to secure a change in America's foreign policy (Pfaff, 2002). The objective was to use force as means of political persuasion. In the attacks of 11 September terrorists used commercial aircraft (and the people onboard) as the vehicle for an attack against the essential political, military, and economic institutions of the United States (Pfaff, 2002). In so doing, the terrorists who orchestrated the attacks on the United States demonstrated the intent and capacity to attack American sovereignty in ways that would have been inconceivable a decade ago.

THE EVOLUTION OF WAR

The French Revolution marked an historic change in inter-state military conflict characterized by a mass war model, which has several hallmark characteristics. First, it is intimately associated with state entities, both in the sense that states lead war efforts and war-making itself is directed almost exclusively at other states. Second, citizens are conscripted or volunteer to take up arms and serve their state for the honor and defense of their nation. Third, state directed military mobilization and war-making is materially and psychologically supported by the citizenry. At the very least, citizen-soldiers show up to fight and do not, in most cases, desert or refuse to fight once inducted. These three elements of the model characterize the way in which military mobilization was organized and war-making orchestrated. They are to be distinguished from another set of three characteristics of the mass war model that focus on the ways in which belligerents define and engage their military opponent. Fourth, the weapons used are generally the same on each side of the conflict (i.e. conventional weaponry). Fifth, one's enemy is clearly defined, both by national allegiance and by the uniform worn. And sixth, the *jus in bello* – the set of rules regulating the actual practice of war on the ground, conventionally embraced and formally articulated in international law – come under pressure from the expanding scale and scope of war as well as the development of increasingly destructive weapons (Best, 1994). These laws and rules provide a structure to military conflict, identifying, among other things, legitimate as opposed to illegitimate targets.

Central to the rules of war are historical protections afforded civilians that crystallized in the doctrine of noncombatant immunity, stipulating that civilians and their property were to be free from the intentional application of coercive force at the hands of the military. However, with the exception of its earliest years, mass war involves an erosion of the distinction between combatant and noncombatant during armed conflict, as militaries come to target not simply their opponent's instrument of war – the armed forces – but their opponent's *capacity* to make war – the resources, political support and even public resolve on which mass war depends (Kestnbaum, 2002).

The advent of nuclear weapons at the end of World War II ushered in a new era of war. In many respects the defining characteristics of war in the nuclear age are similar to those of the mass war era; militaries are associated with states; enemies are clearly defined; laws in the form of international treaties constrain the use of nuclear weapons; and the populations involved show support for the political and military positions of their respective nations. There is, however, at least one major difference. Whereas in the era of mass war a dominant question became how coercive force was targeted, in the nuclear era the weapons themselves are utterly indiscriminate in their effects. Nuclear weapons do not differentiate between soldier and civilian, combatant and noncombatant. The power and indiscriminate nature of these weapons leads to a new form of psychological and symbolic threat over and above their ability to wreak physical destruction. Their existence makes everyone a potential target. It is the psychological and symbolic threats connected to the awesome power of these weapons that ultimately facilitated a successful conclusion to the Cold War. According to Clausewitzian theory (1989), this was possible because in the face of mutual destruction without a realistic hope for victory, militaries and governments will seek to limit the incidence and effects of war.

The type of terrorist conflict in which the U.S. is now engaged appears to be the newest form of war. Terrorism is not new to the United States. There are ample cases of terrorist acts on American soil. Historically, acts of terrorism within U.S. borders have been of a domestic nature. The bombings of abortion clinics, the Unibomber, the Oklahoma City bombing, and the October 2002 Washington D.C. area sniper are examples. The use of terror in these types of domestic circumstances has had rather narrow foci and low levels of coordination. Moreover, they have all been prosecuted as criminal cases. Not until September 2001 did the U.S. consider itself under attack from a terrorist organization such that national security and sovereignty were at stake. Now that such issues are at stake, America finds itself in a state of war.

In the current era of asymmetric terrorist aggression we see a continuation of psychological and symbolic threat identified as characteristic of nuclear-age warfare. However, in terrorist warfare, the distinctiveness of the threat derives primarily from shifts in the characteristics of mass and nuclear warfare rather than from the use of a specific type of weapon. Perhaps the most important element of this shift is that like war in the nuclear era, terrorist war breaks down the barriers between soldier and civilian, producing a socialization of danger. Unlike nuclear war, however, where all members of a society are treated as an undifferentiated enemy subject to lethal attack, terrorist war intentionally targets civilians, and among them, is indiscriminate in its effects. Terrorism stands as a repudiation of the notion of noncombatant immunity, communicating to all members of a society that they too may be its victims.

One of the things that makes this mode of socializing danger possible is that terrorism is characterized by an asymmetry in group identity. Both parties need not be states or state affiliated entities. Generally one side is a state and the other side is a non-state terrorist group. The terrorists do not generally wear uniforms or visually identify themselves as clear enemies of the state. To the extent that terrorists do not affiliate themselves with a state, they are not bound by treaties, alliances, etc.[11] They do not recognize international law. Hence, they have no motivation to abstain from the use of unconventional weapons (e.g. nuclear, chemical, or biological agents) or tactics (e.g. suicide bombings). Indeed, unconventional weapons and tactics add to the psychological and symbolic effect of terrorist attacks.

The motivation fueling terrorist activity is to inflict psychological distress on the target community or society, combined with industrial, economic, and social disruption (Cooper, 2001). Terrorism represents an inversion of Clausewitz's famous dictum that "war is an extension of politics by other means" (Clausewitz, 1989), in which a wide array of military measures are applied to the achievement of specific, refined political objectives. Terrorist war, on the other hand, uses a particular means – indiscriminate attack on civilians – to achieve a diffuse set of political objectives that are necessarily constrained by what is attainable through the sowing of fear and doubt. The bombings of USS Cole and the American embassies in Kenya and Tanzania served notice that the U.S. was a target vulnerable to terrorist attack. The events of 11 September 2001 were a demonstration that the U.S. is vulnerable domestically as well as internationally. While the devastation experienced on that day was physical rather than biological, the subsequent anthrax contaminated mail incidents up and down the eastern seaboard[12] demonstrated that terrorist activity might just as easily be biological in nature. Bioterrorism presents unique challenges to national security and should be viewed as a fundamentally different form of threat to U.S. security than both conventional threats and terrorism using other means.

BIOTERRORISM AS WAR

Bioterrorism is the intentional use of biological agents on humans to induce fear and anxiety in them, often with a desire to exert influence over them. At least 17 nations have, or are suspected to have, biological weapons programs and an additional 12 nations have conducted biological warfare programs in the past (Inglesby et al., 2002; Parker, 2002). This includes only those nations for which we have intelligence. Parker (2002) notes that the General Accounting Office considers bioterrorism to be an increasingly severe threat to U.S. national security, and contends that the coming decade may see an increase in terrorist use of biological weapons.

Of the three classes of asymmetric weapons (chemical, biological, and nuclear) biological weapons hold the most potential for destruction (DiGiovanni, 1999; Parker, 2002). Unique characteristics of biological weapons that make them attractive for terrorist use include: high lethality facilitated by communicability; easy dispersal; easy production of large quantities; stable storage; resistance to environmental degradation, and lack of sensory cues indicating their presence (DiGiovanni, 1999; Parker, 2002). Biological agents may be targeted directly at humans through injection or topical application; deployed against agricultural crops, livestock, poultry, and fish; applied as a contaminant of food or drinking water; disseminated as an aerosol; or introduced through a natural vector such as an insect (Parker, 2002). A small amount of agent can easily be transported and released by a single person. Many very effective biological agents are relatively cheap and easy to produce and acquire (Parker, 2002). It is clear that biological weapons have numerous potential advantages from the view of terrorists given that they are considerably more limited in finances and manpower than most nations.

Biological weapons are also particularly capable of exploiting national security vulnerabilities such as porous borders and dense population centers. It is unlikely that government officials will order the military to seal U.S. borders. Such a task would be virtually impossible, requiring more than the stationing of military personnel at all points of entry. Given the size of U.S. territorial borders, troops and ships would need to be placed every 200 yards in order to ensure that no one got through the security net (Mabry, 1988).

Bioterrorism is unique in its psychological and physical (bodily) destruction given its potentially infectious nature. It is true that not all forms of biological agents available to terrorists are infectious. However, those that are infectious have natural appeal to terrorists. Most disconcerting is the rapidity with which a biological attack could progress. Unlike bombs, gun fire, or nerve gas, contagious biological agents spread well beyond the area in which they are unleashed. Micro-organisms reproduce in hosts at rapid rates; and only a small amount of pathogen is needed to begin a deadly chain reaction. Infection of communicable agents can spread quickly through a target population. Exacerbating the effect, infection could go undetected or misdiagnosed for days or weeks, by a public health infrastructure lacking experience in bioterrorist threats, sufficient surveillance capabilities, and treatment facilities. The result could be a major outbreak of disease before medical, civil, or military authorities were alerted to the danger (CDC, 2000; Parker, 2002).

Thus, the potential impact of a biological agent increases in both numbers and geographic scope well after it is initially released (CDC, 2000). For example, a contagious biological agent released in Chicago could spread regionally or nationally if its symptoms are not apparent for several days. The nature and scale of America's transportation system would greatly facilitate the spread of infectious biological agents. In terms of bodily harm and psychological distress, bioterrorism presents itself as the most potentially damaging form of terrorist threat (DiGiovanni, 1999; Parker, 2002).

Chemical weapons, like biologicals, are relatively cheap and easy to produce and disseminate. These weapons also have the potential to cause great physical harm and psychological distress. However, they have two drawbacks relative to biological weapons for those seeking to inflict maximal devastation. First, the human body shows signs of chemical agent reactivity relatively quickly rather than having long incubation periods prior to detection. Second, most chemical agents lack the capacity, inherent in many biological agents, to spread effectively and efficiently from one human to another. These two considerations do not preclude the likelihood of the use of chemical agents by terrorist groups. The distinction is made here to highlight the destructive capacity of biological weapons as potentially surpassing that of other asymmetric weapons.

The number of civilian casualties that the United States might suffer in the aftermath of a bioterrorist attack could eclipse any previous experience in war (CDC, 2000). However, the explicit focus on civilian targets would merely be an extension of historical trends dating from the end of the nineteenth century and the advent of mass war. "Whether as leaders or led, perpetrators or victims, "the people," have come to play perhaps the central part in armed struggle [in the modern era]. They have been brought into conflict as soldiers, militias, guerillas and paramilitaries. But they have also been drawn in as laborers and manufacturers; volunteers and officials; taxpayers and voters; the families of those lost; the dislocated; *targets and casualties*" (Kestnbaum, 2002, p. 2 *emphasis added*). Indeed, Clausewitz (1989) was correct in observing that in the modern era war again had become the business of the people. Around the turn of the nineteenth century, entire societies became the foundation from which states mobilized in order to wage war (Kestnbaum, 2002; Segal, 1994; Tilly, 1992, 1996). More importantly, however, is the fact that these same popular masses became legitimated as "the enemy against whom such mobilizations were directed" (Kestnbaum, 2002, p. 2).

Rising civilian death tolls in warfare are typically attributed to the historical development of "total war." The deployment of new military technologies and the breakdown of social constraint on the use of weapons of mass destruction against civilian populations are assumed to play central roles. The soldier-to-civilian death ratio has undergone dramatic change from World War I (20:1), to World War II (1:1), to the Korean War (1:5) (Shaeffer, 1989). However, in the modern world this obscures the breakdown of the distinction between noncombatants and combatants. International law distinguishes between noncombatants and combatants as opposed to innocent and guilty or civilians and soldiers. Some civilians, like munitions factory workers who, because they are engaged in an activity that is logically inseparable from war-fighting, have come to be seen as legitimate targets of warfare as much as any uniformed soldier (Pfaff, 2002). Bin Laden has expressed in very clear terms that he makes no distinctions between Americans wearing military uniforms and those that don't. To be American, by definition, is to be the enemy and, thereby, a legitimate target.

CONCLUSION

The socio-political and demographic factors, and the disparities in values and norms identified in much of the pre-September 11 literature on civil-military relations, will continue to be important factors in understanding the civil-military dynamics in the United States. However, these issues may take a back seat to factors directly related to the U.S. war on terrorism and derived from current strategic realities. Legal and popular distinctions between acts of war and criminal activities; criminals, noncombatants, and combatants; the limits to the application of *habeas corpus*; and the proper definition of *posse comitatus* in the war on terrorism will become driving influences in how Americans perceive the threat posed by terrorists. These issues will be critical in defining how the American public views the military, the government and law-enforcement agencies. Additionally, these factors will determine how these groups perceive and receive each other, setting up a complex web of modern civil-military relations.

Decisions yet to be reached about the appropriate functional role of civilian vs. military spheres and the specified division of labor between them; perceptions of cultural, ethnic, and religious components to the current conflict; and the ultimate salience of the bioterrorist threat will also be major factors in understanding civil-military relations in the post-September 11 era. Definitive answers about the proper division of labor between civilian law-enforcement agencies and the military have not yet been issued. Functional clarity depends on how American society in general, and our government and courts particularly, choose to frame the current national security crisis. If terrorist threats are defined as a criminal, one set of scenarios is likely to unfold. If we are at war, we should expect a completely different civil-military dynamic.

The threat of bioterrorist attacks against the U.S. is a real and growing concern. The National Guard and U.S. Coast Guard have numerous historical precedents of supporting civilian agencies during times of crisis. Given their organizational capacity, strategic and tactical training, and specialization in security measures, they present themselves as obvious and attractive agents for dealing with bioterrorist attacks. At the institutional level, the Coast Guard has been relocated from the Department of Transportation to the newly-formed Department of Homeland Defense. In order to best carry out missions to support civilian institutions (e.g. law enforcement, medical professionals), adjustments in organization and training are necessary for both uniformed services.

The future of civil-military relations in America depends largely on factors that have yet to play themselves out. What *is* clear, at this point, is that we are in the midst of a change in American civil-military relations, the permanence of which will coincide with the longevity of the current national security crisis and the attendant socialization of danger that we should expect. Should we completely abandon traditional models of civil-military relations? Certainly not. But to ignore the emergence of the factors identified here, or to blindly attempt to force them into old molds and paradigms, is to ignore an empirical reality, the full understanding of which demands more of our time and much more of our effort.

NOTES

1. It is tempting to frame our analysis in terms of the broader notion of asymmetric warfare, since the arguments we make in this paper may be applied to a wide range of settings, including those in which vastly unequal forces are pitted against one another and one side may make use of irregular fighters employing unconventional tactics. However, this would serve only to shift the emphasis away from our central argument. Terrorism may be a form of asymmetric warfare, but what distinguishes it is the fact that it intentionally targets civilians, and that among civilians, it is indiscriminate in the devastation it wreaks. Terrorism is important because of the way in which it socializes danger, breaking down the barriers between combatant and noncombatant and subjecting all to the worst of harrowing and potentially lethal attacks. It is this socialization of danger produced by terrorism, in turn, that is critical in assessing whether and how civilian and military authorities elect to treat its use against their own societies not as a crime, but as an act of war. Bioterrorism in turn, as we argue below, has unique attributes that distinguish it from other forms of terrorism.

2. The European Union and NATO "create new layers of civil-military relations as they anticipate military cooperation and coordinated civilian control of military activities across national borders" (Burk, 2002, p. 21).

3. For example, Dunlap (2001) cites the Pentagon's recent formation of the Joint Task Force Network Defense, aimed at defending DOD computers from cyberattack, as a move by the military to extend their operational and jurisdictional boundaries in response to the combination of changes in technology and threats to the military and/or state.

4. For a fuller discussion of how the advent of nuclear weapons has limited both time and space in modern warfare, and the implications for how states structure their national security arrangements see McLauchlan (1989).

5. "The Act requires the President to report to the Congress on all unauthorized hostilities, and prohibits their continuance beyond 60 days from initiation, except as provided by the Congress" (Segal, 1994, p. 380).

6. These Congressional Acts placed limitations on "freedom of speech and of the press, and classified certain kinds of statements and actions as seditious and illegal" (Segal, 1994, pp. 377–378).

7. See also Dahl (1989).

8. McVeigh's *criminal* trial and eventual execution for his *crimes* would seem to render a definitive answer to this question. However, in the context of the attacks of 11 September we would like to suggest that the theoretical problem of distinguishing between acts of war and criminal acts is anything but resolved.

9. When terrorist attacks are deemed to threaten civil society or the state directly, it matters not where the military is called upon to operate, nor even the source of the threat to which they respond, but rather the fact that the state or civil society requires protection that allows the military to play an active role.

10. "[A]n enemy combatant who without uniform comes secretly through the lines for the purpose of waging war by destruction of life or property [is] not entitled to the status of prisoner of war, but [they are] offenders against the law of war subject to trial and punishment by military tribunals" (Luban, 2002, p. 11).

11. This point has significant implications for long range resolution possibilities. How does a state, or group of states, negotiate with non-state entities? All formal mechanism of

negotiation at that level are designed for interstate mediation. Moreover, the U.S. has long held the position that it will not negotiate with terrorists.

12. As of the date this article was written it is not known whether this was an international act of terrorism.

ACKNOWLEDGMENTS

This research was funded in part by the Army Research Institute under Contract DASW 0100K16. The views expressed here are those of the authors, and not of the Army Research Institute, the Department of the Army, or the Department of Defense.

REFERENCES

Aron, R. (1968). *Peace and war*. New York: Praeger.

Andreski, S. (1968). *Military organization and society* (2nd ed.). Berkeley: University of California Press.

Bianco, W. T., & Markham, J. (2001). Vanishing veterans: The decline of military experience in the U.S. Congress. In: P. D. Feaver & R. H. Kohn (Eds), *Soldiers and Civilians: The Civil-Military Gap and American National Security* (pp. 275–288). Cambridge, MA: MIT Press.

Burk, J. (2002). Theories of democratic civil-military relations. *Armed Forces & Society*, 29(29), 7–29.

CDC (Centers for Disease Control and Prevention) (2000). Biological and chemical terrorism: Strategic plan for preparedness and response recommendations of the CDC strategic planning workgroup. *Morbidity and Morality Weekly Report*, 49, RR-4.

Clausewitz, C. (1989) [1976]. *On war*. M. Howard & P. Paret (Ed. and Trans.). Princeton: Princeton University Press.

Cohen, E. A. (2001). The unequal dialogue: The theory and reality of civil-military relations and the use of force. In: P. D. Feaver & R. H. Kohn (Eds), *Soldiers and Civilians: The Civil-Military Gap and American National Security* (pp. 429–458). Cambridge, MA: MIT Press.

Comey, J. B., Clement, P. D., Salmons, D. B., Srinivasan, S., Marcus, J. L., & Bruce, E. B. (2002). *Respondents' response to this court's October 21, 2002 order*. Jose Padilla and Donna R. Newman, Next Friend of Jose Padilla (Petitioners) v. George W. Bush, Donald Rumsfeld, John Ashcroft, and Commander M. A. Marr (Respondents): United States District Court for the Southern District of New York.

Cooper, H. H. A. (2001). Terrorism: A problem of definition revisited. *American Behavioral Scientist*, 44(6), 881–893.

Dahl, R. A. (1989). *Democracy and its critics*. New Haven: Yale University Press.

Desch, M. C. (1999). *Civilian control of the military: The changing security environment*. Baltimore: Johns Hopkins University Press.

DiGiovanni, C., Jr. (1999). Domestic terrorism with chemical or biological agents: Psychiatric aspects. *American Journal of Psychiatry*, 156(10), 1500–1504.

Dunlap, C. J. (2001). The thick green line: The growing involvement of military forces in domestic law enforcement. In: P. B. Kraska (Ed.), *Militarizing the American Criminal Justice System* (pp. 29–42). Boston: Northeastern University Press.

Epstein, L., & Walker, T. G. (1995). *Constitutional law for a changing America: Institutional power and constraints.* Washington, DC: Congressional Quarterly Press.

Feaver, P. D. (1995). Civil-military conflict and the use of force. In: D. M. Snider & M. A. Carlton-Carew (Eds), *U.S. Civil-Military Relations: In Crisis or Transition?* (pp. 113–144). Washington, DC: Center for Strategic and International Studies.

Feaver, P. D., & Kohn, R. H. (2001). Conclusion: The gap and what it means for American national security. In: P. D. Feaver & R. H. Kohn (Eds), *Soldiers and Civilians: The Civil-Military Gap and American National Security* (pp. 459–474). Cambridge, MA: MIT Press.

Hammond, M. C. (1997). The Posse Comitatus Act: A principle in need of renewal. *Washington University Law Quarterly, 75*(2), 1–32.

Holsti, O. R. (2001). Of chasms and convergences: Attitudes and beliefs of civilians and military elites at the start of a new millenium. In: P. D. Feaver & R. H. Kohn (Eds), *Soldiers and Civilians: The Civil-Military Gap and American National Security* (pp. 15–100). Cambridge, MA: MIT Press.

Howard, M. (1976). *War in European history.* New York: Oxford University Press.

Huntington, S. P. (1989) [1957]. *The soldier and the state.* Cambridge: The Belknap Press of Harvard University.

Huntington, S. P. (1993). The clash of civilizations? *Foreign Affairs, 72*(3), 22–50.

Huntington, S. P. (1996). *The clash of civilizations and the remaking of world order.* New York: Simon & Schuster.

Inglesby, T. V., O'Toole, T., Henderson, D. A., Bartlett, J. G., Ascher, M. S., Eitzen, E., Friedlander, A. M., Gerberding, J., Hauer, J., Hughes, J., McDade, J., Osterholm, M. T., Parker, G., Perl, T. M., Russell, P. K., & Tonat, K. (2002). Anthrax as a Biological Weapon, 2002: Updated recommendations for management. *Journal of the American Medical Association, 287*(17), 2236–2252.

Jacoby, T. A. (2002). To serve and protect: The changing dynamics of military and policing functions in Canadian foreign policy. Paper presented at the 2002 meeting of the Inter-University Seminar on Armed Forces and Society – Canada. Kingston, Ontario.

Janowitz, M. (1971) [1960]. *The professional soldier.* New York: Free Press.

Jenkins, B. M. (2001). Terrorism: Current and long term threats. In: *Rand Testimony.* Washington, DC: RAND Corporation.

Kestnbaum, M. (2002). Bringing the people into war: A world-historical transformation with global implications. In: S. Douailler, P. Vermeren & J. Riba (Eds), *Philosophie et Globalisation.* Paris: Editions l'Harmattan.

Lasswell, H. (1941). The garrison state. *American Journal of Sociology, 46*(January), 455–468.

Lasswell, H. (1997) [1962]. The garrison state hypothesis today. In: J. Stanley (Ed.), *Essays on the Garrison State* (pp. 77–116). New Brunswick: Transaction.

Luban, D. (2002). The war on terrorism and the end of human rights. *Philosophy of Public Policy Quarterly, 20*(3), 9–14.

Mabry, D. (1988). The U.S. military and the war on drugs in Latin American. *Journal of Interamerican Studies and World Affairs, 30*(2/3), 53–76.

Manicas, P. T. (1989). *War and democracy.* Cambridge: Basil Blackwell.

McIsaac, J. F., & Verdugo, N. (1995). Civil-military relations: A domestic perspective. In: D. M. Snider & M. A. Carlton-Carew (Eds), *U.S. Civil-Military Relations: In Crisis or Transition?* (pp. 21–33). Washington, DC: Center for Strategic & International Studies.

McLauchlan, G. (1989). World war, the advent of nuclear weapons, and global expansion of the national security state. In: R. Shaeffer (Ed.), *War in the World System* (pp. 1–8). New York: Greenwood Press.

McPherson, J. M. (1997). From limited to total war in America. In: S. Forster & J. Nagler (Eds), *On the Road to Total War: The American Civil War and the German Wars of Unification, 1861–1871* (pp. 295–309). Cambridge: Cambridge University Press.

Neely, M. E. (1997). Was the civil war a total war? In: S. Forster & J. Nagler (Eds), *On the Road to Total War: The American Civil War and the German Wars of Unification, 1861–1871* (pp. 29–51). Cambridge: Cambridge University Press.

Nixon, R. (1994). *Beyond peace*. New York: Random House.

Paret, P. (1992). *Understanding war: Essays on Clausewitz and the history of military power*. Princeton: Princeton University Press.

Parker, H. S. (2002). *Agricultural bioterrorism: A federal strategy to meet the threat*. Washington, DC: Institute for National Strategic Studies, National Defense University.

Pfaff, T. (2002). Noncombatant immunity and the war on terrorism. Paper presented at the 2002 meeting of the Inter-University Seminar on Armed Forces and Society – Canada. Kingston, Ontario.

Rakove, J. N. (1996). *Original meanings: Politics and ideas in the making of the constitution*. New York: Alfred A. Knopf.

Roman, P. J., & Tarr, D. W. (2001). Military professionalism and policymaking: Is there a civil-military gap at the top? If so, so does it matter? In: P. D. Feaver & R. H. Kohn (Eds), *Soldiers and Civilians: The Civil-Military Gap and American National Security* (pp. 403–428). Cambridge, MA: MIT Press.

Segal, D. R. (1994). National security and democracy in the U.S. *Armed Forces & Society, 20*(30), 455–468.

Segal, D. R. (1995). U.S. civil-military relations in the twenty-first century: A sociologist's view. In: D. M. Snider & M. A. Carlton-Carew (Eds), *U.S. Civil-Military Relations: In Crisis or Transition?* (pp. 185–200). Washington, DC: Center for Strategic and International Studies.

Shaeffer, R. (1989). Introduction. In: R. Shaeffer (Ed.), *War in the World System* (pp. 1–8). New York: Greenwood Press.

Shanahan, D. J. (2002). The army's role in homeland security. In: W. Murray (Ed.), *Transformation Concepts for National Security in the 21st Century* (pp. 285–309). Carlisle, PA: Strategic Studies Institute.

Snider, D. M., & Carlton-Carew, M. A. (1995). The current state of U.S. civil-military relations: An introduction. In: D. M. Snider & M. A. Carlton-Carew (Eds), *U.S. Civil-Military Relations: In Crisis or Transition?* (pp. 1–20). Washington, DC: Center for Strategic and International Studies.

Stanley, J., & Segal, D. R. (1997). Conclusion: Landmarks in defense literature. In: J. Stanley (Ed.), *Essays on the Garrison State* (pp. 127–134). New Brunswick: Transaction.

Tilly, C. (1992). *Coercion, capital, and European states, AD 990–1992*. Cambridge: Basil Blackwell.

Tilly, C. (1996). The emergence of citizenship in France and elsewhere. In: C. Tilly (Ed.), *Citizenship, Identity, and Social History* (pp. 223–236). Cambridge: Press Syndicate of the University of Cambridge.

Weigley, R. F. (2001). The American civil-military cultural gap: A historical perspective, colonial times to the present. In: P. D. Feaver & R. H. Kohn (Eds), *Soldiers and Civilians: The Civil-Military Gap and American National Security* (pp. 215–246). Cambridge, MA: MIT Press.

INTEGRATION OR DISINTEGRATION? AN EXAMINATION OF THE CORE ORGANIZATIONAL AND MANAGEMENT CHALLENGES AT THE DEPARTMENT OF HOMELAND SECURITY

Nancy M. Dodge, Carlton J. Whitehead and Brian J. Gerber

ABSTRACT

The attacks of September 11th transformed homeland security into a central policy task for governments in the U.S., culminating in the creation of the Department of Homeland Security. Planning and preparation for counter terrorism were no longer secondary priorities. This article seeks to examine some of the salient organizational and management issues that could potentially facilitate or impair DHS's successful integration of its varied 22 agencies, and its subsequent execution of its critical tasks associated with countering terrorism and bioterrorism. Characterizing this change as a type of punctuated equilibrium, this article closes by suggesting that a differentiated network structure offers a potentially powerful mechanism

Bioterrorism, Preparedness, Attack and Response
Advances in Health Care Management, Volume 4, 345–366
© 2004 Published by Elsevier Ltd.
ISSN: 1474-8231/doi:10.1016/S1474-8231(04)04014-5

by which the DHS could proactively and effectively address many of these
leadership, management and organizational challenges.

INTRODUCTION

There is nothing more difficult to carry out, nor doubtful of success, nor more dangerous to
handle than to initiate a new order of things. For the reformer has enemies in all those who
profit by the old order, and only lukewarm defenders by all those who could profit by the new
order. This lukewarmness arises from the incredulity of mankind who do not truly believe in
anything new until they have actual experience with it.

                                                              Niccolo Machiavelli

On November 25th, 2002, President George W. Bush signed into law the *Homeland Security Act of 2002* (Public Law, 107–296) establishing the Department of
Homeland Security (DHS). The new department was created through reorganization of a number of federal agencies possessing security-relevant tasks with the
objective of promoting greater coordination in counterterrorism actions among
those agencies. Creating the new department represented the federal government's
largest reorganization since the Department of Defense was established in 1947
and signifies one of the Bush Administration's most important policy responses to
the events of September 11. Indeed, it has been heralded by the administration as
a major and necessary step in the fight against terrorism.

Specifically, the DHS is charged with the primary missions of "preventing
terrorist attacks within the United States, reducing the vulnerability of the United
States to terrorism at home, and minimizing the damage and assisting in the
recovery from any attacks that may occur [www.whitehouse.gov/deptofhomeland/
analysis/title1.html]." Its operational activities are divided among five major
directorates: border and transportation security, emergency preparedness and
response, science and technology, information analysis and infrastructure protection, and management. At the same time, several other offices/agencies will report
directly to the Secretary rather than to the Under Secretaries heading the several
directorates.[1]

All told, creating the new DHS involved the relocation of twenty-two federal
agencies. Bureaucratic movement on this scale is anything but trivial and carries
with it important complexities. For instance, coupled with the core tasks of
countering terrorism (including acts of bioterrorism) and coordinating emergency
responses to such events, the DHS has also been charged with fulfilling other
significant, but non-security related, functions of the absorbed agencies. As a
result, the DHS is a "blended," extremely complex organization charged with
a series of critical tasks that do not necessarily lend themselves to effective
integration.

An essential question to ask is whether a federal bureaucratic reorganization per se is sufficient to meet the extraordinarily difficult tasks associated with the wide-ranging mission of protecting the country from terrorist acts by not-so-easily identified adversaries.[2] If one accepts the premise that reorganization is indeed necessary to accomplish the stated mission (if perhaps not sufficient), then an equally trenchant issue to consider is whether the organizational and management challenges the new department faces make success in meeting its broad mission likely – or not.

This paper focuses on the latter issue by attempting to identify and outline several salient organizational and management issues that potentially facilitate or impair DHS's successful integration of varied agencies, and its subsequent execution of its critical tasks associated with countering terrorism and bioterrorism. We do so by considering the context of policy change that leads to the creation of the department and by outlining a litany of management challenges for the leadership at DHS: crafting an organizational vision, sustaining policy "urgency," adopting effective change management, fostering a new agency culture, accomplishing information sharing and routine task integration, and identifying relevant external constraints on agency activities. After providing brief discussion of each of those management challenges at the new DHS, we close by suggesting that a differentiated network structure offers a potentially powerful mechanism by which the DHS could proactively and effectively address many of these leadership, management and organizational challenges.

TERRORISM AND BIOTERRORISM IN THE U.S.: RADICAL CHANGE

Following the collapse of the former Soviet Union and with it the end of the Cold War, the American public perceived the nation to be facing a significantly decreased level of threats from abroad. A variety of national polls between 1998 and 2000 suggested a high degree of public indifference to the idea of existing security threats. When asked about *foreign policy* generically, respondents indicated the subject was only minimally important to the activities of the federal government. Indeed, national polling consistently showed that the American public virtually disregarded terrorism as a threat when asked to identify the most important issues facing the country – only about 1% of the public typically named terrorism as most important[3]. However, from the perspective of federal government policy-makers, the importance of the threat of terrorism was not unnoticed, especially given events such as the World Trade Center bombing in 1993, the sarin gas attack in a Tokyo, Japan, subway in 1995, and the Murrah Building bombing in Oklahoma

City in 1995. Recognition of the danger of such threats precipitated President Bill Clinton's Presidential Decision Directive (PDD) 39 which established the Federal Bureau of Investigation (FBI) and the Federal Emergency Management Agency (FEMA) as the lead agencies for domestic counterterrorism efforts in 1995. PDD 39 (also known as the United States Policy on Counterterrorism) is significant in that it represented the first time in American history that the federal government had formally tasked a lead agency to deal with acts of terrorism. In 1998, President Clinton supplemented PDD 39 with a second directive, PDD 62, spelling out with greater detail the roles and responsibilities of federal agencies in the area of counterterrorism.

Awareness of and initial efforts to address terrorist threats to American targets notwithstanding, there was "almost no effective coordination of federal efforts or budgets" in addressing potential terror threats (Cohen & Cook, 2002). Several federal training and grant programs were not efficiently implemented, were hampered by parochial competition over resources, and generally failed to appropriately direct resources to first responders (Smithson & Levy, 2000). Pre-September 11, absent significant public pressure and absent a specific focusing event, orienting the federal bureaucracy (and other levels of government) to address terrorism as a significant security threat can be fairly characterized as a largely incremental or uneven process.

The events of September 11, 2001 undeniably altered the policy priorities of both key government decision-makers and the general public in a dramatic, and anything but incremental, way. Among the many consequences of September 11 was the perceived need to commit greater federal resources to a coherent, coordinated federal counterterrorism response – hence the creation of the DHS. Considering the nature of the change that ultimately brought about a reorganization of that scale, however, has important implications for understanding the likely degree of efficacy for future DHS operations. Or to put it another way, the nature of the change context has implications for the organizational and managerial implications for the navigation of such a radical change.

Some fundamental differences exist among theories of organizational change. Arrow (1997) divides existing models into four primary categories including: (1) robust equilibrium (Bales, 1953), change that is internally drive and represents a homeostatic model; (2) life cycle (Tuckman, 1965), change that is internally driven and follows a fixed progression of incremental change; (3) adaptive matching (Steiner, 1972), change that is externally driven and is basically frictionless; (4) punctuated equilibrium (Gersick, 1991; Gould & Eldridge, 1977), change that is a combination of internal and external forces and can include both step and discontinuous change. In other words, under this type of organizational model, periods of inertia or stability are punctuated by intense periods of discontinuous change (Gersick, 1991) akin to those found in the sciences (Kuhn, 1970) and in nature (Gould & Eldridge, 1977).

Technologically based for-profit firms have been cited by many as the primary exemplar of organizational punctuated equilibrium in that they have cycled through extensive periods of consistent strategies, structures and processes, followed by radical reinvention including the creation of new visions, structures and processes (Christensen, 1997; Tushman, Newman & Romanelli, 1986). Compelling evidence suggests that revolutionary change within organizations holds more promise for organizational success than do less revolutionary changes (Miller & Friesen, 1984; Viranany, Tushman & Romanelli, 1992). A myriad of propositions exist as to strategic drivers of punctuated equilibrium. Organizational inertia linked to entrenched strategies, structures, processes, and culture is frequently cited as a top "driver" of punctuated equilibrium. Others argue that organizational inertia is linked to strategic efforts to reduce agency costs (Galar & McKenzie, 1998).

Accepting that the events of September 11th represent a shock or intervention that interrupted policy stasis at the agencies that now comprise the DHS, the primary tenets of punctuated equilibrium model are worth noting.

As aforementioned through Arrow's (1997) typology of change, some change is characterized as Hassner writes, incremental changes in ideas that in turn elicit incremental changes in behavior, strategies, structure, and processes. In this type of change, there exists a symbiotic relationship between changes in that a small change here may lead to a small change there and a small change here may lead to another small change there. The punctuated equilibrium model, depicts change as being ruled by episodic radical change that penetrates prolonged organizational stasis. What this means in terms organizational and management implications is that change will have substantial and unanticipated influence upon ideas, behavior, strategies, structure and processes. Punctuated equilibrium shocks, caused by both internal and external forces, can drive instantaneous upheavals beyond organizational and managerial expectations. Organizational and managerial contradictions abound as do unpredictable and erratic outcomes and outputs. Thus, forecasting is made difficult due to the unanticipated nature of internal and environmental shocks (Hassner, http://www.stanford.edu/~rony/chess.htm).

Despite the difficult nature of punctuated equilibrium, particularly in terms of the associated organizational and managerial implications for the DHS, certain characteristics of organizational survivability are posited. Carrying out the nature analogy of punctuated equilibrium, some argue that like successful species, successful organizations must be able to respond to environmental change; tolerate stress; compete; exploit new niches; take risks/mutate and develop symbiotic relationships (Price & Evans, 1998). Others argue that radical organizational change cannot occur unless there is power dependence and capacity for action. As Greenwood and Hinings (1996) write,

> power dependence is an existing archetype that gives power to some groups within the orga-
> nization, but marked contexts can shift the balance of power and reconfigure relations within
> the organization. 'Capacity for action' is that ability to manage the transition process from one
> template to another in that . . . different groups within the organizations carry different interests
> and values, and favor the organizational template that better serve their interests and it is only
> when they identify the existing template as one of the causes of their disadvantage, they can
> recognize the possibility of an alternative template.

The primary implication of interpreting the creation of the DHS from a punctuated equilibrium framework is simple: the managerial and organizational issues surrounding radical change add to the complexity of DHS accomplishing an already broadly defined mission of protecting the nation against terrorist attack. We outline several of these issues of managerial and organizational challenges for the DHS below.

ORGANIZATIONAL AND MANAGEMENT ISSUES

Management Issue: Crafting Agency Vision

A central mechanism that builds organizational cohesiveness is its overall vision of what the organization should be trying to accomplish, which in turn guides its behavior as it acts on specific tasks (Kotter, 1995). Some (e.g. Wilson, 1998) contend that vision should be the first step in creating and/or transforming individual, group and organizational change. What is difficult about building a vision for the DHS is the nebulous and expansive nature of the policy problem that DHS was established to address. Indeed, while the Homeland and Security Act is forthright in its mission as outlined above, what is much less clear is what precisely defines acts of terrorism and what exactly constitutes a level or degree of terrorist threat and the national vulnerability to such threats.

Inextricably linked to vision and strategy is the subsequent defining of core processes, procedures and critical tasks. A major challenge is whether or not the DHS will be able to realize its stated functions precisely "at the same time" the concept of terrorism and homeland security is complex, vague and sensitive. The expanse of literature on visions and overarching strategies indicate that visions and overarching strategies should be simple and easy to explain. Verbiage for the DHS's vision is straightforward and even compelling. What is less straightforward is how the DHS's vision should be executed. If Kotter (1995) is correct in his assertion that a vision should "say something that clarifies the direction that an organization needs to move" (p. 63), the DHS's mission statement may miss the mark in that the mission statement does not map a clear direction for the DHS to follow. To identify terrorism prevention, vulnerability reduction, and damage mitigation perhaps gives the DHS a sense of what it is charged with accomplishing, but does not suggest a directive of how to best achieve the "charge."

This issue is critical in a variety of ways. For instance, in naming the FBI and FEMA as the two lead counterterrorism agencies, PDD 39 highlights an important difference in functional roles. FEMA is geared toward consequence management as opposed to crisis management. Consequence management "addresses the effects of an incident on lives and property. It includes measures to protect public health and safety, treat persons injured, mitigate impacts, restore essential government services, and provide emergency relief . . . " (U.S. GAO, 2002, p. 5). FEMA is oriented toward consequence management and is the lead federal agency in this regard in most disaster cases.

However, consequence management should not be conflated with crisis management, which can be defined this way: "crisis management focuses on causes and involves activities to address the threat or occurrence of a terrorist incident. It is predominantly a law enforcement and intelligence function that includes measures to anticipate, prevent, and resolve a threat or act of an incident on lives and property" (U.S. GAO, 2002, p. 5). This distinction is a critical one in terms of thinking about how to manage an administrative response to counteract terrorist threats. Consequence management is likely quite compatible with both the political and management orientation of a decentralized governance model (where significant resources and policy authority is dispersed across the federal system of governments). But on a practical management basis, crisis management likely will necessitate greater centralization in both standard setting and basic coordination of government activity at all levels.

Defining management logic in terms of integrating crisis and consequence management at DHS is a central task the department's leadership must tackle. How it does so also carries implications for whether or not DHS will represent the focal point for a national counterterrorism strategy by effectively integrating subnational government entities toward this mission. That is, the way it allocates resources and distributes authority within the department and without toward state and local governments begs the question of whether we will have a clearly defined domestic security strategy and whether, over time, it will be the job of the DHS to establish an integrative national security strategy at all levels including local, state and federal? At this point, the enabling statute creating DHS and related documents suggest an affirmative answer to both, but as a practical matter the efficacy of those actions depend in large part on how effective the I¹HS is in defining the precise nature of their policy problem and task environm ¿nt.

When implementing strategy and or vision, a clear definition of the policy problem to be addressed is crucial. Defining a policy problem in a helpful manner is one of the top criterions policy analysts employ in their critique of report writing (Susskind et al., 2002). Does the Homeland Security Act adequately define the policy problem of terrorism and to what extent did the authors of the bill understand the nature of terrorism? In order for the problem definition to be helpful, it must clearly define who gains and who loses from the problem's existence and be clear about which stakeholders favor or oppose particular solution or policy options (Susskind et al., 2002). Clearly, the assessment criterion for superior policy report writing is different on numerous levels than writing legislation, but what arguably remains congruent is the need for a clear problem definition in that it is inextricably linked to multiple organizational and management contingency success factors.

For example, problem definition guides "solutions" related to redress of the problem. These solutions can take the form of structures, processes, procedures and critical task definitions. In order for the DHS to effectively and efficiently address the problem of homeland security, there must be systemic agreement on the problem definition. It is easy to see that vision and strategy are closely linked to problem definition and that some form of consensus must be formed.

Management Issue: Creating a Sustainable Sense of Urgency

Kotter (1995) employs the useful metaphor of a "burning platform" to indicate what prompts change in organizations – that is, there needs to exist urgency for change. Organizational change and individual change require high levels of energy, desire, and skill. Change, by definition, implies making either an essential difference that frequently creates a loss of original identity or creates an environment in which the substitution of one thing or a multitude of things [processes, personnel, incentives, etc.] are replaced by other things. Transformation, on the other hand, implies radical difference. Indeed, 9/11 represents a watershed crisis, in that both core operating processes and informational integration between the several relevant bureaucratic agencies were called into question. What is complicated about the transformation and change required by DHS is that it will not only require the loss of original identity, particularly when new processes and tasks are taken on by the absorbed agencies, but also it will also require substitutions, major transformations and possible eradication of processes or tasks. Further complicating matters is the fact that the absorbed agencies will still carry out many of their originally assigned functions. Therein lies another potential challenge in that DHS is simultaneously calling for transformation, assimilation and maintenance.

Maintenance consists of two important dimensions in this context. The readily apparent issue for the various merged agencies is maintenance of core functional tasks prior to reorganization that were not necessarily related to national security generally, and counter-terrorism and bioterrorism specifically. Maintenance also has the dimension of maintaining administrative commitment across a network of linked government actors – federal, state and local – encompassing public service personnel in law enforcement, emergency response, elected officials, and so on. These actors are necessarily linked if the DHS is to succeed in providing effectively coordinated counter-terrorism actions. But this also implies a potential challenge in the maintenance of policy urgency among the highly diverse set of actors and interests, many of whom will likely not perceive the threat of terrorism as an immediate or tangible one.

At a recent conference on the protection of critical infrastructure hosted by the Council of State Governments (CSG) the comments of Nancy Wong, Director of the Office of Planning and Partnership in the Information Analysis and Infrastructure Protection Directorate at the DHS illustrate the point clearly. Echoing the remarks of several state-level officials who head public safety agencies, Wong indicates that "the biggest challenge I have seen personally in state and local governments is that they need [to maintain] a level of interest and attention to homeland security" due to the fact that many governments do not appear to be very close to the terrorist threat "front line." She remarked further that "sustainability is going to be the greatest challenge for us in Homeland Security, as much as it will be for the states" (CSG, 2003).

Management Issue: Building a Management Team Capable of Implementing Changes

An important element of any successful organizational transformation is that the leadership and management provide a clearly defined direction for the changes that must occur (Kotter, 1995). The importance of building an effective team capable, creative, energetic and powerful enough to lead an organizational transformation should not be understated, and the new DHS is no exception. All of the transformation steps (Kotter, 1995) organizational learning strategies (Senge, 1990), and absorptive capacity building strategies (Zahra & George, 2002) would be fruitless if there were not a competent team to initiate and sustain internal agency changes and be capable of responding to external changes to the DHS. Section 102 of the Homeland Security Act, establishes the *Secretary of*

Homeland Security, by Presidential Appointment and confirmed by the Senate, as the head of the DHS. The *Secretary* will have full authority and control over the Department and the duties and activities performed by its personnel (http://www.whitehouse.gov/deptofhomeland/analysis/title1.html). Section 103 creates the personnel structure that will support the *Secretary* including a senior management team of up to 12 confirmed officials. As of this writing, some of these respective individuals have been appointed. Clearly, the Act does not outline all of the personnel required to fulfill the primary missions and responsibilities of DHS, but what is implicit is that the boundaries for teambuilding to meet the mission and functional tasks of DHS will be fluid and perhaps extend beyond the agencies absorbed in DHS.

Sub-challenges related to team and/or coalition building are vast. How big should the core-leadership team be? Sections 102 and 103 outline a head *Secretary* and 12 top senior management members. Will this be too many? Will it be too few? Will this top tier be capable of initiating major transformation? What is needed of this group is a shared sense of urgency, assessment of the policy problem, and a collective view of organizational challenges and potential "fixes." Moreover, there needs to be some basic level of "trust" and open communication the latter of which many scholars posit are the most premier ingredient to forming a "winning" and effective transformation team (Kotter, 1995). Even if the top tiers of DHS management are able to have synergy on these core issues, will they be able to penetrate the DHS system with congruent senses of urgency, assessments and commitment; particularly since the 22 agencies have traditionally have had very disparate levels of urgency, assessments and fixes for issues related to homeland security?

Management Issue: Fostering a New Agency Culture

The integration of the disparate cultures of the 22 agencies previously separate and distinct from one another into a coherent whole is arguably one of the most substantial elements to creating a sustainable and effective DHS. There seems to be more of a consensus than debate on significance of culture within organization transformation. However, switching organizational direction and task is not a benign or simple process, particularly on the scale of DHS with its undeniably complex mission. In such a case, creating a meaningfully coherent sense of a unified organization among its personnel is likely a critical factor to potential success in managing the department.

The mere blending of the institutional norms of the department's newly merged units is a daunting task. Each of those 22 units carry defined task and mission orientations – with attendant cultural ways of accomplishing them. An important issue is whether tasks not originally part of the absorbed units' culture will attract the same energy and resources as tasks that are deeply rooted in agencies' persona? Given the prospect for competing agencies cultures, it is possible that action agendas may be set and manipulated in accordance to cultures that dominate within the DHS. Likewise, the possibility that the absorbed agency units might resist new "cultural" imperatives that appear to be unfamiliar or incompatible with previous lines of thinking? It is also possible that there may be significant cultural domination or imperatives from outside of the DHS. This type of interplay may cause confusion of loyalties and prioritization of mission, processes and activities. Furthermore, even if individuals are diligent in their efforts to conform to the new institutionalisms they may inadvertently neglect, for example, to set items on an agenda or think beyond the familiar. Cultural "baggage" is neither easily unpacked nor integrated. In many ways, starting a new agency and/or department from "ground zero" is less complex than blending existing structures (Crenshaw, 2001).

Part of the reason for this difficulty involves fostering and maintaining organizational "buy-in" at all levels (Kotter, 1995) within the new DHS. This will be a significant challenge. Even if DHS's Secretary, top officials, and management structure clearly define, and act congruent with DHS's vision and build a powerhouse coalition, it is reasonable to ask whether all constituent agencies and all the levels of the organization will follow. This is important because negotiation of the former vision/mission statements of the agencies being absorbed into the DHS's vision statement is a potential source of internal organizational conflict. Will the absorbed agencies still maintain their former vision statements; particularly as they relate to their non-security related functions? Given that not all organizational direction will be focused on homeland security in that the *Act* clearly states that the agencies absorbed into the DHS will continue "to carry out other functions," organizational research suggests that it will be necessary to make clear distinctions between overarching visions and sub-visions that are not related to security. Additionally, organizational change history suggests that there should be a prioritization of the multiple and divergent visions or it should be clearly articulated that security is "*the vision*" and all other is secondary or even tangential. Even if intra-organizational DHS "buy-in" is achieved, external "buy-in" must also simultaneously occur with multiple governmental, private and other units at the federal, state and local level. Thus, both vertical and horizontal integration, in vision, strategy, process and procedures and in general information sharing must occur at all levels.

Even when individuals have a deep sincere desire to transform an agency, it may be that they are unintentionally undermining the organizational mission, goals and activities for a variety of reasons including: they lack the absorptive capacity to transition into their newly assigned tasks; they are confused about how best to prioritize their respective assignments; and their morale is low because of all the agency change, confusion and potential lack of direction. When studying the merger between Custom Services and the Bureau of Narcotics and Dangerous Drugs (BNDD), Rachal (1982) argued the following: "this failed merger should not be attributed to political dynamics, unskilled and incompetent workers or even a cosmetic reorganization, but rather the failure had to do more with the facts that the architects lacked a full understanding of what narcotic agents did and why it was done in a particular manner." Again, individuals may think of themselves as being congruent and capable of carrying out the charge and vision of an organization, but in fact lack the knowledge, skill and in the case of the BNDD, a full understanding of what the organizational "players" need do and how best to do it is missing or severely limited.

Management Issue: Information Sharing and Routine Task Integration

Information sharing represents another serious challenge to the DHS. Developing efficient systems of sharing critical information within DHS as well as with other relevant federal actors and state and local government is essential to the success of DHS. As has been recounted in many different venues, the events of 9/11 suggest communiqué failure was a primary factor that precluded effective prevention of attack occurrence. As elaborated below, communication networks serve to animate organizations. Vertical, lateral, formal and informal communications all influence organization capacity, outcomes, coordinated costs, intraorganizational trust and shared values (Nohria & Ghoshal, 1997).

Establishing streamlined communication systems does not guarantee effective action as a result. The primary issue facing DHS is that in a rapid construction of a blended organization, there exists the real potential for organizational inertia, the ills of which are well documented. Inertia can affect competence in that it is a result of "emergent social and structural processes [that] facilitate convergence on a strategic orientation. A second effect is that it could potentially influence an organization's ability to change and resistance to all but incremental change (Tushman et al., 1985). Neglecting to remove impediments to the new vision or organizational thrust can have far reaching implications (Kotter, 1995). Impediments can be individuals, structures, processes, procedures, and even as suggested above communication flow. Never underestimate the power of an individual, group, or in the case of DHS, agency boycott, with regards to benevolent and or subversive behavior.

Along with the development of informational integration is the challenge of effective coordination of all of the intra-agency functions. Clearly defining core tasks, and processes for disparate constituent agencies is a daunting task – made more so because they must be congruent with the DHS's vision. This affects tasks large and relatively small. For example, narrow job descriptions can significantly undermine productivity while incentive and compensation systems will force individuals to choose between their own self-interest and the new organizational vision (Kotter, 1995).

Who will be responsible for training the individuals on their new tasks? Under the Homeland Security Reorganization Plan the *Director of the Office for Domestic Preparedness* is charged with "providing agency-specific training for agents and analysts within the Department, other agencies, and State and local agencies and international entities" (http://www.whitehouse.gov/deptofhomeland/ analysis/title1.html). Thus, personnel training will need to occur across agencies localities including local, state, national and international sites. Will the trainers of personnel possess the breadth and depth necessary for instruction and mentoring?

Will new tasks that were not originally part of the absorbed agency culture attract the same energy and resources as tasks that are deeply rooted in agencies' persona? Crenshaw (2001) suggests that this is an extremely important concern. Likewise, will there be a "brain drain" from non-security DHS functions to security related functions? Will managers of the absorbed agencies offer out their "inferior" employees as "superior upstarts" and engage in other destructive tactics? Conversely, even if upper agency management have "buy in" to DHS vision and outlined functional tasks and actualize the DHS vision and related processes and tasks as superceding other agency functions, how will mid and lower level employees continue with their non-security tasks given that agency heads may place less emphasis on those tasks and perhaps devote less resources?

Inextricably linked to all of these challenges are resources. As Dependency theorists posit, in order for organizations to survive, they must gather resources (Pfeffer & Salancik, 1978). Organizations depend upon external sources for resources. This outside dependency is characterized as "external dependency." Also, internally, organizations and their respective individuals are struggling for control over resources. Internal resource issues are critical to examine particularly as they relate to individuals core competencies and their ability to retain resources. Thus, the level of analysis of resources needs to be at two levels, internal and external (Pfeffer & Salancik, 1978). Whether one buys into all of the core tenets of organizational dependency theory or not, an analysis of the potential organizational and management transformation challenges would be remiss not to highlight the interplay of resource-related issues to the subsequent development and roll-out of the DHS. Resources, at the two levels of internal and external, do influence organizational structure and behavior (Donaldson, 1996). Moreover, the potential internal and external political struggles, as posited by the Dependency theorists, have already been foreshadowed by the debates surrounding the development of the DHS. It is anticipated that the intensity and scope of these resources related political debates would only escalate as the DHS operates over time.

Management Issue: Managing the External Environment

Accepting that structure does matter, DHS players and stakeholders must also acknowledge that this new organization is more than merely a "black box" filled with lines, titles, processes and specified functions. Frequently, scholars, economists and policy experts highlight the internal complexities of organizations, major challenges and recommend prescriptive measures totally ignoring or negating exogenous variables that significantly influence the internal workings and outcomes of the organization. In order for the DHS success, it must be responsive to external constituencies and concerns. The major challenge of incorporating a or contingency strategy (Donaldson, 1996) is that it is easier posited than realized; particularly when considering the infinitesimal amounts of information that are directly related to homeland security and terrorism in general. How will this information be cataloged, interpreted and incorporated? Technology, markets, and world events, to name a few, will all have some level of interplay with DHS functions and operations.

Any attempt to discuss or forecast potential DHS organizational internal workings and projections of outcomes would be remiss to not highlight the interplay of civil and individual liberties and notions of infringement. Sophisticated security structures and prototypes, such as mechanisms and measures employed by Israel and other nations frequently targeted by terrorist activity, are available for replication and even employed by the U.S. abroad. So why is it that the United States has not taken a bite out of some of this "low hanging fruit" to employ as part of homeland security? Some of this mystery may be enlightened by our constitution, specifically our civil liberties. Thus, when evaluating DHS structures, processes, critical tasks and capacity it must be done in the context of the constraints of civil liberties and other constitutional rights. Civil liberties protections limit search, seizure, and wiretaps, etc. Domestic and local government, although typically not proactive in this area have had, in some cases, more access to certain processes/tactics than traditional national security agencies have had when engaging in certain practices due to legal constraints.

Other major environmental or external factors influencing the development of the DHS and its execution of its functions include: other government agencies, structures, such as Congress and private enterprises at the federal, state and local level. Additionally, international players and stakeholders will also play a major role in influencing the development, implementation and even assessment of DHS.

The importance of private partnerships in providing homeland security has been well documented. The DHS must strategically and methodically identify and path-out its respective interface with private organizations; particularly those organizations that have significant experience in dealing with security related issues and those organizations that will remain instrumental bridges for the carrying out of security and even non-security related functional tasks.

Even if the DHS's structure and its respective processes and players are "correct" or properly positioned, both internal and external DHS stakeholders must perceive the Department as legitimate in order for it to succeed. Thus, perceptions of legitimacy loom as another challenge of the DHS. Why internal legitimacy? Internal legitimacy builds internal cohesiveness among DHS players. Organizational players, like soldiers, perform better with trust in the system (Wilson, 1989). There exist increasing pressures for organizations to adopt similar practices and structures to gain legitimacy and support (DiMaggio & Powell, 1983). Thus, processes, procedures and culture are important mechanisms for building legitimacy into a structure. Again this is easier to posit than it is to realize. Externally, stakeholders must believe in the DHS and have confidence that its respective players and subsequent processes and procedures are congruent with traditional American notions of legitimacy. Already, the *Homeland Security Act* has called into serious question civil liberties and constitutional rights.

REFLECTIONS ON DESIGNING
AND MANAGING THE DHS

Traditionally, organization and management systems are designed to achieve consistency, predictability, stability and efficient use of resources. However, more frequently current organizations are confronted with a dynamic, complex and frequently turbulent set of realities. This is particularly true for the DHS. It is indeed a highly complex organization comprised of a set of diverse agencies with long traditions and different missions most of which must be continued. At the same time, it is asked to solve a highly unstructured, complex, ambiguous and uncertain set of problems inherent in homeland security. Thus, in addition to the attributes of traditional management systems, the design must also support high levels of learning, flexibility and adaptability in order to produce the desired results. In many ways, this poses a truly complex and formidable design problem.

In some regards, multinational firms with many subsidiaries producing varied products/services, operating globally and coping with many different cultures and legal systems, reflect organizing problems facing the DHS. Bartlett and Ghosha (1992) and Nohria and Ghoshal (1997) have researched the design problem of these organizations and have made recommendations for globally operating multinational firms. Nohria and Ghoshal (1997) argue that no single uniform structure can fully address the complexities confronting these firms, and that organizational models that do not attempt to address the unique relationships between subsidiaries and headquarters are not a reflection of reality, should not be employed, and will not lead to a firm's success. In particular, they propose a differentiated network design. At the core of their differentiated network model (DNM) is the contention that organizations should be both appropriately differentiated and integrated in order to tap into and maximize the full value-creating potential for trans-boundary distributed capabilities (Nohria & Ghoshal, 1997). A pictorial view of their proposed model is depicted in Fig. 1.

A differentiated network design allows for the creation of an organization that specifically differentiates between and among the core system and its subsystems

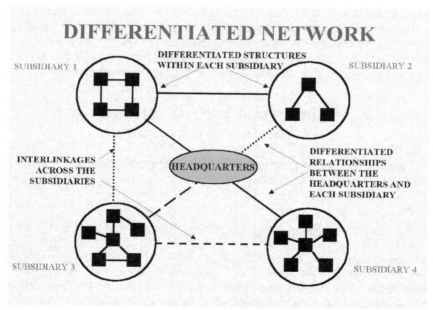

Fig. 1.   Source: Nohria and Ghoshal (1997, p. 14).

while concurrently differentiating between and among various subsystems. This type of differentiated network also allows for different and diverse structures within the network while concurrently allowing for centralization and diffusion of core processes and tasks. The DNM (1997), argue its architects Nohria and Ghosal, is designed to establish a framework that not only examines what "pulls organizations together, but also what pulls them apart" (p. 4).

Based upon the Nohria and Ghoshal research, firms employing a differentiated network approach exhibited superior performance. We believe that the DHS can be effectively conceptualized as a differentiated network and can benefit from a design approach proposed by the DNM. The overriding challenge is how can this complex set of agencies be configured to gain an overall coordinated effort and at the same time perform their varied missions effectively. In the following discussion, the differentiation and integration approaches proposed by the DNM and the potential outcomes will be explored selectively and very briefly. Terminology will be altered to talk about agencies rather than subsidiaries and departments rather than corporate headquarters.

Differentiation Perspective

As the model in Fig. 1 indicates, several key attributes of agencies should be examined and differentiated as the realities of their operations differ. First, relationships between the department and the agencies should be analyzed and differentiated according to their roles in providing homeland security. Some agencies or subsystems in agencies might need to be tightly linked to the department and controlled centrally and others should be more decentralized. The issue is the extent to which authority and decision making should be centralized or localized to the agency in order to best enhance performance. The same analysis should be applied at the intra-agency level.

Second, no single structure should be imposed on all the agencies, but each should analyzed and designed accordingly to best achieve its set of desired outcomes. As Fig. 1 depicts, this can result in a variety of internal configurations of the agencies.

Third, interagency linkages can vary significantly based upon their informational and resource input-output interrelationships (see Fig. 1). If these relationships are tightly coupled, integration mechanisms should be designed to enhance these linkages. Coordination as well as the lack of adequate coordination is expensive. The same analysis should be applied at the intra-agency level.

Fourth, resource configurations across agencies will vary. Some agencies will be more resource rich than others as reflected in their capabilities and expertise to perform certain tasks. Responsibilities and roles should be assigned based upon the resource availability in an agency. Thus, differences in resources among the agencies should influence decisions about allocating roles and responsibilities and creating mechanisms that encourage resource sharing among the agencies (Nohria & Ghoshal, 1997).

In summary, the DNM guidelines for designing complex organizations advocates differentiating structures among and within agencies, relationships between the department and the agencies, the type of linkages needed among the agencies and task assignments based upon resource capabilities.

Integration Perspective

The act of differentiation creates the need for integration, and as differentiation complexity increases, integration demands for coordinating, controlling and governing the system become more challenging. Just as differentiation varies across and within the agencies, integration mechanisms must be used differentially. For example, there needs to be differential use of centralized/decentralized authority and reliance on formalization-the use of standardized policies and systematic rules and procedures.

In addition to the traditional coordination and control mechanism, the DNM puts great emphasis on normative integration (Nohria & Ghoshal, 1997). Shared values, vision and commitment are the core ingredients of normative integration. Integration and control are exercised through socialization processes that create shared understandings, values, visions and commitments. Common understanding and commitment to performance standards and outcomes underpin these processes. Vertical, lateral, formal, and informal informational flows and richness are critical to success of this approach. Many scholars argue, along with Nohria and Ghosal that these kinds of communication flows all influence organizational outcomes such as coordinated costs, combinative capacity, intraorganizational trust and shared values (Arrow, 1997; Edstrom & Galbraith, 1977; Kogut & Zander, 1992; Ring & Van de Ven, 1992).

In summary, by adopting a DNM, an organization adopts the logic not only for a differentiated structural fit, but also, integrative mechanisms through differential use of formal and normative governing mechanisms (Nohria & Ghoshal, 1997).

Performance Outcomes

Strong empirical evidence suggests that organizations, in particular, multinational firms, that employ the DNM approach have enhanced performance. Shared values, a type of dominant normative integration, have, for example, led to higher performance within MNC's (Nohria & Ghoshal, 1997). Also normative integration, both interorganizational and intraorganizational, has a positive effect on the creation, adaptation and diffusion of innovation within organizations (Nohria & Ghoshal, 1997).

Extensive and active interpersonal networks and interpersonal links are important "conduits for information exchange" and are an important outcome of the adopting a DNM. These types of interpersonal networks help to achieve efficiency in innovation; help to build trust; and assist individuals and managers to expand their ties and act as brokers of information. In essence, these types of information exchanges can help fill structural holes (Nohria & Ghoshal, 1997).

One of the most salient attributes of the DNM is its value creation potential as described by Nohria and Ghoshal (1997):

> ...it arises from the ability to create new value through the accumulation, transfer and integration of different kinds of knowledge, resources and capabilities across its dispersed organizational units... Value creation is a form of innovation and processes that utilize diverse resources to create new products, processes and administrative practices. It is exploited both globally and locally (p. 208).

This type of value creation helps to develop a shared code (Moldoveanu et al., 1995; Monteverde, 1995; Nohria & Ghoshal, 1997), facilitates tacit knowledge transfer (Polanyi, 1969), and enhances absorptive capacity (Cohen & Levinthal, 1990). This type of value creation develops value that markets and other external factors are not capable of creating. In other words, value creation allows for purposeful action that markets cannot create or automatically simulate (Nohria & Ghoshal, 1997). Thus, organizational structure, processes, procedures and value creation are not left up to the market or natural evolution, but rather, they are the product of proactive decisions. In summary, it can be said that a differentiated network suggests a proactive management agenda for conceptualizing, actualizing and assessing organizational development in that the types of attributes inherent in DNM must be planned for, stimulated and nurtured.

What Implications Does the DNM have for the DHS?

The suggestion is made that the challenges that await the DHS are complex, but yet not insurmountable. However, the Secretary, senior level management team and other influential stakeholders must be able to effectively manage the different types of relationships (headquarters relationships with agencies; local linkages within each agency; agency relationships between and among other agencies) through differential governing mechanisms while concurrently being creative in resource configuration, normative integration, communication network establishment, capital creation and value creation. This type of shared context creates substantial potential for an organization, represents positive activity and the very real potential for positive management roles.

CONCLUSION

This paper provides a perspective on issues related to developing management and operational strategies for the DHS as it attempts to implement the consolidation of twenty two agencies in the effort to enhance homeland security. Admittedly, our paper does not fully explore all of the potential salient organizational or management issues related to the collapsing of the 22 agencies under the DHS umbrella. However, we hope that it provides useful insight into a currently evolving process and organization as well as insight into strategies for maximizing the effectiveness of this large and important undertaking by the federal government.

We suggest that a differentiated network model might offer viable mechanisms to both understand and address the institutional and managerial changes that are emerging with the formation of the DHS. This model provides structure for development of governing mechanisms thereby facilitating a more efficient and innovative approach that ultimately may allow the DHS to fully achieve its new objectives while at the same time preserving the quintessential operations of the separate agencies. Strategies such as core autonomous teams may be of benefit in facilitating informational and managerial flow within organizations, previously monistic in nature, now operating under a new holistic directive. Additionally, the DNM creates a sustainable structure both horizontally and vertically allowing the organization to be fluid in nature while at the same time providing for clear lines of command and control as new threats are encountered or as a better understanding of the terroristic threat is attained.

Our hope is that this article will stimulate further research and management practices aimed at examining both the negative and positive implications of the DHS organizational restructuring. This challenge remains more than strictly an academic undertaking as the ability of the DHS to be successful will have a great impact on our everyday lives.

NOTES

1. The U.S. Coast Guard, U.S. Secret Service, Bureau of Citizenship and Immigration Services, Office of State and Local Government Coordination, Office of Private Sector

Liaison, and Office of Inspector General will all operate within the department, but not as subunits of any of the five directorates.

2. It is worth noting that the public position of the Bush Administration – literally up to the point when the White House announced it supported creating a Department of Homeland Security – was opposition to the creation of a new department on the grounds that it was not necessary and would represent an unproductive expansion in the size of the federal government.

3. Cohen and Cook (2002) provide a discussion of these national poll results.

REFERENCES

Arrow, H. (1997). Stability, bistability, and instability in small group influence patterns. *Journal of Personality and Social Psychology, 72*, 75–85.

Bales, R. F. (1953). The equilibrium problem in small groups. In: T. Parsons, R. E. Bales & E. Shils (Eds), *Working Papers in the Theory of Action* (pp. 111–161). New York: Free Press.

Bartlett, C., & Ghosha, S. (1992). *Transnational management, text, cases and readings in cross-border management*. Homewood, IL: Irwin.

Christensen, C. (1997). *The innovator's dilemma: When new technologies cause great firms to fail*. Harvard Business School Press.

Cohen, D., & Cook, A. (2002). Institutional redesign: Terrorism, punctuated equilibrium, and the evolution of homeland security in the United States. Paper presented at the annual meeting of the American Political Science Association. Boston, MA, August 29–September 1.

Cohen, W., & Levinthal, D. (1990). Absorptive capacity: A new perspective on learning and innovation. *Administrative Science Quarterly, 35*, 128–152.

Council of State Government (2003). *Homeland security teleconference series: Bridging the public and private gap – infrastructure security in the states*. http://www.csg.org/CSG/Policy/public+safety+and+justice/homeland+security/infrastructure+security+teleconference+. htm (March 23).

Crenshaw, M. (2001). Counterterrorism policy and the political process. *Studies in Conflict and Terrorism, 24*, 329–337.

DiMaggio, P., & Powell, W. (1983). The iron cage revisited: Institutional isomorphism and collective rationality in organizational fields. *American Sociological Review, 48*, 147–160.

Donaldson, L. (1996). *For positivist organization theory: Proving the hard core*. London: Sage.

Edstrom, A., & Galbraith, J. (1977). Transfer of managers as a coordinated control strategy in multinational corporations. *Administrative Sciences Quarterly, 22*, 24–263.

Galar, R. B., & McKenzie, R. (1998). *Punctuated life cycles of firms: A principal/agent perspective on the evolution of firms*. Http://www.gsm.uci.edu/mckenzie/lifecyc.htm.

Gersick, C. (1991). Revolutionary change theories: A multilevel exploration of the punctuated equilibrium paradigm. *Academy of Management Review, 16*(1), 10–36.

Gould, S., & Eldridge, N. (1977). Punctuated equilibria: The tempo and mode of evolution reconsidered. *Paleobiology, 3*, 115–151.

Greenwood, R., & Hinings, C. (1996). Understanding radical organizational change: Bringing together the old and new institutionalism. *Academy of Management Review, 21*, 1022–1054.

Kogut, B., & Zander, U. (1992). Knowledge of the firm and the evolutionary theory of multinational corporation. *Journal of International Business Studies, 24*, 625–645.

Kotter, J. P. (1995). Leading change: Why transformation efforts fail. *Harvard Business Review* (March–April).

Kuhn, T. (1970). *The structure of scientific revolution* (2nd ed.). Chicago: University of Chicago.

Miller, D., & Friesen, P. (1984). *Organizations: A quantum view*. Englewood Cliffs, NJ: Prentice-Hall.

Moldoveanu, M., Nohria, N., & Stevenson, H. (1995). The path dependent evolution of organizations. Harvard Business School Working Paper, Series #596–005.

Monteverde, L. (1995). *Applying resource based strategic analysis.: Making the model more accessible to source-based strategic analysis*. Paper No. 95–1, Department of Management and Information Systems, St. Joseph's University.

Nohria N., & Ghoshal, S. (1997). *The differentiated network: Organizing multinational corporations for value creation*. San Francisco, CA: Jossey-Bass.

Pfeffer, J., & Salancik, G. (1978). *The external control of organizations*. New York: Harper & Row.

Polanyi, M. (1969). Objectivity in science – a dangerous illusion. *Scientific Research* (28 April).

Price, I., & Evans, L. (1998). Punctuated equilibrium: An organic model for the learning organisation, *Journal of the European Foundation for Management Development, 93*, 1. http://members.aol.com/ifprice/peqform.html.

Rachal, P. (1982). *Federal narcotics enforcement: Reorganization and reform*. Boston, MA: Auburn House.

Ring, P., & Van de Ven, A. (1992). Structuring cooperative relationships between organizations. *Strategic Management Journal, 13*, 483–498.

Senge, P. (1990). *The fifth discipline: The art and practice of the learning organization*. New York: Doubleday.

Smithson, A., & Levy, L. (2000). *Ataxia: The chemical and biological terrorism threat and the U.S. response*. Washington, DC: Henry L. Stimson Center.

Steiner, J. (1972). *Group process and productivity*. New York: Academic Press.

Susskind, L., Moorman, W., & Gallager, K. (2002). *Transboundary environmental negotiation: New approaches to global cooperation*. San Francisco, CA: Jossey-Bass.

Tuckman, B. (1965). Developmental sequence in small groups. *Psychological Bulletin, 63*, 384–399.

Tushman, M., Michael, L., & Romanelli, E. (1985). Organizational evolution: A metamorphosis model of convergence and reorientation. In: L. L. Cummings & B. Staw (Eds), *Research in Organizational Behavior* (Vol. 7, pp. 171–222). Greenwich, CT: JAI Press.

Tushman, M., Newman, W., & Romanelli, E. (1986). Conveyance and upheaval: Managing the steady pace of organizational evolution. *California Management Review, 29*(1), 29–44.

U.S. General Accounting Office (2002). *Testimony before the select committee on homeland security*. House of Representatives, Homeland Security: Critical Design and Implementation Issues, Statement of David Walker, Comptroller General of the United States (July 17).

Virany, B., Tushman, M., & Romanelli, E. (1992). Executive succession and organization outcomes in turbulent environments: An organizational learning approach. *Organization Science, 3*(4), 72–92. http://www.whitehouse.gov/deptofhomeland/analysis/title1.html.

Wilson J. (1989). *Bureaucracy*. New York: Basic Books.

Zahra, S., & George, G. (2002). Absorptive capacity: A review, reconceptualization, and extension. *Academy of Management Review, 27*(2), 185–203.